MCQ'S ON FORESTRY

MCQ'S ON FORESTRY

Nilay Kumar
Assistant Professor (Sr. Scale) in the Department of Floriculture
(Ornamental and Medicinal Plants)
College of Horticulture and Forestry
Central Agricultural University, Pasighat, Arunachal Pradesh

Sunandani Chandel
M.Sc. Forestry, Forest Products and Utilization
Navsari Agricultural University
Navsari, Gujarat

New Delhi – 110 034

NEW INDIA PUBLISHING AGENCY
101, Vikas Surya Plaza, CU Block, LSC Market
Pitam Pura, New Delhi – 110 034, India
Email: info@nipabooks.com
Web: www.nipabooks.com

For customer assistance, please contact

Phone: + 91-11-27 34 17 17 Fax: + 91-11- 27 34 16 16
E-Mail: feedbacks@nipabooks.com

ISBN: 978-93-89907-31-5

Composed and Designed by NIPA.

Dedicated to
Our Beloved
Parents
and
Family Members

Dr. Y.S. Parmar University of
Horticulture and Forestry
Nauni- 173 230, Solan
Himachal Pradesh

Foreword

The field of Forestry has seen a tremendous expansion in last few decades. This discipline deals with the art and science for the scientific management of the forests for sustenance. In the present era of climate change and increasing environmental issues such as global warming, loss of biodiversity, etc., the rising demand of the hour is to strengthen this field of elegance. It is one of the best tools for conserving all the life forms present on this earth. The professional undergraduate and postgraduate courses have found place both at ICAR and ICFRE levels. Due to the increasing competition in the field of forestry several competitive examinations are conducted and there is urgent need to develop competitive reading material.

The main aim of this book is to provide all the important information related to forestry generally required for various competitive examinations. This book covers all the important fact and figures of the subject in brief to make it easy and understandable. This book entitled "MCQ's on Forestry" shall be of great support to the students appearing in the examinations conducted by various institutes like ICAR, ICFRE, etc. This book will definitely help the students in achieving their goals in forestry and others competitive examinations such as JRF, SRF, ARS, RO, ACF held at state and national levels. The authors have put lot of the efforts to provide all the important information at a single juncture.

The tremendous efforts taken by the authors in writing this book are commendable and appreciable. I hope this book will serve its purpose in enlightening the knowledge and understanding of students in the field of forestry.

I congratulate and compliment the authors for bringing out this wonderful publication.

Nauni, (Solan)

(Dr Kulwant Rai Sharma)
Dean, College of Forestry

Preface

"A forest is not a resource for us, it is life itself. It is the only place for us to live" so is true the death of the forest is the end of our lives. Thus, the increasing significance of the forest in human lives has intended the students to take up professional forestry courses in the last few decades. The last few decades we have witnessed enormously growing competition in the field of forestry. In this era of immense competition, availability of all the important forestry related study material to students at a single stage is a boon.

The book has put together all the important forestry related information to our aspiring graduates and postgraduate forestry students appearing for various state level and national level exams such as ICAR-JRF, SRF, ARS, NET, RO, ACF, IFS prelims. These multiple-choice questions will help the students to understand the deep perception of the subject and would add to their knowledge. The book covers all the core subjects of the forestry along with some of the untouched topics such as Rangeland Management, Forest Engineering, Forest Economics, Forest Tree Seed Technology, Statistics, Tree Physiology, Forest Soils, Forest Protection and General Agriculture.

The book covers multiple choice questions of all the subjects related to the forestry and efforts are made to cover the important topics from the examination point of view. It includes original previous year ICAR-JRF question paper.

The book is written to acquaint all the UG and PG forestry students with the important information asked in the competitive exams. It is sincerely hoped that the students of forestry will find this book quite useful and prove beneficial to them. This book incorporates all the crucial and important information of forestry subjects in a single volume to save the time of students in search of material in other books.

Authors convey their gratitude to the entire management of New India Publishing Agency, New Delhi for taking all efforts in the publishing of this book.

Authors are highly thankful to all those persons especially research scholars as well as various books, periodicals, magazines, newsletters, journals etc., that helped in the preparation of this book. We especially acknowledge the crucial role of external reviewers and offer our profound thanks for their generous contributions, which certainly improved the quality of this publication.

Being first version of this book, there could be some errors for which authors/ publishers look valuable suggestions and information from all the readers/ scholars/experts/scientist for further improvement of this book and their proposal will be immensely acknowledged.

Nauni (Solan)

Nilay Kumar &
Sunandani Chandel

Syllabus for JRF Exam

SUBJECT CODE 09: MAJOR SUBJECT GROUP - FORESTRY/ AGROFORESTRY AND SILVICULTURE

(**Sub-Subjects**: **9.1**: Forest Production & Utilization, **9.2**: Silviculture & Agroforestry, **9.3**: Tree Physiology and Breeding, **9.4**: Agroforestry, **9.5**: Forest NRM/Forest Mgmt. & Utilization, **9.6**: Plantation Technology, **9.7** Wild Life Sc. Forestry, **9.8** Wood Science)

UNIT-I: Importance of Agriculture/Forestry/Livestock in national economy. Basic principles of crop production, important rural development programmes in India, elementary principles of economics and agri-extension, organizational set up of agricultural research, education and extension in India, major diseases and pests of crops, elements of statistics.

UNIT-II: Forest- importance, types, classification, ecosystem, biotic and abiotic components, ecological succession and climax, nursery and planting technique, social forestry, farm forestry, urban forestry, community forestry, forest management, silvicultural practices, forest mensuration, natural regeneration, man-made plantations, shifting cultivation, taungya, dendrology, hardwoods, softwoods, pulp woods, fuel woods, multipurpose tree species, wasteland management. Agroforestry-importance and land use systems, forest soils, classification and conservation, watershed management, forest genetics and biotechnology and tree improvement, tree seed technology, rangelands, wildlife-importance, abuse, depletion, management, major and minor forest products including medicinal and aromatic plants, forest inventory, aerial photo interpretation and remote sensing, forest depletion and degradation – importance and impact on environment, global warming, role of forests and trees in climate mitigation, tree diseases, wood decay and discolouration, tree pests, integrated pest and disease management, biological and chemical wood preservation, forest conservation, Indian forest policies, Indian forest act, forest engineering, forest economics, joint forest management and tribology.

Syllabus for SRF Exam

FORESTRY/AGROFORESTRY

Unit 1: Forests and Forest Polic

Forests-extent, basis for classification and distribution in India; Geographical distribution and salient features of major world forest types; Phylogeographical regions and vegetation of India; Role of forests in national economy-productive, protective and ameliorative, tribal and rural livelihoods; Forest types of India: distribution and types; Succession, climax and retrogression; Concepts of biomass, productivity, energy flow and nutrient cycling in forest ecosystem; Migration and dispersal mechanism. National Forest Policy 1894, 1952 and 1988; Indian Forest Act, 1927; Forest Conservation Act,1980 and Wildlife Protection Act, 1972; Amendments 1991, 2003 and 2006, Biological Diversity Act, 2002, The Scheduled Tribes and Other traditional forest dwellers (Recognition of Forest Rights) Act, 2006. National Agroforestry Policy, 2014.

Unit 2: Silviculture

Definition, object and scope of silviculture; Site factors-climatic, edaphic, physiographic, biotic and their influence on forest vegetation; Forest regeneration: natural and artificial; Silvicultural systems-high forest and coppice systems; Seed collection, processing, storage, viability and pre- treatment; Seed dormancy and methods for breaking dormancy; Seed testing and germination tests; Seed certification and ISTA Rules; Forest nursery-need, selection and preparation of site, layout and design of nursery beds; Types of containers; Root trainers; Growing media and sowing methods; Management of nursery-shading, watering, manuring, fertilizer application, weed control, insect pest and diseases control; Planting techniques: site selection, evaluation and protection; Soil working techniques for various edaphic and climatic conditions; Planting patterns; Plant spacing, manure and fertilizer application, irrigation/moisture conservation techniques; Choice of species. Afforestation on difficult sites: saline-alkaline soils, coastal sands, lateritic soils, wetlands, ravines and sand dunes, dry and rocky areas, cold desert; Tending operations - weeding, cleaning, climber cutting, thinning - mechanical, ordinary, crown and selection thinning, improvement felling, pruning and girdling; Silviculture of important tree species- *Populus, Eucalyptus, Dalbergia, Acacia, Tectona, Shorea, Prosopis, Casuarina, Pinus, Gmelina, Azadirachta, Diospyros, Pterocarpus, Anogeissus, Santalum, Quercus and Albizia, bamboos, Melia dubia, Ailanthus excelsa, Simarouba and Karanja.* Plantation forestry- industrial and energy plantations.

Unit 3: Forest Biology and Tree improvement

Tree improvement: nature and extent of variations in natural population; Natural selection; Concept of seed source/provenance; Selection of superior trees; Seed production areas, exotic trees, land races; Collection, evaluation and maintenance of germplasm; Provenance testing. Genetic gains; Tree breeding: general principles, mode of pollination and floral structure; Basics of forest genetics-inheritance, Hardy weinberg Law, genetic drift; Aims and methods of tree breeding. Seed orchard: types, establishment, planning and management, progeny test and designs; Clonal forestry-merits and demerits; Techniques of vegetative propagation, tissue culture; Role of growth substances in vegetative propagation.

Unit 4: Forest Mensuration

Forest mensuration-definition, object and scope; Measurement of diameter, girth, height, stem form, bark thickness, crown width and crown length; Measurement methods and their principles. Measurement and computation of volume of logs and felled/standing trees; Construction and application of volume tables; Biomass measurement; Growth and increment; Measurement of crops; Forest inventory: kinds of enumeration, sampling methods, sample plots and aerial photo interpretation; Geographic information systems and remote sensing-concept and scope.

Unit 5: Social forestry and Agroforestry

Social forestry, community forestry and farm forestry; Concept and definition of agroforestry, Benefits and constraints of agroforestry; Historical development of agroforestry and overview of global agroforestry systems. Classification of agroforestry systems: structural, functional, socio-economic and ecological; Diagnosis and design of agroforestry system; Land capability classification and land use; Criteria of an ideal agroforestry design, productivity, sustainability and adoptability; Multipurpose tree species and their characteristics suitable for agroforestry.

Unit 6: Agroforestry management

Plant management practices in agroforestry; Tree-crop interactions: ecological and economic; Concept of complementarity, supplementarity and competition; Productivity, nutrient cycling and light, water and nutrient competition in agroforestry; Concept of allelopathy and its impact on agroforestry; Agroforestry practices and systems in different agro-ecological zones of India.

Unit 7: Wood Science and Forest Products

Logging and ergonomics- wood anatomy, wood seasoning and preservation techniques; Forest Products and utilization-Manufacturing and utilization of wood products (timber and composite wood) and non-wood forest products such as fibers, flosses, dyes, gums, resins & tannins, medicinal plants, essential oils, edible fruits, spices, bamboo and canes.

Unit 8: Forest Degradation and Protection

Extent and causes of land denudation; Effects of deforestation on soil erosion, land degradation. Wastelands: their extent, characteristics and reclamation; Watershed management and its role in social, economic and ecological development; Forest fires: causes, types, impacts and control measures; Major forest pests, diseases and weeds and its management.

Unit 9: Forest Management and Forest Economics

Forest management: definition and scope; Concept of sustained yield and normal forest; Rotation; Estimation of growing stock, density and site quality; Management of even aged and uneven aged forest; Regulation of yield in regular and irregular forests by area, volume, increment and number of trees; land equivalent ratio; Working plan; Joint forest management; Conservation and management of natural resources including wildlife; Forest evaluation; Internal rate of return, present net worth and cost benefit analysis. Ecosystem services. Economic evaluation of agroforestry systems: cost benefit analysis and land equivalent ratio.

Unit 10: Wildlife

Wildlife biology, ornithology, herpetology, wildlife management-Population estimation in wildlife-census methods, Man-animal conflicts and management strategies.

Unit 11: Forest Genetic Resources and Ecotourism

Role of green revolution in forest conservation in India. *In-situ* and *ex-situ* conservation of forest genetic resources-Sacred groves; Urban forestry-Choice of species, design, development and management, Eco-tourism.

Unit 12: Climate change and mitigation

Climate change: greenhouse effect, sources and sinks of greenhouse gases, major greenhouses gases; Global climate change-its history and future predictions; Impact of climate change on agriculture, forestry, wildlife, water resources, sea level; Livestock, fishery and coastal ecosystems; International conventions on climate change; Global warming: effect of enhanced CO2 on productivity; Ozone layer depletion; Disaster management, floods, droughts, earthquakes; Tsunami, cyclones and landslides; Agroforestry-environmental conservation- carbon sequestration.

Unit 13: Statistics

Statistics: definition, object and scope; frequency distribution; mean, median, mode and standard deviation, introduction to correlation and regression; Experimental designs: basic principles, completely randomized, randomized block, latin square and split plot designs.

AGRICULTURAL METEOROLOGY

Unit 1: General Meteorology

Scope and importance of meteorology; Layers of atmosphere and their characters; Laws of radiation: Planck's law, Stephan-Boatman law, Wein's displacement law; Kirchoff's law, Beer's law and Lambert's, Cosine law, Solar constant, length of day; Atmospheric and astronomical factors affecting depiction of solar radiation; Ozone hole; Direct and diffuse radiation; Albedo, Heat transfer, convection, conduction and radiation; Concepts of latent and sensible heat; temperature inversion Radiant flux and flux density; Atmospheric motion balanced forces; Gas laws, pressure gradient, isobars, hydraulic equation and its application; Carioles force, geotropic, gradient and cyclostrophic winds; three dimensional wind circulation; Pressure systems; Cyclones and anticyclonic motions: trough, ridge and col; Thermal wind; Contour charts, Concepts of specific heat at constant volume and pressure; First and second laws of thermodynamics, vapor pressure, specific humidity, relative humidity, mixing ratio, absolute humidity and dew point temperature; Vapour pressure deficit; Psychometric equation, entropy, T-phi gram; Vertical stability of atmosphere, virtual temperature and potential temperature; Moist and dry adiabatic processes; Clouds their description and classification; Condensation process-artificial rain making; Bergeron-Findeison theory; coalescence theory; Forms and types of precipitation Dew, frost, fog, mist, haze thunderstorm and hail; Air masses and fronts; Extra tropical cyclones; Land and sea breeze; Mountain and valley winds; Wind rose, Tropical cyclones and their structures; Extreme events, Tornados, water spout, thunder storm, dust storm, Avalanche, blizzard etc., Weather variables and their measurements; Different types of observatories and instruments; Units for measurements of momentum, force work, power, surface tension, pressure, temperature; Thermal-conductivity and diffusivity, resistance, radiation light intensity and water vapour.

Unit 2: General Climatology

Elements of weather and climate; Climatic controls, factors affecting climate Seasonal distribution of radiation, rainfall. temperature sunshine, wind pressure over India; Climatic classification-Koppen and Thornwaite; Hargreaves and Trolls classification; Climatology principles of weather phenomena occurring in four main seasons of India; Mechanism of Indian monsoons; Climatic variability, recent trends, Mitigation and adaptation strategies, factor affecting rainfall distribution, cyclones and cyclonic tracks over the Indian region; North western disturbances and monsoon breaks; Drought climatology, rainfall and its variability, atmospheric and agricultural droughts intensity, duration, beginning and end of drought and wet spells; Moisture availability indices; Heat and cold waves; Continental, polar maritime and monsoon climates, El-Niño, La Nino and their impact on Indian rainfall systems.

Unit 3: Agricultural Climatology

Meaning and scope; Effect of thermal environment, Diurnal variation; on growth and yield of crops; Cardinal temperatures; Thermoperiodism, photoperiodism; Vont Hoff's law, phenology of crops; Heat unit concept, thermal time and thermal use-efficiency and their applications; Length of growing period determination. Contingency planning far different weather aberrations, and extreme weather events; Meteorological factors associated with incidence and development of crop pests and disease, pest and disease out breaks etc.; Effect of climate on humans and animals, warm and cold season indices for comfort zones, role of weather in animal disease and protection against weather hazards.

Unit 4: Micrometeorology

Concept of micro, meso and macro meteorology; Micrometeorological processes near bare ground and crop surfaces; Shearing stress, molecular and eddy diffusion, forced and tree convection; Boundary layer, frictional velocity, roughness length and zero plane displacement; Micrometeorology of crops, rice and wheat; Day and night radiation, humidity, temperature, wind and CO2 profiles in crop canopies; Richardson number, Reynolds analogy, exchange coefficients, fluxes of momentum, water vapors, CO2 and heat; Inversion and its effect on smoke plume distribution; Windbreaks and shelterbelts, different methods on modification of field microclimate; Frost protection, hail suppression spectral properties of vegetation; Light interception by crop canopies as influenced by leaf area index, leaf arrangement and leaf transmissibility, extinction coefficient and radiation use-efficiency; Microclimate of field crops, forest and orchards etc.

Unit 5: Evapotranspiration

Hydrological cycle and concept of water balance, concepts of evaporation. evapotranspiration, potential and actual evapotranspiration, consumptive use, different approaches of ET determination direct and empirical methods, energy balance and Bowen's ratio methods, water balance single and multilayered soil methods, aerodynamic, eddy correlation and combination approaches, field lysimetric approaches and canopy temperature based methods; Advantages and limitations of different methods; Water use and water use-efficiency, dry matter production and crop yield functions; Irrigation scheduling based on ET; Advective energy determination and its effect on water use by crops; Physiological variation in relation to crop growth and development.

Unit 6: Crop Weather Modeling

Concepts of mechanistic and deterministic models; General features of dynamical and statistical modeling techniques; Crop weather models and their use in crop yield assessments; Crop weather analysis models, empirical, statistical models, and different crop growth simulation models for yield assessment; concepts for crop growth and yield; Advantages and limitations of modeling, climatic change, greenhouse effect, CO2 increase, global warming and their impact on agriculture.

Unit 7: Weather Forecasting for Agriculture

Methods and types of weather forecasting,: Short, medium and long range weather forecasting; Crop and pest weather calendars Monsoon onset and rainfall forecasts; Weather forecasting and agro-advisories; Use of satellite cloud imageries in weather forecasting; Synoptic charts and synoptic approach to weather forecasting, use of medium, long range and vegetative indices based agro meteorology forecasts for monitoring crop prospects and crop yield forecasts; Meteorological satellites for weather forecasts; Forecast of Indian monsoon rainfall; Early warning systems for agriculture operation forecasts.

ENVIRONMENTAL SCIENCES

Unit 1

Definition and scope of environmental science and its interrelationship with other sciences and agriculture; Segments of atmosphere: hydrosphere, Lithosphere and biosphere; Components of environment-biotic, abiotic and social; Ecological Foot prints. Natural resources: land, soil, water and forest and; present status-Land degradation-Wasteland: their extent, characteristics and reclamation; water conservation : watershed management and rain water harvesting-Major river projects and its impacts; Mineral resources-Environmental effects of mining; Food resources-problems; Ecology concepts-types-habitat ecology, systems ecology, synecology, autecology; Ecosystem: Structure- Functions; Population-characteristics and measurement; Communities- habitats, niches, biomes, population dynamics, species and individual in the ecosystem; Recent trends in ecology; Characteristic features-structure and function of forest, grassland, plantation, desert; Aquatic and agro-ecosystem. Energy flow in ecosystems and environment; Energy exchange and productivity-food chains and food webs-ecological pyramids; Ecological succession-types and causes. Biogeochemical cycles; nutrient cycles and recycle pathways.

Unit 2

Biodiversity concepts, levels and types, Values and Significance of biodiversity; Theories on biodiversity; Agro-biodiversity-Transgenic crops and animals-Impact on Environment. Plant genetic resources, exploration and collection; Biogeographical zones of India; Biodiversity hot spots in India and world; Loss of biodiversity-Causes-Crop domestication, plant introductions -exotics and invasive plants-IUCN clauses and concept of threatened and endangered species; Methods of conservation: *in-situ* and *ex*-situ, national parks, wildlife sanctuaries, biosphere reserves; National and global conservation measures-institutions and conventions-Indian Biodiversity Act 2002; World heritage sites; Wetlands-Mangroves-Ramsar convention.

Unit 3

Environmental Pollution - Point and non-point sources-Atmosphere-stratification-Composition of air; Air pollution: sources and classification-Criteria pollutants-Indoor and out-door air pollution; Types-primary and secondary pollutants-Thermal Inversion-Air pollution Episode-Air Quality standards-Greenhouse gases-Global warming-Ozone depletion-Acid rain-Impacts on Environment-Effects of air pollutants on vegetation, animals and human health; mitigation measures for combating air pollution; Automobile pollution, Noise pollution-source and effects, Disasters and their management: floods, droughts, earthquakes; Tsunami, cyclones and landslides; Adaptation and mitigation strategies of climate change-Carbon sequestration and clean development mechanism. National and international laws and policies on air pollution. Environmental treaties; Role of NGO's in environmental protection; Corporate Social Responsibility (CSR) of industries in environmental protection; Advance tools for ecosystem analysis-Remote Sensing (RS) and Geographic Information Systems (GIS). EIA and Environmental Auditing.

Unit 4

Urban and Industrial wastewater - Pollution of ponds, lakes, rivers and ground water. Impacts of water pollutants on Environment-Effluent Treatment Processes-Energy production recycling of treated waste water and value addition to wastes-Permissible limits. Soil pollution-sources-Organic and inorganic contaminants, Xenobiotics and their effect on agriculture: Heavy metals and pesticides-Effects of pollutants on soil health and productivity; Radioactive pollutants- Impacts; Remediation of contaminated soil-Microbial, chemical ameliorants, phytoremediation and Nano-remediation; Permissible limits of organic and inorganic pollutants.

Unit 5

Solid waste-sources, Categories, hazardous and non-hazardous, impact on Environment Management strategies-5R concepts-Thermal conversions-Pyrolysis-Gasification-Incineration; Biodegradation of organic wastes-Composting, Vermicomposting, Mushroom production, SCP; Energy recovery-biogas, landfill, etc. E-waste-impacts and resource recovery; Solid waste management rules in India.

Unit 6

Energy-Types of renewable sources of energy; Solar energy: Energy transfer and applications- Solar thermal system and their applications Wind energy-Types Geothermal and tidal energy; Bio-energy from biomass. Liquid fuels from petro crops, Concepts of producer gas; types of gasifiers; Briquetting of agro-wastes for fuel; Potential of renewable energy sources in India, Integrated rural energy programme; Nuclear energy.

Unit 7

Frequency distribution, mean, median, mode and standard deviation; Normal, binomial and poisson distribution; Correlation-partial and multiple; Regression coefficients and multiple regression. Tests of significance F and Chi-square ($\chi 2$) tests; Experimental designs-basic principles, completely randomized, randomized block, Latin square and split plot designs.

AGRICULTURAL PHYSICS

Unit 1: Basic Physics

Conservation of mass, energy and momentum; Forces in nature; Measurement of heat, specific heat, transfer of heat; Huygen's principle, reflection, refraction, diffraction, polarization, interference and scattering of light waves; Optics theory, principles of optical instruments; Change of phase and polarization, equation of state, Laws of thermodynamics; Free energy, Entropy and concept of negative entropy; Vont Hoff's law; Cathode rays; Radio activity, alpha, beta, and gamma- rays, detection and measurement of radiation; Properties of X-rays; Bragg's law; Nuclear fission, fusion, nuclear reactions, neutron moderation, nuclear energy, atomic power; Radioactivity and its applications in agriculture.

Unit 2: Soil Physics

Factors and processes of soil formation; Physical, physicochemical and biological properties of soils; Soil water retention and movement under saturated and unsaturated conditions; Infiltration, redistribution and evaporation of soil water; Field water balance and water use efficiency; Soil aeration; Thermal properties of soil and heat transport; Influence of soil water, temperature and aeration on crop growth and their management; Soil erosion and control; Soil physical constraints and their management.

Unit 3: Radiation Physics

Basics of Electromagnetic spectrum and its interaction with matter; Laws of radiation, scattering, reflection, transmission, absorption, emission, diffuse and specular radiations; Radiation units, flux, intensity, emittance, inter conversion of radiometric units; Energy balance of land surfaces.

Unit 4: Plant Biophysics and Nano Technology

Introduction and scope of biophysics; Structure and properties of water; Experimental techniques used for separation and characterization of bio-molecules sedimentation, ultra- centrifugation, diffusion, osmosis, viscosity, polarization and electrophoresis, chromatography; Fiber physics; Basic Spectroscopic techniques, UV-Visible, IR, NMR, EPR spectroscopy, X-ray diffraction; Chlorophyll fluorescence; Nanostructures, Properties and characterization of nano-materials; Nano-biology, hazards of nano-material; Applications of nanotechnology in agriculture.

Unit 5: Remote sensing

Electromagnetic radiation, and interactions with the matter, remote sensing system-active and passive, sensor and platform; Radiometric quantities; Spectral signatures of natural targets and its physical basis, spectral indices; Satellite characteristics, spatial, spectral, radiometric and temporal resolutions; Air borne remote sensing; Imaging and non-imaging systems; Multispectral, hyperspectral, thermal and microwave remote sensing; Digital image processing; National and International satellite systems for land, weather, ocean and other observations; Applications of remote sensing in agriculture.

Unit 6: Geo-informatics

Basic concepts and principles: Hardware and software requirements; Common terminologies of geographic information system (GIS); Maps and projections, principles of cartography; Basic geodesy: Geiod/Datum/Ellipsoid; Cartographic projections, coordinate systems, types and scales; Accuracy of maps; Raster and Vector data model; DBMS; Geostatistical analyses; Spatial interpolation-Thiessen polygon; Inverse square distance; Digital Elevation Model; Principles of GPS; DGPS; Errors in GPS data and correction; GPS constellations; Geo-informatics application in agriculture and natural resource management.

Unit 7: Atmospheric physics

Weather and climate: Atmosphere and its constituents; Meteorological elements and their measurements; Heat balance of the earth and atmosphere; Climatic classification systems; climatology of India, agro-ecological regions; Monsoon, western disturbances, cyclones, droughts; Wind system, precipitation, cloud, pressure pattern. Atmospheric stability; Weather forecasting: numerical weather prediction; El Nino, La Nina and ENSO; Climate change, global warming, impacts of climate change on agro-ecosystems; Physiological response of crop plants to weather (light, temperature, CO2, moisture and solar radiation); Heat units, thermal time and thermal use-efficiency and their applications; Micro, Meso and Macro climates; Exchange of mass, momentum and energy between surface and atmosphere, exchange coefficients; Richardson number & Reynold's analogy; Boundary layer; Eddy covariance techniques; Wind profile; Modification of microclimate; Radiation distribution within the plant canopy; Concept of evapotranspiration: potential, reference and actual evapotranspiration, crop coefficient; Measurement of evapotranspiration.

Unit 8: Mathematical Modeling of soil-plant-atmosphere system

Applications of matrices: Differentiation and integration; Numerical modeling: finite difference and finite element; Spatial statistics: Variogram and interpolation techniques; Surface modeling; Root water uptake models; Simulation models for water, heat, and solute movement in two and three dimensional porous media; Fundamentals of dynamic simulation, systems, models and simulation; Mechanistic, stochastic and deterministic models; Model calibration, validation and sensitivity analysis; Crop weather models and its use in crop yield estimation; Advantage and limitations of modeling.

Contents

Glossary

Acre: The acre is a unit of land area used in the imperial and US customary systems. One acre is equal to 43,560 square feet or is equivalent to 0.4047 hectares or 160 square rods.

Amphibian: Any of a class of vertebrates that regulate their body temperature externally; lay shell-less eggs in wet areas; live in water during early development and live both in water and on land as adults; and use lungs, gills and their skin for breathing. Most have four legs and smooth, moist skin without scales.

Angiosperm: A plant that has true flowers and bears its seeds in fruits In temperate zones, many angiosperms are deciduous trees, while in tropical zones, many are evergreen trees Examples include oaks, willows, maples and birches.

Annual Ring: The combination of one earlywood layer (light coloured) and one latewood layer (dark coloured) seen in a cross-section of a tree. One annual ring usually represents one year of growth.

Artificial Regeneration: The growth of new trees through seeding and planting.

Bark: The tough exterior covering of a woody root or stem that protects the tree from injury caused by insects and other animals, by other plants, by disease and by fire.

Best Management Practices: Procedures employed during harvesting and/ or timber stand improvement activities that reduce erosion and prevent or control water pollution.

Biltmore Stick: A stick similar to a yardstick in appearance, but usually about 25 inches long. One side is scaled to read a tree's diameter by holding the stick horizontally at arm's length and against the tree at breast height. A Merritt hypsometer runs along one edge of the stick and is scaled to read a tree's height from 66 feet away from the tree's base. These two measurements are then used to find the tree's volume according to the volume table printed on one face of the stick.

Biodiversity: The variety of life forms in a given area; can be categorized in terms of number of species, variety of plant and animal communities, genetic variability or some combination of these categories.

Bird: Any of a class of vertebrates that regulate their body temperature internally, have bodies that are covered almost entirely with feathers and have forelimbs modified as wings that enable most to fly.

Board Foot: A unit of measure equal to a board that is 1 inch thick, 12 inches long and 12 inches wide, or 144 cubic inches.

Bole: The main trunk of a tree.

Broadleaf: A class of trees that have broad, flat leaves of many different shapes; most are deciduous; also called hardwood because most broad-leaved trees have harder wood than do conifers. Examples include oak, hickory, maple and ash.

Buffer Strip: A narrow zone or strip of land, trees or vegetation bordering an area. Common examples include visual buffers, which screen the view along roads, and streamside buffers, which are used to protect water quality. Buffers may also be used to prevent the spread of forest pests.

Cambium: A thin layer of specialized cells within a tree's trunk that divide to produce new inner bark cells to the outside and new sapwood cells to the inside. The narrow band of cells that is responsible for the tree's growth in circumference.

Canopy: The "roof" of the forest formed by the crowns of the tallest trees.

Carrying Capacity: The maximum number of healthy wildlife that a given habitat or area can support without degradation of the habitat.

Cellulose: The scientific name for wood fibre.

Chain: A distance of 66 feet.

Clear-cut: A harvesting and regeneration method that removes all trees within a given area. Most commonly used in pine and hardwood forests that require full sunlight to regenerate and grow efficiently.

Clinometer: An instrument that is held at eye level to read stump height and merchantable or total height when standing 50 and 66 feet from the base of the tree. The difference between the two readings yields the height.

Competition: The struggle between trees to obtain sunlight, nutrients, water and growing space every part of the tree, from the roots to the crown, competes for space and food.

Conifer: A class of trees that are evergreen, have needle or scalelike foliage and conelike fruit; often called softwood. Examples include pine, hemlock, cedar and cypress.

Conservation: Planned management and wise use of natural resources for present and future generations.

Cord: A standard unit of measure equivalent to 128 cubic feet of round or split wood. A standard cord measures 4 feet by 4 feet by 8 feet A face cord or short cord is 4 feet by 8 feet by any length of wood under 4 feet.

Cover: (a) Any plant that intercepts rain drops before they reach the soil or that holds soil in place; (b) hiding place or vegetative shelter for wildlife from predators or inclement weather.

Crown: The branches and foliage at the top of a tree.

Cruise: A survey or inventory of forestland to locate timber and estimate its quantity by species, products, size, quality or other characteristics.

Deciduous: A group of trees that lose all of their leaves every year.

Decomposition: The process by which organic material such as leaves and branches are broken down by bacteria, fungi, protozoans and the many different kinds of animals that live in the soil.

Dendrology: The study of trees; tree identification.

Diameter at Breast Height (DBH): Tree diameter measured at 45 feet above ground level.

Diameter Tape: A steel measuring tape that has a scale calibrated to read a tree's diameter when wrapped around the tree's circumference.

Earlywood: Wood cells produced at the beginning of a tree's growing season that are generally light in colour. Also called springwood.

Ecology: The science or study of the relationships between organisms and their environment.

Ecological Succession: The gradual change of plant and animal communities over time.

Ecosystem: A loosely defined area consisting of numerous habitats.

Edge: The transition between two different types or ages of vegetation.

Endangered Species: Any species that has been classified by the US Fish and Wildlife Service or a state wildlife agency as being in danger of extinction throughout all or a significant portion of its range. A species is endangered when the total number of remaining members may not be sufficient to reproduce enough offspring to ensure survival of the species.

Environment: The sum of all external living and non-living conditions and influences that affect the development and survival of an organism.

Erosion: The wearing away or removal of land or soil by the action of wind, water, ice or gravity.

Even-Aged Management: A forest management method used to produce stands that are all the same age or nearly the same age by harvesting all trees in an area at one time or in several cuttings over a short time. This management method is commonly applied to shade-intolerant conifers and hardwoods.

Evergreen: A group of trees that do not lose all of their leaves every year but go through a gradual replacement by dropping only their oldest leaves each year. Instead of being bare in winter, these trees have leaves all year.

Foliage: The leaves of a tree or other plant.

Forage: Vegetation such as leaves, stems, buds and some types of bark, that can be eaten for food and energy.

Forb: Any herb other than grass.

Forest Floor: The lowest level of the forest that is made up of tree seedlings, dead leaves and needles, grasses, ferns, flowers, fungi, and decaying plants and logs.

Forest Management: Caring for a forest so that it stays healthy and vigorous and provides the products and values the landowner desires.

Forest Stewardship Plan: A written document listing activities that enhance or improve forest resources (wildlife, timber, soil, water, recreation and aestheticson private land over a five-year period).

Forest Type: A designation or name given to a forest based on the most abundant tree type or types in the stand; groups of tree species commonly growing in the same stand because their environmental requirements are similar. Examples of North Carolina forest types include (a) pine (b) mixed hardwood (c) cypress, tupelo and black gum and (d) oak and hickory.

Forestry: The art and science of managing forests to produce various products and benefits including timber, wildlife habitat, clean water, biodiversity and recreation.

Fuel Loading: A build-up of easily ignited leaves, pine straw, branches and trees on the forest floor.

Group Selection: (a) The removal of small groups of trees to regenerate shade-intolerant trees in the opening (usually at least 1/4 acre) (b) specific type of selective cutting.

Gymnosperm: A plant whose seeds are not enclosed in flowers. Most gymnosperms produce their seeds on the surface of the scales of female cones and are pollinated by wind. Conifers are the most common type of gymnosperm.

Habitat: An area in which a specific plant or animal naturally lives, grows and reproduces; the area that provides a plant or animal with adequate food, water, shelter and living space.

Hardwoods: Trees with broad, flat leaves as opposed to coniferous or needled trees. Wood hardness varies among the hardwood species, and some are actually softer than some softwoods.

Heartwood: The central core of a tree, which is made up of dense, dead wood and provides strength to the tree.

High-Grading: A harvesting technique that removes only the biggest and most valuable trees from a stand and provides high returns at the expense of future growth potential. Poor quality, shade-loving trees tend to dominate in continually high-graded sites.

Hypsometer: Any device used for measuring tree height.

Increment Borer: A hollow auger-like tool with a screw bit used to remove core samples from trees.

Latewood: Wood cells produced at the end of the growing season that make up the darker section of an annual ring. Also called summerwood.

Limiting Factor: Any requirement for wildlife survival that is in limited supply.

Mammal: Any of a class of higher vertebrates whose bodies are covered with hair, who give birth to live young, nourish their young with milk from mammary glands, regulate their body temperature internally, have four types of well-developed teeth and typically have four well-developed legs with toes that have nails, claws or hoofs.

Mast: Fruits or nuts used as a food source by wildlife. Soft mast includes most fruits with fleshy coverings, such as persimmon, dogwood seed or black gum seed. Hard mast refers to nuts such as acorns and beech, pecan and hickory nuts.

Merritt Hypsometer: A scale that measures the number of 16-foot logs in a tree.

Multiple-Use Management: The management of land or forest for more than one purpose, such as wood production, water quality, wildlife, recreation, aesthetics and clean air.

Natural Regeneration: The growth of new trees in one of the following ways without human assistance: (a) from seeds carried by wind or animals, (b) from seeds stored on the forest floor, or (c) from stumps that sprout.

Naval Stores: Products such as turpentine, pitch and rosin that come from pine trees and are used in the construction and maintenance of wooden sailing vessels.

Phloem: The part of a tree that carries sap from the leaves to the rest of the tree. Also called inner bark.

Photosynthesis: The process by which a plant or tree combines water and carbon dioxide with energy from the sun to make glucose and oxygen.

Plant Succession: The progression of plants from bare ground to mature forest.

Prescribed Burning: The practice of using regulated fires to reduce or eliminate material on the forest floor, for seedbed preparation or to control competing vegetation. Prescribed burning simulates one of the most common natural disturbances Also called controlled burning.

Pulpwood: Wood used in the manufacture of paper, fibre board or other wood fibre products. Pulpwood-sized trees are usually a minimum of 4 inches in diameter.

Reforestation: Re-establishing a forest by planting or seeding an area from which forest vegetation has been removed.

Release: To free a tree from competition with its immediate neighbours by removing the surrounding trees. This occurs naturally and artificially.

Renewable Resource: A naturally occurring raw material or form of energy that has the capacity to replenish itself through ecological cycles and sound management practices.

Reptile: Any of a class of vertebrates that regulates its body temperature externally, has dry, glandless skin covered with scales, breathes through lungs and lays large eggs that develops on land.

Resin: A group of sticky liquid substances secreted by plants that appear on the plant's external surface after a wound.

Roots: The underground portion of a tree that helps anchor the tree in the ground and absorbs water and nutrients from the soil.

Rotation: The number of years required to establish and grow trees to a specified size, product or condition of maturity. A pine rotation may range from as short as 20 years for pulpwood to more than 60 years for sawtimber.

Salvage Cut: The harvesting of dead or damaged trees, or the harvesting of trees in danger of being killed by insects, disease, flooding or other factors in order to save their economic value.

Sawtimber: Wood of large enough size to be used to produce lumber for construction and furniture.

Sedimentation: The deposition or settling of soil particles suspended in water.

Seed Tree Cut: A harvesting method in which a few scattered trees are left in the area to provide seeds for a new forest stand. Selection of seed trees is based on growth rate, form, seeding ability, wind firmness and future marketability This harvesting method produces an even-aged forest.

Selective Cutting: The periodic removal of individual trees or groups of trees to improve or regenerate a stand.

Shade-Intolerant Species: Trees that require full sunlight to thrive and cannot grow in the shade of larger trees.

Shade-Tolerant Species: Trees that have the ability to grow in the shade of other trees and in competition with them.

Shelterwood Cut: Removing trees in the harvest area in a series of two or more cuttings so that new seedlings can grow from the seeds of older trees. This method produces an even-aged forest.

Silviculture: The art, science and practice of establishing, tending and reproducing forest stands of desired characteristics. It is based on knowledge of species' characteristics and environmental requirements.

Site Index: A relative measure of forest site quality based on the height (in feet of the dominant trees at a specific age (usually 25 or 50 years, depending on rotation length). Site index information helps estimate future returns and land productivity for timber and wildlife.

Snag: A standing dead or dying tree.

Softwood: A tree belonging to the order Coniferales. Softwood trees are usually evergreen, bear cones and have needles or scale like leaves. Examples include pines, spruces, firs and cedars.

Species: A group of related organisms having common characteristics and capable of interbreeding. Loblolly and Virginia pine are common tree species that can interbreed.

Stand: A group of trees that are sufficiently the same in species composition and arrangement of age classes and condition so that they can be managed as a unit.

Streamside Management Zone (SMZ): An area adjacent to a stream in which vegetation is maintained or managed to protect water quality.

Suppression: The process by which a tree loses its vigour due to inadequate light, water and nutrients.

Thinning: A tree removal practice that reduces tree density and competition between trees in a stand. Thinning concentrates growth on fewer, high-quality trees, provides periodic income and generally enhances tree vigor. Heavy thinning can benefit wildlife through the increased growth of ground vegetation.

Threatened Species: Any species that has been classified by the US Fish and Wildlife Service or a state wildlife agency as likely to become endangered within the foreseeable future throughout all or a significant portion of its range. A threatened species has declining or dangerously low populations but still has enough members to maintain or increase numbers.

Transpiration: The loss of water through leaves.

Tree Calliper: A metal or wooden device consisting of an arm and two prongs, one of which is free to slide along a graduated scale on the arm. The prongs are placed against opposite sides of a tree to read its diameter on the scale.

Turpentine: A distilled chemical produced from tapping into a living pine and harvesting the sap.

Understory: The area below the forest canopy that comprises shrubs, snags and small tree. Because the understory receives little light, many of the plants at this level tolerate shade and will remain part of the understory Others will grow and replace older trees that fall.

Wildlife: A broad term that includes non-domesticated vertebrates, especially mammals, birds and fish.

Wood: The solid interior of a tree.

Wood Chemicals: Chemical that are found naturally in the various parts of a tree.

Xylem: The part of a tree that transports water and nutrients up from the roots to the leaves Older xylem cells become part of the heartwood. Also called sapwood.

1

Glimpse of Forestry

Chronological events in Indian Forestry:

Year	Chronological event
1790	Tipu sultan introduced *Eucalyptus* in Nandi hills, Mysore from Australia.
1800	Commission on Forestry appointed to enquire into the availability of "Teak" in Malabar region. Based on their report felling of Teak below 21 inches is prohibited.
1806	Capton Watson appointed as "First Conservator of Forest".
1828	Wasteland is declared as Government properties.
1842	Establishment of "Nilambur Teak Plantation" by "Conolly" (Collector of Madras).
1843	Afforestation work started and Champell introduces *Eucalyptus pinnata* at Wellington, Madras in this year.
1855	Charter of Indian Forest – Memorandum submitted by MC Clelland to Lord Dalhousie.
1856	"Cleghom" appointed as first regular conservator of Forest and "Dietrich Brandis" appointed as Superintendent of Pegu forest by Lord Dalhousie in this year.
1858	Introduction of *Acacia* and *Eucalyptus* in Nilgiris, Madras.
1864	Introduction of Scientific Forest Management as well as appointment of first IG of Forest in India (Dietrich Brandis).
1865	Indian Forest Act passed.
1866	Establishment of Changamanga fuel wood plantation.
1867	IFS (Indian Forest Services) Promulgated.
1869	Re-organization of Forest Services and created post of CF, DCF, and ACF.
1870	Introduction of Wattles at Ooty and Kodaikanal.
1871	First Working Plan set up at Nilambur and Cattle Trespass Act passed.
1875	"The Indian Forest" Journal issued by Dr. Schlich and Baden Powell.
1878	First Forest school started at Dehradun and "Gambel" also started research work in NWFP (Non-Wood Forest Products).
1879	Elephant Preservation Act passed.
1882	Madras Forest Act passed.
1883	Establishment of "Bombay Natural History Society" during this year and also appointment of "Schlich" as second Inspector General of Forest.
1884	Forest School renamed as "Imperial Forest School" and *Eucalyptus grandis* was Introduced in Kerela.

Year	Chronological event
1885	Training of Forest Officers of India started at Cooper's Hill (England).
1890	A study on problems of Indian agriculture started by Dr. Voelkar.
1894	First Indian Forest Policy passed.
1895	Completion of First Working Plan at Nilambur.
1898	"Vedanthangal" declared as First Wildlife Sanctuary in India.
1901	Establishment of "Forest Grass Museum" at Madras.
1905	World Forestry Congress was formed and Systematic research started on "Paper and Pulp" in India.
1906	Imperial Forest School rename as "Imperial Forest Research" (Currently known as "Forest Research Institute, FRI) and "Ruther" was appointed as First Chief Conservator.
1909	Royal commission on Decentralization set up.
1912	"Wild Animal and Preservation Act" passed.
1914	"Destructive Insect-Pest Act" passed.
1919	Appointment of Forest Researchers.
1927	"Indian Forest Act" passed.
1928	Establishment of "Royal Commission on Agriculture" in India.
1929	"Ornithological Survey" conducted in India.
1932	Plantation of "Teak stump" started at Madras.
1934	Establishment of "New Forest Botanical Garden".
1936	First National Park of India named as "Hailey".
1938	Madras Forest College Closed.
1939	Research on "Forest Soil" started.
1945	Madras Forest College re-opened.
1947	"Import and Export Control Act" passed.
1948	Forest Colleges handed to Government of India and Central Board of Forestry was setup during this year.
1950	Van-Mahotsav (Tree planting festival) started by "K.M. Munshi".
1952	Second Forest Policy passed and establishment of "Indian Board for Wildlife".
1954	"Grassland Survey" conducted in India.
1955	Hill Area (Preservation) Act passed.
1957	Hailey National Park renamed as "Jim Corbett National Park".
1958	Establishment of "Wildlife Preservation Society of India".
1959	First Logging center established at "Batote" and Tree Improvement work started at FRI, Dehradun.
1960	Prevention of Cruelty to Animals Act passed.
1961	Forest seed testing started at FRI, Dehradun.

Year	Chronological event
1963	"Aerial Survey of Kulu Forest" conducted by G.A. Jones.
1965	"Forest type" Map published by Pre-Investment Survey of Forest Resources.
1966	Seed Act passed.
1968	Forest Types of India classified by Champion & Seth (Based on ecosystem approach) and the termed "Social Forestry" was coined by "Westoby".
1970	Gujarat started the first separate "Social Forestry Wing" and "Project Hangul" started in J&K.
1972	"Wildlife Protection Act" passed and also "Project Lion" started at Gir sanctuary.
1973	"Project Tiger" started, "Chipko movement" started by Sundar Lal Bahuguna at Uttarakhand and UNESCO started MAB (Man & Biosphere) programme in this year.
1975	"Crocodile Breeding & Management Project" launched.
1976	"Forest and Wildlife" included in concurrent list in 7th schedule under 42nd Amendment. India became the member of CITES in this year. NCA also published its report.
1978	Established of "Central Crocodile Breeding Institute" at Hyderabad.
1979	Social Forestry Project started.
1980	Forest (Conservation) Act passed.
1981	Air (prevention & control) Act passed.
1982	IIFM (Indian Institute of Forest Management) established at Bhopal and also WII (Wildlife Institute of India) started at Dehradun.
1983	First National Wildlife Action Plan adopted.
1984	MOEF (Ministry of Environment & Forest) created. Now known as MOEF&CC (Ministry of Environment, Forest & Climate Change).
1985	B.Sc. (Forestry) course started in State Agricultural Universities and NWDB (National Wasteland Development Board) was constituted.
1986	Environment (Protection) Act passed.
1987	FSI (Forest Survey of India) published its first report.
1988	Third National Forest Policy passed.
1989	Establishment of "Biotechnological Centre for Tree Improvement" at Tirupati.
1991	Establishment of "Centre for Social Forestry and Eco rehabilitation" Ibex and Musk deer project started.
1992	Project Elephant started.
1993	The Great Indian Bustard Project started.
1995	Siberian Crane Project started at Bharatpur.
1996	Seed Act passed.
1998	National Zoo Policy adopted.
1999	Creation of National Policy and a macro level Action Strategy on Biodiversity by Ministry of Environment and Forest.

Year	Chronological event
2000	Kanchenjunga (Sikkim) declared as the 12th Biosphere Reserve in the country.
2001	S.K. Pandey became the Director General of Forest.
2002	PM of India releases the National Wildlife Action Plan (2002-2016) and also Biological Diversity Act was passed.
2003	Establishment of National Biodiversity Authority at Chennai.
2004	CAMPA (Compensatory Afforestation Management Planning Authority) notified by MOEF (Now MOEF&CC).
2006	National Bamboo Mission started and also FRA (Forest Right Act) was passed in this year.
2007	Government of India constituted "Wildlife Crime Control Bureau".
2008	Implementation of Forest Right Act (FRA).
2010	National Tribunal Act passed and Wetland (Conservation and Management) rules constituted.
2011	Green India Mission launched.
2012	Land Acquisition Bill passed.
2013	Criminal Law (Amendment) Act passed.
2014	3rd meeting on "World Congress in Agroforestry" held at New Delhi and National Agroforestry Policy was passed in this year.
2015	World Sustainable Development Summit (WSDS) at New Delhi.
2016	India's first repository on Tiger opened at WII.
2017	Commonwealth Forestry Conference held at Dehradun.
2018	ICFRE Sign two pacts to spread awareness on forests and environment.
2019	International conference on Forestry Food and Sustainable Agriculture ICFFSA at Pune, India.

First in Indian Forestry

Events	Year	Name/Place/Act/Policy
First Conservator of Forest	1806	Watson
First Zoo in India	1854	Calcutta Zoo, West Bengal
First Regular Conservator of Forest	1856	Cleghorn
First Inspector General of Forest	1864	Dietrich Brandis
First National Forest Act	1865	Indian Forest Act
First Special Assistant Conservator of Forest	1867	Ribbon Troup and W. Schlich
First Director of the Forest School, Dehradun	1878	F. Bailey
First Forest Policy	1894	Indian Forest Policy

Events	Year	Name/Place/Act/Policy
First Working Plan of India	1895	Nilambur district
First Wildlife Sanctuary	1898	Vedanthangal, Madras
First Forest Entomologist	1901	Lefroy
First Chief Conservator of Forest	1906	Ruther
First President of FRI and College, Dehradun	1906	S. Eward Wilmont
First Silviculturalist of India	1909	Troup
First National Park	1936	Hailey National Park (Now "Jim Corbett National Park")
First Indian, Inspector General of Forest	1949	M.D. Chaturvedi
First Forest Movement	1973	Chipko movement
First Biosphere Reserve in India	1986	Nilgiri Biosphere Reserve

ICFRE Institutes

Name of Institutes	Established	City
Arid Forest Research Institute (AFRI)	1988	Jodhpur
Forest Research Institute (FRI)	1906	Dehradun
Himalayan Forest Research Institute (HFRI)	1977	Shimla
Institute of Forest Biodiversity (IFB)	2012	Hyderabad
Institute of Forest Genetics and Tree Breeding (IFGTB)	1988	Coimbatore
Institute of Forest Productivity (IFP)	1993	Ranchi
Institute of Wood Science and Technology (IWST)	1938	Bangalore
Rain Forest Research Institute (RFRI)	1988	Jorhat
Tropical Forest Research Institute (TFRI)	1988	Jabalpur

ICFRE Advance Research Centres

Name	Established	City
Advanced Research Centre for Bamboo and Rattan (a unit of RFRI)	2004	Aizawl
Centre for Forestry Research and Human Resource Development (satellite centre of TFRI)	1995	Chhindwara
Centre for Social Forestry and Eco-Rehabilitation (a centre of ICFRE)	1992	Allahabad
Centre for Forest Based Livelihood and Extension	2013	Agartala

Forest Institutes and Authorities in India

S.No.	Institute/Authority	Acronym	Locations	State
1.	Arid Forest Research Institute	AFRI	Jodhpur	Rajasthan
2.	Biotechnological Centre for Tree Breeding	BIOTRIM	Tirupati	Andhra Pradesh
3.	Bombay Natural History Society	BNHS	Mumbai	Maharashtra
4.	Central Arid Zone Research Institute	CAZRI	Jodhpur	Rajasthan
5.	Central Zoo Authority	CZA	Delhi	Delhi
6.	Directorate of Forest Education	DFE	Dehradun	Uttarakhand
7.	Forest Research Centre	FRC	Hyderabad	Andhra Pradesh
8.	Forest Research Institute	FRI	Dehradun	Uttarakhand
9.	Forest Survey of India	FSI	Dehradun	Uttarakhand
10.	Himalayan Forest Research Institute	HFRI	Shimla	Himachal Pradesh
11.	Indra Gandhi National Forest Academy	IGNFA	Dehradun	Uttarakhand
12.	Indian Board for Wildlife	IBWL	New Delhi	Delhi
13.	Indian Council of Forest Research and Education	ICFRE	Dehradun	Uttarakhand
14.	Institute of Forest Genetics and Tree Breeding	IFGTB	Coimbatore	Tamil Nadu
15.	Indian Institute of Forest Management	IIFM	Bhopal	Madhya Pradesh
16.	National Research Centre for Agroforestry	NRCAF	Jhansi	Uttar Pradesh
17.	Zoological Survey of India	ZSI	Kolkata	West Bengal
18.	Tropical Forest Research Institute	TFRI	Jabalpur	Madhya Pradesh
19.	Advance Center of Social Forestry and Rehabilitation	ACSFER	Allahabad	Uttar Pradesh
20.	Biotechnological Center for Tree Improvement		Tirupati	Andhra Pradesh
21.	Central Pollution Control Board	CPCB	New Delhi	New Delhi
22.	Indian Institute of Biodiversity	IIB	Itanagar	Arunachal Pradesh
23.	Wildlife Institute of India	WII	Dehradun	Uttrakhand

Important Forest Trees

S.No.	Common name	Scientific name	Family
1.	Silver fir	*Abies pindrow*	Pinaceae
2.	Australian wattle	*Acacia auriculiformis*	Mimosaceae
3.	Khair	*Acacia catechu*	Mimosaceae
4.	Babool	*Acacia nilotica*	Mimosaceae
5.	Himalayan maple	*Acer oblongum*	Aceraceae
6.	Pink cedar	*Acrocarpus fraxinifolius*	Leguminaceae
7.	Haldu	*Adina cordifolia*	Rubiaceae
8.	Bel, God fruit	*Aegle marmelos*	Rutaceae
9.	Mahaneem, Tree of heaven	*Ailanthus excelsa*	Simaroubaceae
10.	Womens tounge tree	*Albizzia lebbeck*	Mimosaceae
11.	Alder	*Alnus nepalensis*	Betulaceae
12.	Devil's tree	*Alstonia scholaris*	Legumniosae.
13.	Kaju	*Anacardium occidentalis*	Anacardiaceae
14.	Custard apple	*Annona squamosa*	Annonaceae
15.	Safed donkra, Axelwood	*Anogeissus latifolia*	Combretaceae
16.	Dhonkra	*Anogeissus pendula*	Combretaceae.
17.	Kadamba	*Anthocephalus cadamba*	Rubiaceae.
18.	Jackfruit	*Artocarpus heterophyllus*	Moraceae
19.	Neem	*Azadirachta indica*	Meliaceae
20.	Bamboo, Green gold	*Bambusa bambos*	Poaceae
21.	Kachnar	*Bauhinia variegata*	Caesalpinaceae
22.	Semul, Kapok	*Bombax ceiba*	Bombacaceae
23.	Dhak, Flame of Forest	*Butea monosperma*	Legumniosae
24.	Bottle brush	*Callistemon lanceolatus*	Myrtaceae
25.	Kadaya, Fishtail palm	*Caryota urens*	Arecaceae
26.	Amaltas, Indian laburnum	*Cassia fistula*	Caesalpinaceae
27.	Cassod/Beet tree	*Cassia siamea*	Caesalpinaceae
28.	Casuarina, Beef wood	*Casuarina equisetifolia*	Casuarinaceae
29.	Deodar	*Cedrus deodara*	Pinaceae
30.	Indian kapok	*Ceiba pentandra*	Bombaceae
31.	Khirak, European nettle	*Celtis australis*	Cannabaceae
32.	Silk cotton tree	*Cochlospermum religiosum*	Bixaceae
33.	Coconut tree	*Cocus nucifera*	Arecaceae
34.	Cupress	*Cupresses torulosa*	Cupreseaceae

S.No.	Common name	Scientific name	Family
35.	Sago palm	*Cycus revoluta*	Cycadaceae
36.	Rosewood	*Dalbergia latifolia*	Paplionaceae
37.	Shisham	*Dalbergia sissoo*	Paplionaceae
38.	Red gulmohar	*Delonix regia*	Caesalpinaceae
39.	Solid bamboo	*Dendrocalamus strictus*	Poaceae
40.	Tendu, Bedi leaf	*Diospyros melanoxylon*	Ebenaceae
41.	Gurjan	*Dipterocarpus turbinatus*	Dipterocarpaceae
42.	Dodonea	*Dodonea viscosa*	Sapindaceae
43.	Safeda, Eucalyptus	*Eucalyptus globulus*	Myrtaceae
44.	Mysoore gum	*Eucalyptus tereticornis*	Myrtaceae
45.	Bargad	*Ficus bengalensis*	Moraceae
46.	Anjir	*Ficus carica*	Moraceae
47.	Indian rubber plant	*Ficus elastica*	Moraceae
48.	Gular	*Ficus glomerata*	Moraceae
49.	Pepal	*Ficus religiosa*	Moraceae
50.	Butter tree	*Garcinia indica*	Clusiaceae
51.	Wintergreen	*Gaultheria fragrantissima*	Ericaceae
52.	Ghamar	*Gmelina arborea*	Lamiaceae
53.	Silver oak	*Grevillea robusta*	Proteaceae
54.	Bhimal	*Grewia optiva*	Tiliaceae
55.	Dhaman tree	*Grewia tiliifolia*	Malvaceae
56.	Indian elm	*Holoptelia integrifolia*	Ulmaceae
57.	Malawar wood	*Hopea parviflora*	Dipterocarpaceae
58.	Blue gulmohar	*Jacaranda mimosifolia*	Bignoniaceae
59.	Ratanjot, Biodiesel plant	*Jatropha curcus*	Euphorbiaceae
60.	Walnut	*Juglan regia*	Juglandaceae
61.	Indian ash tree	*Lannea coromandelica*	Anacardiaceae
62.	Subabool, Ipil-Ipil	*Leucaena leucocephala*	Mimosaceae
63.	Mahua	*Madhuca indica*	Sapotaceae
64.	Mahua	*Madhuca longifolia*	Sapotaceae
65.	Drek	*Melia azedarach*	Meliaceae
66.	Champa	*Michelia champaca*	Magnoliaceae
67.	Drumstick	*Moringa oleifera*	Moringaceae
68.	Mulberry	*Morus alba*	Moraceae
69.	Indian trumpet	*Oroxylum indicum*	Bignoniaceae
70.	Yellow gulmohar	*Peltophorum ferruginium*	Caesalpinaceae

S.No.	Common name	Scientific name	Family
71.	Amla	*Phyllanthus emblica*	Euphorbiaceae
72.	Spruce	*Picea smithiana*	Pinaceae
73.	Chilgoza/Neoza pine	*Pinus gerardiana*	Pinaceae
74.	Chir pine	*Pinus roxburghii*	Pinaceae
75.	Blue pine	*Pinus wallichiana*	Pinaceae
76.	False ashok	*Polyalthia longifolia*	Annonaceae
77.	Karanj	*Pongamia pinnata*	Fabaceae
78.	Himalayan popular	*Populus ciliata*	Salicaceae
79.	Popular	*Populus deltoides*	Salicaceae
80.	Black popular	*Populus nigra*	Salicaceae
81.	Khejri	*Prosopis cineraria*	Mimosaceae
82.	Bengal kino, Bijasal	*Pterocarpus marsupium*	Dipterocarpaceae
83.	Red oak	*Quercus rubra*	Fagaceae
84.	Black locast, Kashmir kikar	*Robinia pseudocasia*	Fabaceae
85.	Salix	*Salix alba*	Salicaceae
86.	Chandan	*Santalum album*	Santalaceae
87.	Ashok	*Saraca asoca*	Fabaceae.
88.	Sequoia	*Sequoia sempervirens*	Pinaceae
89.	Sterculia	*Sterculia urens*	Sterculiaceae
90.	Jamun	*Syzygium cumini*	Myrtaceae
91.	Imli	*Tamarindus indica*	Caesalpinaceae
92.	Teak	*Tectona grandis*	Lamiaceae
93.	Arjuna	*Terminalia arjuna*	Combretaceae
94.	Bahera	*Terminalia bellirica*	Combretaceae
95.	Harar	*Terminalia chebula*	Combretaceae
96.	Toon, Red cedar	*Toona ciliata*	Meliaceae
97.	Elm	*Ulmus* species	Ulmaceae
98.	Ber	*Ziziphus mauritiana*	Rhamnaceae

Indian Forest Act, 1927

Chapters	Particulars
Chapter 1	Preliminary
Chapter 2	Reserve Forest
Chapter 3	Village Forest
Chapter 4	Protected Forest
Chapter 5	Control over forest and lands not being the property of government

Chapters	Particulars
Chapter 6	Duty on timber and other forest produce
Chapter 7	The control of timber and other forest produce in transit
Chapter 8	Collection of drift and stranded timber
Chapter 9	Penalties and procedure
Chapter 10	Cattle-trespass
Chapter 11	Forest officers
Chapter 12	Subsidiary rules
Chapter 13	Miscellaneous

State Animals, Birds, Trees and Flowers of India

State/UT	Animal	Bird	Tree	Flower
Andhra Pradesh	Blackbuck	Rose ringed Parakeet	Neem	Jasmine
Arunachal Pradesh	Mithun	Great hornbill	Hollong	Lady slipper orchid
Assam	One-horned rhinoceros	White-winged wood duck	Hollong	Foxtail orchids
Bihar	Ox	House sparrow	Peepal	Kachnar
Chhattisgarh	Wild buffalo	Hill myna	Sal	*
Delhi	Nilgai	House sparrow	*	*
Goa	Gaur	Flame throated bulbul	Matti	*
Gujarat	Asiatic lion	Greater flamingo	Mango	Marigold
Haryana	Blackbuck	Black francolin	Peepal	Lotus
Himachal Pradesh	Snow leopard	Western tragopan	Deodar	Rhododendron
Jammu & Kashmir	Hangul	Black necked crane	Chinar	Lotus
Jharkhand	Elephant	Koel	Sal	Palash
Karnataka	Elephant	Indian roller	Sandal	Lotus
Kerala	Elephant	Great hornbill	Coconut	Golden shower tree
Madhya Pradesh	Swamp deer	Asian paradise flycatcher	Banyan	Parrot tree
Maharashtra	Giant squirrel	Yellow-footed green pigeon	Mango	Jarul
Manipur	Sangai	Mrs. Hume's pheasant	Toon	Shirui lily

State/UT	Animal	Bird	Tree	Flower
Meghalaya	Clouded leopard	Hill myna	White teak	Lady slipper orchid
Mizoram	Serow	Mrs. Hume's pheasant	Iron wood	Red vanda
Nagaland	Mithun	Blyth's tragopan	Alder	Rhododendron
Odisha	Sambar	Indian roller	Banyan	Lotus
Puducherry	Squirrel	Asian koel	Vilva tree	Cannonball
Punjab	Blackbuck	Northern goshawk	Sheesham	*
Rajasthan	Camel	Great Indian bustard	Khejri	Rohira
Sikkim	Red panda	Blood pheasant	Rhododendron	Noble orchid
Tamil Nadu	Nilgiri tahr	Emerald dove	Palmyra palm	Kandhal
Telangana	Deer	Indian roller	Jammi	Tangedu
Tripura	Phayre's langur	Green imperial pigeon	Agar	Nageshwar
Uttarakhand	Musk deer	Himalayan monal	Burans	Brahm kamal
Uttar Pradesh	Swamp deer	Sarus crane	Ashok	Brahm kamal
West Bengal	Fishing cat	White-breasted kingfisher	Chatian	Shephali

Endangered Species

Common name	Scientific name
Snow Leopard	*Panthera uncia*
Great Indian bustard	*Ardeotis nigriceps*
Dolphin	Delphinus delphis
Hangul	*Cervus elaphus hanglu*
Nilgiri tahr	*Nilgiritragus hylocrius*
Dugongs	*Dugong dugon*
Asian wild buffalo	*Bubalus arnee*
Brow antlered deer	*Rucervus eldii*
Vulture	*Gyps indicus*
Malabar civet	*Viverra civettina*
Great one- horned rhinoceros	*Rhinoceros unicornis*
Asiatic lion	*Panthera leo persica*
Swamp deer	*Rucervus duvaucelii*

2

Agroforestry

1. Agroforestry started in Burma
(a) 1890 (c) 1850
(b) 1856 (d) 1886

2. "Agroforestry" is also known as
(a) Ancient land use practice
(b) Integrating practice of farming
(c) Diverse practice of farming
(d) All of these

3. Agroforestry helps in
(a) Improving socio economic status
(b) Proper land use management
(c) Increasing productivity
(d) All of the above

4. Alley cropping is also known as
(a) Intercropping
(b) Hedge row intercropping
(c) Mixed cropping
(d) None of the above

5. The ideal spacing between the tree rows of Alley cropping is
(a) 4-8 m (c) 5-10 m
(b) 1-3 m (d) 10-15 m

6. Crop grown between *lantana camara* rows is
(a) Alley cropping
(b) Hedge row intercropping
(c) Agroforestry
(d) Both (a) and (b)

7. Which of the following is an example of Agrisilviculture?
(a) Ground nut and *sesbania*
(b) Ground nut and sal
(c) Pulses and *Gmelina arborea*
(d) All of the above

8. The" WAC" is the new name given to
(a) AICRP (c) CRIDA
(b) ICRAF (d) CAFRI

9. ICRAF was founded in
(a) 1890 (c) 1978
(b) 1987 (d) 1956

10. Shifting cultivation is also known as
(a) Slash and burn (c) Dahiya
(b) Podu (d) All of these

11. Optimum size of area in shifting cultivation for 3–5 family members
(a) 1–2.5 ha (c) 5–6 ha
(b) 0.2–0.4 ha (d) 0.5–1 ha

12. Jhumming in north east referred as
(a) Taungya
(b) Shifting cultivation
(c) Tree farming
(d) None of these

13. Taungya is derived from
(a) Latin
(b) Greek
(c) Burmese
(d) Germany

14. Taungya was first started in
(a) North Bengal (c) Java
(b) Myanmar (d) India

15. In India, Taungya was introduced by
(a) Brandis (c) Troupe
(b) Gayer (d) Lefroy

16. In British period, Taungya was first done under which tree crop
(a) Sal
(b) Teak
(c) Gmelina
(d) All of the above

17. In India first plantation under taungya was done in
(a) North Bengal (c) UP
(b) West Bengal (d) Assam

18. Area allotted in village taungya
(a) 1–2 ha
(b) 5–6 ha
(c) 0.2–0.4 ha
(d) 0.5–1 ha

19. Raised nursery beds are preferred in
(a) High rainfall areas
(b) Low rainfall areas
(c) Adequate rainfall areas
(d) Dry areas

20. ____ beds are preferred for dry areas
(a) Flat
(b) Raised
(c) Sunken
(d) None of the above

21. Nursery beds should be oriented in
(a) N-S (c) W-E
(b) S-N (d) E-W

22. In hedge row inter cropping the trees should be oriented in
(a) E-W (c) W-E
(b) N-S (d) S-W

23. Photo insensitive plants are known as
(a) Long day plants
(b) Short day plants
(c) Day neutral plants
(d) None of the above

24. Agro-ecological zones in India are
(a) 10 (b) 9 (c) 12 (d) 15

25. Which of the following tree species are used for charcoal making?
(a) *Pinus roxburghii*
(b) *Quercus leucotrichophora*
(c) *Jatropa curcus*
(d) All of these

26. IDRC was commissioned by
(a) John Bene
(b) Nair
(c) Brandis
(d) Beneetal

27. Agro-climatic regions in India are
(a) 12 (b) 8 (c) 10 (d) 15

28. In 1984, Nair classified the agro-forestry systems into
(a) 2 parts (c) 3 parts
(b) 4 parts (d) 5 parts

29. The most acceptable classification of agroforestry systems based on
(a) Socio-economic classification
(b) Functional classification
(c) Ecological classification
(d) Structural classification

30. Silvicultural practice of Protein bank comes under
(a) Agrisilviculture
(b) Agrisilvipasture
(c) Silvopasture
(d) Silvipastoral

31. The key feature of D & D are
(a) Flexibility
(b) Speed
(c) Repetition
(d) All of the above

32. The major attributes of Agroforestry is
(a) Productivity
(b) Flexibility
(c) Speed
(d) Repetition

33. Lopping is generally done in
(a) Agrisilviculture
(b) Silvipastoral
(c) Home garden
(d) None of these

34. Operations in agrisilviculture
(a) Thinning
(b) Lopping
(c) Pollarding
(d) Cutting

35. Home garden is also known as
(a) Home stead plantations
(b) Pekar angan
(c) Multitier system
(d) All of the above

36. Home garden is dominating in
(a) Space
(b) Time
(c) Both (a) and (b)
(d) None of the above

37. Based on spatial arrangement, the agro-forestry system is classified into how many categories
(a) 2 (b) 3 (c) 4 (d) 6

38. Based on temporal arrangement the agroforestry system is classified into how many categories
(a) 3 (b) 5 (c) 4 (d) 6

39. Which of the following is not the spacial arrangement of agroforestry system?
(a) Taungya
(b) Closely planted trees in strips
(d) Home-garden
(d) Scattered tree in pasture land

40. Space and time dominating is referred to
(a) Improved fallow species in shifting cultivation
(b) Taungya
(c) Home garden
(d) Hedge row intercropping

41. Tea crop under shade tree is an example of
(a) Concomitant (c) Interpolated
(b) Intermittent (d) Coincident

42. Arecanut with betelvine is an example of
(a) Interpolated (c) Intermittent
(b) Overlapping (d) Separate

43. Which of the following is time-dominating
(a) Homegarden (c) Overlapping
(b) Separate (d) Coincident

44. Cocunut with annual crops is an example of
(a) Interpolated (c) Intermitant
(b) Coincident (d) Concomitant

45. Which of the following is space dominating?
(a) Home-garden
(b) Coconut with annual crops
(c) Alley cropping
(d) Both (b) and (c)

46. Mixed farming is prevalent in
(a) Urban areas
(b) Villages

(c) Rain fed areas
(d) Both (b) and (c)

47. Which is the richest food grain production tract in India?
(a) Indo-gangetic plains
(b) Brahmaputra plains
(c) Central plateau
(d) Peninsular India

48. Hills generally have
(a) Low organic matter, low pH
(b) High organic matter, low pH
(c) Low organic matter, high pH
(d) High organic matter, high pH

49. Most common type of erosion found in hilly areas
(a) Rill (c) Sheet
(b) Gullied (d) Ravine

50. Aqua-silviculture is also known as
(a) Mesomorphic (c) Hydramorphic
(b) Xeromorphic (d) Hydromorphic

51. The father of modern agroforestry is
(a) PK Nair (c) Brandis
(b) Lefroy (d) Gayer

52. In shifting cultivation, the area is clear felled during
(a) November-December
(b) June -July
(c) March -April
(d) January-February

53. The host tree species for lac insect is/are
(a) *Ziziphus mauritiana*
(b) *Acacia catechu*
(d) *Butea monosperma*
(d) All of the above

54. The scientific name of silkworm is
(a) *Bombax mori* (c) *Bambaux morii*
(b) *Laccifer lacca* (d) *Apis mellifera*

55. The strains for the lac insect is/are
(a) Kusumi (c) Both (a) and (b)
(b) Rangeeni (d) None of these

56. In Sunderban deltas, commonly practised agro-forestry system is
(a) Silvipisciculture
(b) Agrisilviculture
(c) Aquasilviculture
(d) None of the above

57. Shelterbelt is also known as
(a) Protection belt
(b) Multi row windbreak
(c) Wind breakage
(d) Both (a) and (b)

58. Windbreak is also known as
(a) Narrow shelterbelt
(b) Wind belt
(c) Wind breakage
(d) All of the above

59. Windbreak and shelterbelt provide protection against
(a) Snow
(b) Sun
(c) Wind
(d) Both (b) and (c)

60. Which one is of long duration?
(a) Squalls
(b) Dust
(c) Cyclonic storm
(d) All of the above

61. Shifting sand dunes are prevalent in
(a) Arid zone
(b) Hills
(c) Plains
(d) All of the above

62. ……….. protected area lies on wind-ward side of the shelterbelt
(a) 1/4 (b) 3/4 (c) 2/3 (d) 1/3

63. Windbreak velocity depends upon
(a) Density
(b) Crown spread
(c) Height
(a) All of the above

64. Maximum area is protected by windbreak and shelterbelt when direction of wind flow is
(a) 180° (c) 90°
(b) 80° (d) 150°

65. In shelterbelt, the ratio of height to width should be
(a) 1:10 (c) 1:20
(b) 1:2 (d) 1:30

66. The term "Social forestry" was introduced by
(a) Troup (c) Brandis
(b) Tansley (d) Westoby

67. The tree crop interactions is/are
(a) Positive
(b) Negative
(c) Neutral
(d) Both (a) and (b)

68. The D & D approach is developed by
(a) CIFOR (c) IUFRO
(b) ICRAF (d) ICFRE

69. The key features of D & D is/are
(a) Speed
(b) Repetition
(c) Flexibility
(d) All of the above

70. The suitable criteria for good agroforestry design is/are
(a) Productivity
(b) Adoptability
(c) Sustainability
(d) All of the above

71. A forest composed entirely of one species usually to the extent not less than 80% called as
(a) Pure forest
(b) Normal forest
(d) Mixed forest
(d) Evenaged forest

72. The man first appeared in
(a) Palaeolithic age
(b) Neolithic age
(c) Eolithic age
(d) None of the above

73. The alley cropping system was developed by
(a) lCRAF (c) IGFRI
(b) IITA (d) CAZRI

74. The headquarter of IITA is situated in
(a) Nigeria (c) Tokyo
(b) Nairobi (d) Brazil

75. The rate of infiltration of soil water is
(a) High in forest soils
(b) High in agriculture lands
(c) Both (a) and (b)
(d) None of the above

76. The first taungya cultivation in North Bengal was raised for
(a) Teak (c) Pongamia
(b) Sal (d) Grewia

77. In Karnataka, taungya cultivation is practiced for raising which crops
(a) Teak and Populus
(b) Teak and Sandalwood
(c) Teak and Bombax
(d) None of the above

78. In taungya system, main emphasis is/are given to
(a) Growth and yield of crop

(b) Regeneration of forest
(c) Both (a) and (b)
(d) None of the above

79. The important varieties of *Morus alba*
(a) L-146
(b) K2
(c) RSS-135
(d) All of the above

80. In temperate areas, tassar silk worm feeds on
(a) *Morus alba*
(b) *Syzygium cumini*
(c) *Quercus leucotrichophora*
(d) *Shorea tolura*

81. Which of the following are the main host for tassar silk worm?
(a) *Terminalia alata*
(b) *Terminalia arjuna*
(c) *Morus alba*
(d) All of the above

82. Which of the following are main host for lac insect?
(a) *Butea monosperma*, *Schleichera oleosa*, *Acacia catechu*
(b) *Ziziphus* species, *Butea monosperma*, *Shorea tolura*
(d) *Butea monosperma*, *Schleichera oleosa*, *Ziziphus* species
(d) All of the above

83. Land use efficiency of any farming practise can be calculated using
(a) L = C/2F
(b) L = (C + F)/F
(c) L = (C + F)/C
(d) L = (2F + C)/F

84. Species widely used for tank shore afforestation
(a) *Acacia nilotica*
(b) *Eucalyptus* species
(c) *Casuarina equisetifolia*
(d) All of the above

85. The woodlots are primarily managed for
(a) Timber production
(b) Fuelwood production
(c) Wood production upto rotation age
(d) All of the above

86. The usual spacing for the woodlots is
(a) 3m × 3m (c) 5m × 5m
(b) 1m × 1m (d) 3m × 4m

87. The MPT found in arid region is/are
(a) *Prosopis cineraria*
(b) *Acacia nilotica*
(c) *Ziziphus nummularia*
(d) All of the above

88. In block planting usual spacing followed for *Populus deltoids* is/are
(d) 4m × 5m (c) 1m × 1m
(b) 5m × 5m (d) Both (a) and (b)

89. Tree species commonly found in Konkan area is/are
(a) *Mangifera indica*
(b) *Terminalia arjuna*
(c) *Bambusa arundinaria*
(d) All of the above

90. Home gardens are prevalent in
(a) Low rainfall and humidity areas
(b) High rainfall and humidity areas
(c) Low rainfall and high humidity areas
(d) High rainfall and low humidity areas

91. System mostly prevalent in Kerala and Karnataka
(a) Agrisilvipastoral

(b) Silvipastoral
(c) Shifting cultivation
(d) Agrisilviculture

92. In Home gardens usually crop intensity is/are
(a) Low
(b) Medium
(c) High
(d) Both (a) and (b)

93. Which of the following is a good fodder tree for arid regions?
(a) *Acacia catechu*
(b) *Prosopis cineraria*
(c) *Ziziphus nummularia*
(d) All of the above

94. The Kancha system of forest grazing prevalent in
(a) Himachal Pradesh
(b) Kerala and Tamil Nadu
(c) Gujarat
(d) Andhra Pradesh and Karnataka

95. The usual lopping season comes during which month
(a) November-December
(b) April-May
(c) June-September
(d) September-November

96. Species suitable for planting central row in wind break
(a) *Acacia nilotica*
(b) *Azadirachta indica*
(c) *Eucalyptus* species
(d) All of the above

97. Species suitable for flank row in wind break is/are
(a) *Acacia leucophloea*
(b) *Casuarina equisetifolia*
(c) *Prosopis chilensis*
(d) All of the above

98. Which of the following crop is/are deep rooted?
(a) *Cajanus cajan*
(b) Sugarcane
(c) Cotton
(d) All of the above

99. Which of the following plants grow well under shade?
(a) Tea
(b) Cassava
(c) Cotton
(d) All of the above

100. Which of the following has low evapotranspiration rate?
(a) *Eucalyptus cameldulensis*
(b) *Eucalyptus globulus*
(c) *Eucalyptus tereticornis*
(d) None of the above

101. The water consumption per plants is highest in
(a) *Dalbergia sissoo*
(b) *Pongamia pinnata*
(c) *Eucalyptus* species
(d) *Acacia auriculiformis*

102. Non-leguminous nitrogen fixing tree in Himalayas is/are
(a) *Casuarina equisetifolia*
(b) *Alnus* species
(c) Both (a) and (b)
(d) None of the above

103. Sticky hairs on fruit is/are
(a) *Plumbago* species
(b) *Tribulus* species
(c) *Chrysopogon* species
(d) All of the above

104. Hooked spines on fruit is
(a) *Achyranthus* species
(b) *Chrysopogon* species

(c) *Medicago* species
(d) All of the above

105. The water use efficiency is usually expressed in terms of
(a) Transpiration ratio
(b) Transpiration pull
(c) Both (a) and (b)
(d) None of the above

106. LCC necessarily put trees under
(a) IV and V
(b) VII and VIII
(c) V, VI, VII
(d) All of the above

107. Plants propagated through root suckers is/are
(a) *Dalbergia sissoo*
(b) *Diospyros melanoxylon*
(c) Both (a) and (b)
(d) None of the above

108. The cutting back of the stem in order to obtain a flush of shoots usually above the height of browsing animals is known as
(a) Pollarding
(b) Coppicing
(c) Both (a) and (b)
(d) None of the above

109. The pollarding system is usually employed for
(a) *Hardwickia binata*
(b) *Grewia optiva*
(c) *Morus alba*
(d) All of the above

110. Optimum height for pollarding is
(a) 1m (b) 3m (c) 2m (d) 5m

111. In *Adina cordifolia* and *Eucalyptus* species the sowing is done by
(a) Broadcasting
(b) Dibbling
(c) Drill sowing
(d) Both (a) and (c)

112. Transplanting or pricking out is commonly done for
(a) *Cedrus deodara*
(b) *Pinus* species
(c) *Eucalyptus* species
(d) All of the above

113. Which of the following are raised by stem and branch cutting?
(a) *Populus* species
(b) *Salix* species
(c) *Ficus* species
(d) All of the above

114. Bamboos are propagated through
(a) Rhizomes (c) Root suckers
(b) Stem cutting (d) Shoot cutting

115. Optimum age of deodar planting stock is
(a) 9 months (c) 4.5 years
(b) 2 years (d) 6 years

116. The casualty replacement of planting stock is known as
(a) Beating up
(b) Blanking
(c) In-filling
(d) All of the above

117. The growth in grasses is/are due to
(a) Apical meristem
(b) Intercalary meristem
(c) Lateral meristem
(d) All of the above

118. The most important tending operation in Agroforestry is
(a) Pruning (c) Cleaning
(b) Thinning (d) Felling

119. In shifting cultivation areas, the land man ratio is usually
(a) Narrow (c) Constant
(b) Wide (d) Both (a) and (b)

120. Multipurpose trees means
(a) Many products derived from same tree
(b) Many products derived from few trees of some species
(c) Many trees produce different products
(d) All of the above

121. *Casuarina equisetifolia* fixes atmospheric nitrogen in association with
(a) Frankia (c) Rhizobium
(c) Azolla (d) Nitrobacter

122. National Wasteland Development Board (NWDB) was established in
(a) 1995 (c) 1985
(b) 1970 (d) 1895

123. Toxicity in fodder is due to
(a) Glycoside (c) Saponin
(b) Tannin (d) None of these

124. Allelochemicals are released by
(a) Leaching
(b) Volatizations
(c) Exudations
(d) All of the above

125. Allelochemicals secreted by plants are mainly due to
(a) Primary metabolites
(b) Secondary metabolites
(c) Tertiary metabolites
(d) All of the above

126. NCA emphasis on agroforestry education during
(a) 4th plan (c) 6th plan
(b) 7th plan (d) 9th plan

127. The basic procedures of D & D consist of _____ stages.
(a) Four (c) Six
(b) Five (d) Seven

128. D & D approach can be applied at
(a) Macro level
(b) Micro level
(c) Meso level
(d) All of the above

129. All India Coordinated Research on Agroforestry was started by
(a) ICFRE (c) ICRAF
(b) ICAR (d) IARI

130. What does SALT stand for
(a) Sloping Agroforestry Land Technology
(b) Sloping Agriculture Technology
(c) Sloping Agriculture Land Technology
(d) Sloping Agriculture Land use Technology

131. The distinctive arrangement of components in space and time is
(a) Agroforestry practice
(b) Agroforestry system
(d) Both (a) and (b)
(d) None of the above

132. National Agroforestry Policy was formulated in
(a) 2016 (c) 2010
(b) 2015 (d) 2014

133. National Agro-forestry day is celebrated on
(a) 5th June
(b) 8th May
(c) 21th March
(d) 22th April

134. Alley cropping was developed by
(a) PK Nair (c) BT Kang
(b) Brandis (d) Troup

135. D & D is developed by
(a) JB Raintree
(b) KB Raintree
(c) BT Kang
(d) PK Nair

136. Shelterbelts should have ___ shape
(a) Square (c) Triangle
(b) Rectangular (d) Circular

137. The term social forestry was first used in India by
(a) Royal Commission on Agriculture
(b) National Commission on Agriculture
(c) Indian Council of Agriculture and Research
(c) None of the above

138. Social forestry was started in
(a) 1952 (c) 1956
(b) 1976 (d) 1950

139. First social forestry wing created in
(a) West Bengal (c) Rajasthan
(b) Gujarat (d) Kerala

140. Social forestry was started in
(a) Fifth five-year plan
(b) Third five-year plan
(c) Sixth five-year plan
(d) Fourth five-year plan

141. Important and best shade tree used for coffee plantations
(a) *Grevillea robusta*
(b) *Acrocarpus fraxinifolius*
(c) *Calliandra* species
(d) None of the above

142. The term social forestry was first used in India
(a) 1975 (c) 1976
(b) 1959 (d) 1978

143. The oldest known agroforestry system is
(a) Alley cropping
(b) Home garden
(c) Taungya
(d) Shifting cultivation

144. Mimosine is present in
(a) *Pongamia* (c) Neem
(b) *Leuceana* (d) *Robinia*

145. Shading effect of tree can be overcome by
(a) Canopy management
(b) Root management
(c) Water management
(d) None of the above

146. *Acacia tortilis* was introduced in India by
(a) CAFRI (c) CAZRI
(b) ICRISAT (d) ICRAF

147. The fourth world congress of Agroforestry 2019 held at
(a) Montpellier, France
(b) Bonn, Germany
(c) Montreal, Canada
(d) Cannum, Mexico

148. The land use factor (L) in shifting cultivation where permanent cultivation is reached (Zero fellow phase) through tree improvement fallow management will be
(a) = 2 (b) = 1 (c) > 10 (d) < 1

149. Most common tree used for tea and coffee plantations
(a) Grevillea (c) Sissoo
(c) Teak (d) Gamhar

150. Indian Journal of Agroforestry is published by
(a) Indian Society of Agroforestry
(b) Agroforestry Society of India
(c) Association of Agroforestry in India
(d) Indian Agroforestry Association

151. Coffee based Agroforestry system belongs to which silvicultural system
(a) Agrisilviculture system
(b) Silvipastoral system
(c) Agrisilvipastoral system
(d) Other systems

152. Juglone is an Allelochemicals released from
(a) Mango (c) Apple
(b) Walnut (d) Neem

153. *Ailanthus excelsa* is the major fodder tree species of which of the state
(a) Himachal Pradesh
(b) Assam
(c) Rajasthan
(d) Punjab

154. Agroforestry system followed in the Kilimanjaro hilly tact of China is
(a) Kihamba-Chagga
(b) Satoyama
(c) Parkland
(d) Pekarangan

155. Major agroforestry system found in Africa is
(a) Parkland
(b) Chagga
(c) Kihamba- Chagga
(d) Satoyama

156. The *Populus deltoides* based Agro-forestry system is found in
(a) J & K
(b) Punjab
(c) Himachal Pradesh
(d) Haryana

157. Agroforestry was recognized as a GHG mitigation strategy as per
(a) 4th assessment of IPCC, 2007
(b) 3rd assessment of IPCC, 2007
(c) Bali action plan
(d) Copenhagen summit 2009

158. The First National Seminar on agroforestry was held in
(a) 1947 (c) 1979
(b) 1950 (d) 1982

159. Multiple cropping is raising of two or more crops on the
(a) Same land
(b) Different land
(c) Mixed land
(d) Dry land

160. First project on Agroforestry in India was initiated in
(a) 1990 (c) 1972
(b) 1983 (d) 1980

161. Which of following species is not applicable in Agroforestry system for plain areas?
(a) *Acacia auriculiformis*
(b) *Mesua ferrea*
(c) *Mangifera indica*
(d) *Pinus* species

162. Poplar plantation raised under Agroforestry systems used as a source of raw materials for
(a) Match Industry
(b) Rubber Industry
(c) Board Industry
(d) All of the above

163. Planting trees as alley cropping under agroforestry is best suited for
(a) Desert area
(b) Sloping terrain
(c) Plain areas
(d) Rocky areas

164. In shallow soil, where terraces are not possible, the system practised is
(a) Agri-horti system
(b) Strip system
(c) Silvipasture
(d) Agriculture

165. Which fodder tree species is/are widely used in arid zone of the Central India
(a) *Prosopis juliflora*
(b) *Prosopis cineraria*
(c) *Ailanthus excelsa*
(d) All of the above

166. Practice of growing leguminous trees and collecting fodder from them is called
(a) Protein bank
(b) Pastoral system
(c) Live fence
(d) Agri-pastoral

167. *Casuarina equisetifolia* was introduced in Karawar coast during
(a) 1883 (c) 1845
(b) 1963 (d) 1663

168. *Robinia pseudocacia* is a valuable fodder tree species due to high content of
(a) Nitrogen (c) Crude protein
(b) Calcium (d) Both (b) and (c)

169. Which of the following species are suitable in coastal areas?
(a) *Pongamia glabra*
(b) *Eucalyptus* species
(c) *Acacia auriculiformis*
(d) *Azadirachta indica*

170. Which of the following grows best in black cotton soil?
(a) Sal (c) Teak
(b) Karanj (d) Oak

171. The commonly followed agroforestry system in Java is/are
(a) Kebun-Talun
(b) Parkland
(c) Perka rangam
(d) Both (a) and (c)

172. In Tree Patta Scheme, the area should not be more than
(a) 1 ha (c) 3 ha
(b) 7 ha (d) 10 ha

173. The major donor for the "Social forestry" programme is/are
(a) World bank
(b) CIDA
(c) SIDA
(d) All of the above

174. Present number of coordinating centres for "AICRP" on Agroforestry" is
(a) 35 (b) 37 (c) 50 (d) 27

175. The Alder based agroforestry system is commonly found in
(a) N India (c) N-E India
(b) S-W India (d) N-W India

176. Cultivation factor is calculated by
(a) L = F/(C + F) (c) L= (2F + C)/C
(b) L = C/(C + F) (d) L= F/(2C + F)

177. The term "Allelopathy" is derived from
(a) French word (c) Latin word
(b) Arabic word (d) Greek word

178. The native of *Grewia optiva* is
(a) Mexico (c) America
(c) India (d) China

179. The tree to tree interface can be studied by
(a) Parallel rows of tree and crops
(b) Concentric rows of crop
(c) Simple design
(d) All of the above

180. The ideal spacing of energy plantation for *Prosopis juliflora* is
(a) 0.5 × 1 m (c) 2 × 3 m
(b) 1.3 × 1.3 m (d) 5 × 2 m

181. Species used for the life fence
(a) *Grewia optiva*
(b) *Sesbania grandiflora*
(c) *Euphorbia* species
(d) All of the above

182. Which of the following is generally used in the multi-storeyed cropping?
(a) Areca-nut (c) Coconut
(b) Bamboo (d) Mango

183. Species not suitable for dry land agroforestry system?
(a) *Shorea robusta*
(b) *Ziziphus nummularia*
(c) *Terminalia arjuna*
(d) *Acacia tortilis*

184. Best species for fixing sand
(a) *Casuarina equisetifolia*
(b) *Crotolaria burhia*
(c) *Prosopis juliflora*
(d) *Acacia catechu*

185. Which of the following is the most suitable agroforestry system for wasteland?
(a) Agrisilviculture
(b) Hortisilviculture
(c) Agrihorticulture
(d) Silvipasture

186. If the land is managed by the owner itself for satisfying his basic needs than it comes under
(a) Subsistence agroforestry system
(b) Intermediate agroforestry system
(c) Commercial agroforestry system
(d) Social forestry

187. Which of the following is the best species for sand dune stabalization?
(a) *Acacia auriculiformis*
(b) *Gmelina arborea*
(c) *Acacia senegal*
(d) *Acacia tortolis*

188. Which of the following are the clones of poplar?
(a) G-48
(b) Kranti
(c) Bahar
(d) All of the above

189. Poplar is popularly grown in northern India because of
(a) Fast biomass growth
(b) Easy propagation
(c) No allelopathic effect
(d) All of the above

190. Which of the following tree is becoming an aggressive weed in several states of India?
(a) *Acacia tortolis*
(b) *Prosopis juliflora*
(c) *Eucalyptus teriticornis*
(d) None of the above

191. Poplar was introduced in India by
(a) WIMCO (c) CAZRI
(b) CAFRI (d) CRIDA

192. Which of the following is considered as the best fodder during the lean months of green fodder?
(a) *Grewia optiva*
(b) *Robinia pseudoacacia*
(c) *Leucaena leucocephala*
(d) *Morus alba*

193. The native place of *Leucaena leucocephala* is
(a) Indonesia (c) Brazil
(b) Mexico (d) Australia

194. Poplar in north India is extensively grown with
(a) Rice (c) Bajra
(b) Maize (d) Wheat

195. In India research work on agroforestry started during
(a) 1940 (c) 1950
(b) 1960 (d) 1970

196. The silkworm rears on the leaves of
(a) *Terminalia alata*
(b) *Morus alba*
(c) *Terminalia tomentosa*
(d) All of the above

Answer Keys

1. (b)	**2.** (d)	**3.** (d)	**4.** (b)	**5.** (a)	**6.** (d)	**7.** (a)	**8.** (b)	**9.** (c)	**10.** (d)
11. (a)	**12.** (b)	**13.** (c)	**14.** (b)	**15.** (a)	**16.** (b)	**17.** (a)	**18.** (c)	**19.** (a)	**20.** (c)
21. (d)	**22.** (a)	**23.** (c)	**24.** (b)	**25.** (a)	**26.** (a)	**27.** (d)	**28.** (b)	**29.** (d)	**30.** (d)
31. (d)	**32.** (a)	**33.** (b)	**34.** (c)	**35.** (d)	**36.** (c)	**37.** (b)	**38.** (d)	**39.** (a)	**40.** (c)
41. (d)	**42.** (b)	**43.** (b)	**44.** (c)	**45.** (d)	**46.** (d)	**47.** (a)	**48.** (b)	**49.** (c)	**50.** (d)
51. (a)	**52.** (a)	**53.** (c)	**54.** (a)	**55.** (c)	**56.** (a)	**57.** (d)	**58.** (d)	**59.** (d)	**60.** (c)
61. (a)	**62.** (a)	**63.** (d)	**64.** (c)	**65.** (a)	**66.** (d)	**67.** (d)	**68.** (b)	**69.** (d)	**70.** (d)
71. (a)	**72.** (c)	**73.** (b)	**74.** (a)	**75.** (a)	**76.** (b)	**77.** (b)	**78.** (b)	**79.** (d)	**80.** (c)
81. (a)	**82.** (d)	**83.** (b)	**84.** (a)	**85.** (d)	**86.** (b)	**87.** (d)	**88.** (d)	**89.** (d)	**90.** (b)
91. (a)	**92.** (c)	**93.** (b)	**94.** (d)	**95.** (d)	**96.** (d)	**97.** (d)	**98.** (d)	**99.** (d)	**100.** (a)
101. (c)	**102.** (b)	**103.** (a)	**104.** (d)	**105.** (a)	**106.** (d)	**107.** (c)	**108.** (a)	**109.** (d)	**110.** (c)
111. (a)	**112.** (d)	**113.** (d)	**114.** (a)	**115.** (c)	**116.** (d)	**117.** (b)	**118.** (a)	**119.** (b)	**120.** (a)
121. (a)	**122.** (c)	**123.** (b)	**124.** (d)	**125.** (b)	**126.** (b)	**127.** (b)	**128.** (d)	**129.** (b)	**130.** (c)
131. (a)	**132.** (d)	**133.** (b)	**134.** (c)	**135.** (a)	**136.** (c)	**137.** (b)	**138.** (d)	**139.** (b)	**140.** (a)
141. (a)	**142.** (c)	**143.** (d)	**144.** (b)	**145.** (a)	**146.** (c)	**147.** (a)	**148.** (b)	**149.** (a)	**150.** (b)
151. (a)	**152.** (b)	**153.** (d)	**154.** (a)	**155.** (a)	**156.** (b)	**157.** (b)	**158.** (c)	**159.** (a)	**160.** (b)
161. (d)	**162.** (a)	**163.** (b)	**164.** (c)	**165.** (d)	**166.** (a)	**167.** (d)	**168.** (d)	**169.** (a)	**170.** (c)
171. (d)	**172.** (a)	**173.** (d)	**174.** (b)	**175.** (c)	**176.** (a)	**177.** (d)	**178.** (b)	**179.** (d)	**180.** (b)
181. (d)	**182.** (c)	**183.** (a)	**184.** (b)	**185.** (d)	**186.** (a)	**187.** (c)	**188.** (d)	**189.** (d)	**190.** (b)
191. (a)	**192.** (c)	**193.** (b)	**194.** (d)	**195.** (b)	**196.** (b)				

3

Silviculture

1. The theory and practise of raising forest crop is known as
(a) Forestry
(b) Silvics
(c) Silviculture
(d) All of the above

2. Virgin forest is also known as
(a) Primeval forest
(b) Aged forest
(c) New forest
(d) All of the above

3. The practice of forestry with object of obtaining maximum return per unit area is
(a) Commercial forestry
(b) Intensive forestry
(c) Both (a) and (b)
(d) None of the above

4. The second largest tree in world
(a) *Pseudosuga taxifolia*
(b) *Cedrus deodara*
(c) *Sequoia sempervirens*
(d) None of the above

5. In India the second oldest known tree is
(a) *Tectona grandis*
(b) *Shorea robusta*
(c) *Cedrus deodara*
(d) *Dalbergia latifolia*

6. *Populus nigra* and *Cupressus sempervirens* have branching angle
(a) 20–30° (c) 60–70°
(b) 70–80° (d) 40–50°

7. The colour of new leaves of *Quercus incana* is
(a) Brown (c) Pink
(b) White (d) Red

8. The striking change in leaf colour before fall is known as
(a) Autumn tints
(b) Winter tints
(c) Summer tints
(d) All of the above

9. On lower altitudes, Chir pine species behave as
(a) Evergreen
(b) Deciduous
(c) Both (a) and (b)
(d) None of the above

10. Epicormic branches originate from
(a) Adventitious bud
(b) Dormant bud
(c) Axillary bud
(d) Both (a) and (b)

11. Which species is known as “Paragon of Indian Timber”?
(a) Teak (c) Arjun
(b) Pine (d) Mahogany

12. Which of the following decreases basal volume?
(a) Buttressing
(b) Fluting
(c) Lignotubers
(d) None of these

13. Epicormic branches are caused due to
(a) Drought
(b) Excessive light
(c) Fire
(d) All of the above

14. Fluting in Teak occurs due to
(a) Unsuitable site
(b) Insect attack
(c) Epicormic branches
(d) All of the above

15. Mycorrhizae forms the association of
(a) Bacteria with roots of higher plants
(b) Fungus with roots of higher plants
(c) Virus with roots of higher plants
(d) Protozoa with roots of higher plants

16. Ectotrophic mycorrhizae found in
(a) Basidiomycetes
(b) Actinomycetes
(c) Zygomycetes
(d) All of the above

17. Lignotubers are found in
(a) *Eucalyptus* species
(b) *Podocarpus* species
(c) *Dipterocarpus* species
(d) None of the above

18. Vigrous height growth and absence of dead bark seen in
(a) Seedling (c) Sapling
(b) Pole (d) Tree

19. Fast growth is indicated by rings which is wider than
(a) 5 mm (c) 10 mm
(b) 15 mm (d) 20 mm

20. Which of the following is/are polygamous species?
(a) *Sterculia*
(b) *Garuga*
(c) *Rhus*
(d) All of the above

21. In conifers, pollination occurs through
(a) Wind (c) Birds
(c) Insects (d) Water

22. Pneumatophores found in
(a) Littoral and swamp forest
(b) Mangrove forest
(c) Rain forest
(d) None of the above

23. In *Pinus roxburghii* the interval between pollination of flowers and ripening of seed is
(a) 15–17 months (c) 6–7 months
(b) 12–13 months (d) 24–26 months

24. The pollination through bats is known as
(a) Chiropterophily
(b) Zoophily
(c) Entomophily
(d) Anemophily

25. 3/4th of the tallest trees whose leading shoots are not free are
(a) Dominants (c) Dominated
(b) Suppressed (d) Codominants

26. In forest natural regeneration can be obtained by
(a) Root suckers
(b) Coppice
(c) Seeds
(d) All of the above

27. Factors of locality are also known as
(a) Site factors

(b) Habitat factors
(d) Both (a) and (b)
(d) None of the above

28. Pneumatophore helps in
(a) Photosynthesis (c) Digestion
(b) Respiration (d) Both (b) and (c)

29. Irregular involutions and swelling on the bole just above the basal swell is known as
(a) Buttressing (c) Lignotubers
(b) Swelling (d) Fluting

30. Which light is effective for photosynthesis?
(a) Red (c) Green
(b) Blue (d) Both (a) and (b)

31. The plants growing in UV light is
(a) Dwarf (c) Etiolated
(b) Normal (d) Irregualr

32. The stunted growth of vegetation in alpine region is attributed to
(a) Preponderance of UV rays
(b) Slow water absorption
(c) Desiccating winds
(d) All of the above

33. Root pruning is beneficial for
(a) *Eucalyptus*
(b) *Albizzia*
(c) Both (a) and (b)
(d) None of the above

34. Photosynthesis used light upto
(a) 3–4% (c) 1–2%
(b) 5–8% (d) 10–15%

35. The relative length of day and night to which plant is exposed is called as
(a) Photoperiod
(b) Photoperiodism
(c) Phototropism
(d) All of the above

36. Species capable of persisting and developing under shade is
(a) Shade demander
(b) Shade bearer
(c) Light demander
(d) None of the above

37. Woody perennial plant having height not more than 6m is
(a) Herb (c) Shrub
(b) Pole (d) Tree

38. Upper branchy part of a tree above the bole is known as
(a) Crown (c) Bole
(b) Stem (d) Root

39. Conifers are also known as
(a) Gymnosperms (c) Softwoods
(b) Evergreens (d) All of these

40. The crown of *Phoenix, Cocos,* and *Borasus* are
(a) Spherical (c) Conical
(b) Cylindrical (d) Flat topped

41. Crown of .... is not affected by the situation in which they grow
(a) *Palms*
(b) *Phoenix*
(c) *Borasus*
(d) All of the above

42. Teak behave as
(a) Evergreen
(b) Deciduous
(c) Both (a) and (b)
(d) None of the above

43. Which of the following are deciduous species.?
(a) *Lannea coromandelica*
(b) *Garuga pinnata*
(c) *Holoptelia integrifolia*
(d) All of the above

44. Which of the following species are evergreen
(a) Mango
(b) Himalayan cypress
(c) Rohini
(d) All of the above

45. Adventitious roots which arise from the butt end of the tree are known as
(a) Prop roots
(b) Buttresses
(c) Stilt roots
(d) Aerial roots

46. The appearance of characteristic periodic events in life cycle of an organism is known as
(a) Physiognomy (c) Growth
(b) Development (d) Phenology

47. What is gregarious flowering?
(a) General flowering with one or few years over considerable area of all or most of the whole of certain species that do not flower annually
(b) Flowering of one or few culms in a clump or few clumps in a locality
(c) General flowering followed by death of plant
(d) None of the above

48. Which of the following are the annual flowering species of bamboo?
(a) *Arundinaria wightiana*
(b) *Bambusa lineata*
(c) *Oclendra stridula*
(d) All of the above

49. Forest is derived from ......word
(a) Latin (c) German
(c) Persian (d) French

50. Forest regenerated through seed is known as
(a) Coppice forest
(b) Seedling forest
(c) High forest
(d) Low forest

51. In Indian forest act, Protected forest comes under
(a) Chapter V (c) Chapter III
(b) Chapter II (d) Chapter IV

52. Normal and Abnormal forest are classified on the basis of
(a) Composition
(b) Growing stock
(c) Age
(d) All of the above

53. The top height of tree is generally calculated from mean diameter of
(a) 100 trees (c) 200 trees
(b) 250 trees (d) 500 trees

54. 42% incoming radiation reflected back is known as
(a) Albido
(b) Radiation
(c) Albino
(d) Insolation

55. Leaf mosiac is common feature of
(a) Tropical areas
(b) Alpine regions
(c) Temperate regions
(d) Arid regions

56. Which of the following is light demander species. in Tropical wet evergreen forest?
(a) *Tectona grandis*
(b) *Callophyllum inophyllum*
(c) *Toona ciliata*
(d) *Artocarpus* species

57. The solar radiations greater than 720 millimicron in
(a) UV rays (c) Visible rays
(b) Infrared (d) X rays

58. One degree increase in latitude decreases the temperature by
(a) 2°C (c) 0.55°C
(b) 1°C (d) 0.2°C

59. Every 270m increase in altitude decreases temperature by
(a) 0.55°C (c) 4°C
(b) 2°C (d) 1°C

60. The optimum air temperature for plant respiration is
(a) 0°C–40°C (c) Upto 25°C
(b) 27°C–35°C (d) All of these

61. Optimum soil temperature for plant species range from
(a) 35°–40°C (c) 27°–35°C
(b) 40°–50°C (d) 20°–30°C

62. On the basis of temperature the forest of India are classified into......... types.
(a) Six (c) Sixteen
(b) Five (d) Four

63. Chilling injury is caused to plants in tropical region when temperature falls below
(a) 0°C (c) –5°C
(b) 5°C (d) –2°C

64. Dehradun valley is the typical example of
(a) Convection frost
(b) Advective frost
(c) Pool frost
(d) Hoar frost

65. Which type of the frost has the most deleterious effect on vegetation?
(a) Pool frost
(b) Hoar frost
(c) Advective frost
(d) Radiation frost

66. Area where frost is more frequent and intense than the adjoining area called as
(a) Frost pocket
(b) Frost hole
(c) Frost locality
(d) All of the above

67. Mineral reduces frost hardiness in tree species
(a) Potassium
(b) Phosphorus
(c) Nitrogen
(d) Carbon

68. Root pruning is beneficial for
(a) *Eucalyptus*
(b) *Albizzia*
(c) Both (a) and (b)
(d) None of these

69. Species susceptible to snow break
(a) *Pinus wallichiana*
(b) *Cedrus deodara*
(c) *Abies pindrow*
(d) *Picea smithiana*

70. Except TN and Kashmir most of the country receives rainfall from
(a) Retreating NE monsoon
(b) Westerly depressions
(c) Advancing SW monsoon
(d) Cyclonic depressions

71. Kashmir and Great Himalayan range receives maximum precipitation through
(a) Retreating NE monsoon
(b) Westerly depressions
(c) Advancing SW monsoon
(d) Nor's Westers

72. In Deccan plateau the Nor's westers and thunderstorms bring.
(a) Mango showers
(b) April rains
(c) Summer showers
(d) All of the above

73. Which of the following is frost hardy?
(a) *Dalbergia sissoo*
(b) *Dalbergia latifolia*
(c) *Morus alba*
(d) All of the above

74. Drought sensitive species is/are
(a) *Mangifera indica*
(b) *Shorea robusta*
(c) *Anogeissus latifolia*
(d) All of the above

75. *Manilkara hexandra* and *Diospyros* are commonly found in
(a) Montane wet temperate forest
(b) Himalyan moist temperate forest
(c) Tropical dry evergreen forest
(d) None of the above

76. *Tenstroemia gymnanthera*, *Meliosma* and *Eurya* are commonly found in
(a) Montane wet temperate forest
(b) Moist evergreen forest
(c) Dry evergreen forest
(d) Himalayan moist temperate forest

77. A dry month receives rainfall less than
(a) 30mm (c) 50mm
(b) 100mm (d) 20mm

78. A day receiving rainfall less than 2.5mm is known as
(a) Crop rainy day
(c) Trace day
(c) Rainy day
(d) None of these

79. .... is known as the lower limit of permanent snow on mountains.
(a) Snow line (c) Timber line
(c) Tree line (d) Any of these

80. Freezing of falling rain drops into pellets less than 2cm in size is called
(a) Dew (c) Snow
(b) Hail (d) Frost

81. The period between two gregarious flowerings over the same year of a species is often known as
(a) Physiological cycle
(b) Reproductive year
(c) Physiographic cycle
(d) None of the above

82. Shola forest is present in
(a) Western Himalayas
(b) Khasi hills
(c) Anamudi hills
(d) Nilgiri hills

83. In Dehradun, average annual dewfall is
(a) 50 mm (c) 20 mm
(b) 35 mm (d) 60 mm

84. Ecological index is given by
(a) Gindel (c) Moor
(b) Koppen (d) Lang

85. .... classified the world climate into five principal climatic groups.
(a) Moor (c) Koppen
(b) Lang (d) Seth & Khan

86. The Mean annual rainfall in Tropical Savanna is
(a) More than 500mm
(b) Less than 500mm
(c) 250-500mm
(d) 0-250mm

87. The strong noisy winds blowing up and down the valley in Northern

India are locally known as

(a) Squalls (c) Nor's westers
(b) Dadu (d) None of these

88. Generally heavy rainfall in mountains regions is experienced on

(a) Leeward side
(b) Windward side
(c) Both (a) and (b)
(d) None of these

89. The relation of a site to weather conditions especially Sun and wind is known as

(a) Aspect (c) Exposure
(b) Radiation (d) Solar flux

90. Generally the evaporation from forest floor is..............more than in open

(a) 10–80% (c) 10–20%
(b) 2–5% (d) 10–50%

91. Generally wind velocity in forest is reduced..............than in open

(a) 20–60% (c) 10–20%
(b) 20–30% (d) 20–50%

92. The direction towards which slope faces is known as

(a) Slope (c) Exposure
(b) Aspect (d) Altitude

93. In India, greater insolation is experienced on .... aspects.

(a) North-East (c) East-West
(b) South-North (d) South-West

94. Generally the climate of a small area differs significantly from general climate is known as

(a) Bioclimate
(b) Microwatershed
(c) Microclimate
(d) Macroclimate

95. Generally the poleward aspect are...........as compared to equators

(a) Warmer and moist
(b) Hot and dry
(c) Cool and dry
(d) Cool and moist

96. Humus mixed with mineral soil is known as

(a) Mor (c) Sour humus
(b) Mull (d) Raw humus

97. Mull and Mor are found in........... horizon.

(a) A°° (c) A°
(b) A (d) B

98. Illuviation is common in which soil horizon

(a) B (b) C (c) A° (d) A

99. Accumulation of salts takes place due to

(a) Excessive evaporation
(b) High water table
(c) Basin shape topography
(d) All of the above

100. pH of Saline-Alkaline soil is

(a) < 7.5 (c) > 7.5
(b) > 8.5 (d) > 9

101. Conifers have greater tendency to form

(a) Lateritic Fe pan
(b) Podsolic Fe pan
(c) Silica pan
(d) Plough pan

102. .... is commonly known as Plough pan

(a) Kankar pan
(b) Lateritic Fe pan
(c) Silica pan
(d) Clay pan

103. The soil particles < 0.002 mm is known as
(a) Clay (c) Sand
(b) Silt (d) Gravel

104. Soil particles belonging to soil group 'silt' having particle size
(a) <0.002 mm (c) 0.002–002mm
(b) 0.02–02 mm (d) 0.2–2.0 mm

105. Fine textured soils have
(a) High moisture content
(b) High nutrient status
(c) Both (a) and (b)
(d) None of the above

106. .... soils support pioneer and hardy species
(a) Clayey (c) Latretic
(b) Sandy (d) Loamy

107. Useful soil moisture for plant growth is
(a) Capillary water
(b) Hygroscopic water
(c) Gravity water
(d) Ground water

108. Process of soil formation in tropical forest is
(a) Laterization
(b) Podsolization
(c) Mineralisation
(d) None of these

109. Plants growing in moist sites is known as
(a) Hydrophyte
(b) Mesophyte
(c) Hygrophyte
(d) Xerophytes

110. Plants growing in extremely dry conditions is known as
(a) Thalophytes (c) Thalophytes
(b) Xerophytes (d) Mesophytes

111. Relative proportion of sand, silt, clay is called
(a) Soil texture
(b) Soil aggregation
(c) Soil structure
(d) None of the above

112. 'Ortstein' is generally found in
(a) A° Horizon (c) C Horizon
(b) A Horizon (d) B Horizon

113. Properties of soil cannot be altered
(a) Soil structure (c) Both (a) and (b)
(b) Soil texture (d) None of these

114. CEC of soil is measured as
(a) Mili eq/100 g soil
(b) Mili eq/1000 g soil
(c) Mili eq/100 g oven dry soil
(d) Mili eq/1000 g oven dry soil

115. Application of fertilizers is inefficient in
(a) Sandy soil (c) Clayey soil
(b) Loamy (d) Clayey loam

116. Sal grows best on soil having pH
(a) 6.5–7.6 (c) 9
(b) 7 (d) 4.5–5.5

117. Teak grows best on soil having pH
(a) 9–9.5 (c) 6–8
(b) 6.5–7.6 (d) 4–5

118. Nitrogen is best available when pH
(a) 6–8 (c) less than 7
(b) more than 7 (d) above 8

119. Tree species that grow up to 9.8 pH
(a) Neem
(b) Terminalia
(c) Pongamia
(d) All of the above

120. Nitrifying organisms inhibited in
(a) pH < 5.5
(b) pH > 5
(c) pH < 7
(d) pH > 9

121. Damping off in nursery can be controlled by maintaining pH
(a) 7 (b) 6 (c) 9 (d) 5.5

122. Increase in silica-sesquioxide ratio results in
(a) Low CEC
(b) Low moisture retentivity of soil
(c) Both (a) and (b)
(d) None of the above

123. Annual production of leaf litter found maximum in
(a) Tropical deciduous forest
(b) Tropical rain forest
(c) Tropical thorn forest
(d) Sub-Tropical rain forest

124. .... constitutes the 25% of dry weight of the plant.
(a) P (b) K (c) N (d) C

125. The free-living nitrogen fixing bacteria
(a) *Azotobacter* (c) *Clostridium*
(b) *Rhizobium* (d) Both (a) & (c)

126. Nitrogen fixing bacteria resides on leaves of
(a) *Pongamia glabra*
(b) *Pavetta indica*
(c) *Ixora parviflora*
(d) All of the above

127. The bacteria involved in conversion of ammonical form of 'N' to nitrate form of 'N' is
(a) Nitrobacter
(b) Nitosomonas
(c) Nitrosococcus
(d) *Bacterium denitrificans*

128. The bacteria involved in conversion of Nitrites into Nitrates is
(a) Nitrosomonas (c) Nitrosococcus
(b) Nitrobacter (d) Rhizobium

129. Lower the C:N ratio the greater amount of Nitrogen will be available to plants in the form of
(a) Nitrites (c) Nitrates
(c) Ammonia (d) None of these

130. Optimum C:N ratio for availability of minerals is
(a) 1:10 (c) 17:1
(b) 33:1 (d) 10:1

131. Total stem parasites found in which species
(a) *Cuscuta reflexa*
(b) *Cassytha filiformis*
(c) Both (a) and (b)
(d) None of the above

132. The *Santalum album* derive its nutrients from host through
(a) Houstoria (c) Parasite
(a) Epiphytes (d) Both (a) and (b)

133. In early stages of plant life
(a) Shoot growth = Root growth
(b) Shoot growth < Root growth
(c) Shoot growth > Root growth
(d) All of the above

134. Partial stem parasites is/are
(a) *Viscum* species
(b) *Loranthus* species
(c) *Arceuthobium* species
(d) All of the above

135. Examples of the Epiphytes is/are
(a) *Ficus religiosa*
(b) *Tinospora cordifolia*

(c) *Ficus bengalensis*
(d) All of the above

136. Large woody plants that climbs up trees in tropical areas is called as
(a) Climber (c) Liana
(b) Creepers (d) Twinner

137. Examples of the climbers is/are
(a) *Dioscorea deltoidea*
(b) *Bauhinia vahlii*
(c) *Butea superba*
(d) All of the above

138. Examples of the abnoxious weed is/are
(a) *Latana camara*
(b) *Eupatorium odoratum*
(c) Both (a) and (b)
(d) None of the above

139. Which of the following is responsible for die back of shoot?
(a) Copper (c) Calcium
(b) Zinc (d) Sulphur

140. Micro nutrient is
(a) Calcium (c) Manganese
(b) Magnesium (d) Sulphur

141. Intervenal necrosis occurs due to deficiency of
(a) Manganese (c) Boron
(b) Zinc (d) Magnesium

142. Saline-alkaline soils are deficient in
(a) Nitrogen (c) Iron
(b) Calcium (d) Phosphorus

143. Forest type is a category of forest defined with reference to
(a) Geographical location
(b) Climatic factors
(c) Composition and Condition
(d) All of the above

144. The general appearance of forest community is termed as
(a) Physiography (c) Physiognomy
(b) Structure (d) Floristics

145. Baurn-blanquet, Beard, Fosberg and Webb classified the forest types on the basis of
(a) Vegetation (c) Ecosystem
(b) Climate (d) Bioclimate

146. .... Classified the forest types on the basis of climate.
(a) Seth and khan
(b) Gaussen
(c) Champion and Seth
(d) Schimper

147. Emberger's formula was given by
(a) Shanbhag
(b) Paterson
(c) Seth and Khan
(d) Champion and Seth

148. Champion and Seth classified the forest types on the basis of
(a) Climate (c) Vegetation
(b) Ecosystem (d) Bioclimate

149. Champion and Seth classified the forest types of India into ....major groups.
(a) Four (c) Five
(b) Seven (d) Sixteen

150. Champion and Seth classified the Indian forest into
(a) Five (c) Seven
(b) Sixteen (d) Ten

151. Good seed year in *Abies pindrow* comes after
(a) 10 years (c) 4–5 years
(b) 5-6 years (d) 3-8 years

152. The regeneration crop obtained from seed is known as
(a) Seedling crop (c) Coppice forest
(b) High forest (d) Low forest

153. In *Acacia arabica* seed is dispersed through
(a) Birds (c) Animal
(b) Water (d) Wind

154. Seeds of .... lie dormant on forest ground for whole year.
(a) Sal
(b) *Junipers macropoda*
(c) *Acrocarpus fraxinifolius*
(d) *Fraxinus floribunda*

155. In *Junipers macropoda*, germination is delayed due to
(a) Immature embryo
(b) After-ripening
(c) Permeability to water
(d) Low viability

156. The viability of Sal seeds is for
(a) One week
(b) One month
(c) Three weeks
(d) Four months

157. The concept of the Germinative energy was given by
(a) Baldwin (c) Troupe
(b) Holmes (d) Schimper

158. Which one is irrespective of time
(a) Germinative capacity
(b) Germinative energy
(c) Both (a) and (b)
(d) None of the above

159. The Germinative capacity of teak is
(a) 10% (c) 80%
(b) 90% (d) 50%

160. Which of the following statement is true?
(a) Germinative capacity is higher than plant percent
(b) Germinative capacity is lower than plant percent
(c) Both (a) and (b)
(d) None of the above

161. Seeds of ____ require light for germination
(a) *Cassia fistula*
(b) *Albizzia procera*
(c) Both (a) and (b)
(d) None of the above

162. The ratio of CO_2:O_2 in Sal mortality is
(a) 1:3 (c) 2:5
(b) 2:8 (d) 1:10

163. The harmful weeds for Sal regeneration
(a) *Clerodendron*
(b) *Strobilanthes*
(c) *Parrotia*
(d) *Viola canescens*

164. *Viola canescens* is an indicator species for
(a) Teak
(b) Deodar and Kail
(c) Sal
(d) Deodar

165 Good indicator for Sal regeneration
(a) *Flamengia* species
(b) *Narenga* species
(c) Both (a) and (b)
(d) None of these

166. Dying back occurs in
(a) Sal
(b) *Terminalia tomentosa*

(c) *Boswellia serrata*
(d) All of the above

167. Sampling intensity for regeneration survey is
(a) 10–20% (c) 5–10%
(b) 2–4% (d) 15–20%

168. The regeneration survey is carried out in quadrate of size
(a) 2 m × 2 m (c) 1 m × 1 m
(b) 5 m × 5 m (d) 10 m × 10 m

169. In regeneration survey, 'u' indicates
(a) Sub-whippy seedling
(b) Woody shoot browsed by animals
(c) Unestablished whippy seedling
(d) Unestablished sub-whippy seedling

170. In steep slopes the regeneration is protected against fire till it reaches
(a) 2 m (b) 5 m (c) 7 m (d) 3 m

171. In Spruce, good seed year comes after
(a) 10 years (c) 4–5 years
(b) 8 years (d) 6 years

172. Pollarding is done in
(a) *Salix alba*
(b) *Hardwickia binata*
(c) *Grewia optiva.*
(d) All of the above

173. Lopping is done in
(a) *Quercus incana*
(b) *Morus* species
(c) *Butea monosperma*
(d) All of the above

174. Natural seedling obtained from the forest is known as
(a) Wildling (c) Seedling
(b) Sapling (d) Both (a) and (c)

175. Indicators for lime rich soil
(a) *Ixora parviflora*
(b) *Cupressus torulosa*
(c) *Acacia leucophloea*
(d) Both (a) and (b)

176. Indicators for soil with high concentration of soluble salts
(a) *Prosopis spicigera*
(b) *Prosopis juliflora*
(c) *Soymida febrifuga*
(d) Both (a) and (b)

177. Indicators for clayey soil liable to water logging
(a) *Desmostica bipinata*
(b) *Vetiveria zizanioides*
(d) *Chloroxylon* species
(d) None of the above

178. Indicator for stiff kankar clayey soil
(a) *Capparis* species
(b) *Chloroxylon swietinia*
(c) *Acacia leucophloea*
(d) All of the above

179. Indigenous fast growing species is/are
(a) *Acrocarpus fraxinifolius*
(b) *Populus ciliata*
(c) *Toona ciliata*
(d) All of the above

180. Example of exotic fast growing species
(a) *Populus deltoidea*
(b) *Populus ciliata*
(c) Both (a) and (b)
(d) None of the above

181. Dibling is common in
(a) *Quercus* species
(b) *Juglans regia*

(c) Chir pine
(d) All of the above

182. The time of neem seed collection is
(a) January-February
(b) May-June
(c) June-July
(d) March-April

183. The best example of collection of fallen seeds from ground is
(a) Teak and *Gmelina arborea*
(b) Sal and *Artocarpus chaplasa*
(c) Oak
(d) All of the above

184. Generally seed collection from standing tree is done in which species
(a) *Dalbergia sissoo*
(b) *Morus* species
(c) *Acacia catechu*
(d) *Acrocarpus* species

185. Large amount of mortality in Sal seedlings occur in monsoon due to
(a) Weed growth
(b) Moisture scarcity
(c) Accumulation of CO_2
(d) All of the above

186. Heavy growth of weeds like *Strobilanthus* and *Petalidium* is harmful for the regeneration of
(a) Teak (c) Sal
(b) Deodar-kail (d) Chirpine

187. As a general rule if seeds are larger in size, the rate of germination is
(a) Lower (c) Higher
(b) Medium (d) Both (a) and (b)

188. Easily attacked by bacteria and fungi in seed is/are
(a) Protien
(b) Enzymes
(c) Carbohydrates
(d) Both (a) and (c)

189. Seed production is higher in
(a) Warmer climate
(b) Cool climate
(c) Both (a) and (b)
(d) None of the above

190. Autumn tints is associated with change in
(a) Flower colour (c) Fruit colour
(b) Leaf colour (d) All of these

191. Soil preferred for nursery preparation is
(a) Clayey soil (c) Clayey loam
(b) Silt loam (d) Sandy loam

192. In Stool coppice, new shoots arise from
(a) Axillary bud
(b) Adventitious bud
(c) Both (a) and (b)
(d) None of the above

193. Which species. coppice fairly?
(a) Neem
(b) *Cedrus deodara*
(c) *Hardwickia binata*
(d) *Shorea robusta*

194. Restocking of felled or otherwise cleared woodland by artificial means
(a) Reforestation (c) Afforestation
(b) Deforestation (d) None of these

195. Establishment of forest on abandoned area by artificial means
(a) Reforestation (c) Plantation
(b) Afforestation (d) Deforestation

196. Reforestation is done to
(a) Supplement natural regeneration
(b) Introduce exotics

(c) Change composition of crop
(d) All of the above

197. A specie having height of 60 cm/year with minimum yield of 10 m^3/ha/yr in short rotation of 10-15 years is called
(a) Slow growing
(b) Medium growing
(c) Fast growing
(d) None of the above

198. In high rainfall areas ______ is done
(a) Ridge sowing
(b) Ridge ditch sowing
(c) Trench sowing
(d) None of the above

199. In dry or low rainfall areas ____ is done
(a) Ridge ditch sowing
(b) Trench sowing
(c) Ridge sowing
(d) All of the above

200. A nursery which is maintained without irrigation is known as
(a) Wet nursery
(b) Temporary nursery
(c) Permanent nursery
(d) Dry nursery

201. A nursery should be .... of plantation area.
(a) 2.5% (b) 10% (c) 5% (d) 4%

202. The standard size of nursery bed is about
(a) 12.2 m × 1.2 m (c) 10 m × 1 m
(b) 12.2 m × 1 m (d) 10 m × 1.2 m

203. In hills the width of nursery bed for deodar is
(a) 1.8 m (b) 0.9 m (c) 2 m (d) 1 m

204. Generally nursery beds for protection against Sun and frost are oriented in
(a) North-south (c) East- West
(b) West-East (d) South-East

205. In dry and arid area, Nursery bed should be
(a) Raised (c) Flat
(b) Sunken (d) None of these

206. In high rainfall areas, nursery bed should be
(a) Raised
(b) Flat
(c) Sunken
(d) Any of these

207. Seeds of *Adina* and *Eucalyptus* before sowing are mixed with
(a) Sand
(b) Ash
(c) Fine soil
(d) All of the above

208. Quantity of seed required for nursery bed is given by
(a) W=(A*P*100)/(D*N)
(b) W=(P*D*100)/(A*N)
(c) W=(A*D*100)/(P*N)
(d) P=(W*D*100)/(A*N)

209. Seeds are protected using
(a) Red lead
(b) Kerosene oil
(c) Camphor
(d) All of the above

210. Aldrex is a/an
(a) Pesticide (c) Insecticide
(b) Nematicide (d) None of these

211. The most common nursery disease
(a) Damping off

(b) Powdery mildew
(c) Wilting
(d) Root rot

212. In Nursery, damping off is controlled by
(a) Acetic acid
(b) Formaldehyde
(c) Copper sulphate
(d) All of the above

213. An area where plants are raised for eventual planting out is
(a) Forest (c) Plantation
(b) Nursery (d) None of these

214. In western countries, seeds of oak and chestnut transported through packing in
(a) Gunny bags
(b) Powdered dry charcoal
(c) Moss
(d) None of the above

215. Quantity of seed required for drill sowing is
(a) Two times (c) Four times
(b) Six times (d) Eight times

216. Quantity of seed required for broadcasting is
(a) Three times
(b) Four times
(c) Six times
(d) Eight times

217. Transferring of nursery stock from the seed bed directly to the planting site is known as
(a) Lining out
(b) Planting out
(c) Pricking out
(d) Transplanting

218. If seedlings are transferred or shifted from seedling bed to transplant bed or nursery container is known as
(a) Transplanting
(b) Lining out
(c) Pricking out
(d) All of the above

219. Different plant containers used in nursery is
(a) Dona
(b) Moss cylinders
(d) Baskets
(d) All of the above

220. In Ootacamund, mossing is used for which species
(a) *Ficus bengalensis*
(b) *Shorea robusta*
(c) *Eucalyptus globulous*
(d) *Eucalyptus hybrid*

221. In Nursery, polythene bags used are of following gauge
(a) 150
(b) 200
(c) 250
(d) All of the above

222. The proprietary product, ADCO is added to vegetable waste in the ratio
(a) 1:30 (c) 1:20
(b) 1:40 (d) 1:10

223. In India, green manuring is done using
(a) *Crotolaria juncea*
(b) *Sesbania aculeata*
(c) *Cannabis sativa*
(d) All of the above

224. Scale of plantation map is
(a) 1:5000 (c) 1:50000
(b) 1:250 (d) 1:1000

225. Special pig and porcupine-proof cum game-proof fence is common in
(a) Temperate forest
(b) Tropical forest
(c) Both (a) and (b)
(d) None of the above

226. Game proof fence provides protection against
(a) Wild animals
(b) Domestic animals
(c) None of these
(d) None of the above

227. "Method steppique" is used for
(a) Deep consolidated soil
(b) Shallow soil
(c) Soil forming hard pan
(d) All of the above

228. A type of pit used for planting in clayey and saline-alkaline soil in all rainfall classes is
(a) Saucer pit (c) Ordinary pit
(b) Ring pit (d) Ridge pit

229. The most common type of pit used in sandy soils is
(a) Ordinary pit (c) Saucer pit
(b) Ring pit (d) Both (a) and (b)

230. Ridge ditch sowing is generally preferred in
(a) Sloping areas
(b) Plain areas.
(c) Waterlogged areas
(d) Dry areas

231.................. is suitable for salt impregnated areas.
(a) Double trench
(b) Shelfed trench
(c) Trench pit
(d) Trench ridges

232. "Shallow trench ridge" is known as
(a) Trench mound
(b) Wurli
(c) Methode steppique
(d) None of the above

233. The most suitable soil preparation technique used in stony detrital slopes, low water holding capacity and high surface run-off is
(a) Shelfed trench (c) Trench mound.
(b) Double trench (d) Wurli

234. Pre-monsoon planting is done in
(a) *Toona ciliata*
(b) Deodar
(c) *Morus alba*
(d) *Dalbergia* species

235. Monsoon planting is generally done in
(a) Deodar
(b) Silver fir
(c) Spruce
(d) All of the above

236. Winter planting is done in
(a) *Toona ciliata*
(b) *Dalbergia latifolia*
(c) Chukrasia
(d) All of the above

237. The approximate age of *Morus alba* seedlings for planting is
(a) 1 year
(b) 9 months
(c) 2–3 years
(d) 5 months

238. The stump planting is common in
(a) Teak (c) Sal
(b) Poplar (d) Deodar

239. Naked root planting is successfully adopted in
(a) Sal (c) Conifers
(b) Bamboos (d) Teak

240. In Willows and poplar usual length of cuttings is
(a) 2.0–2.5 cm (c) 4.0–5.0 cm
(b) 6.0–8.0 cm (d) 2.5–3.0 m

241. Planting of stem cuttings is common in
(a) *Ficus* species
(b) *Bombax ceiba*
(c) *Boswelia* species
(d) All of the above

242. The most commonly used pattern of planting in forest is
(a) Triangular planting
(b) Quincunx planting
(c) Rectangular planting
(d) Circular planting

243. Calculate the number of plants required for 20 hectares of plantation, planted at the spacing of 2 m × 2 m with a plant in the centre of square
(a) 32,000 (c) 1,00,000
(b) 50,000 (d) 57,750

244. Calculate the number of plants required for 10 hectares of plantation in which plants are spaced 2.5 m apart in triangular pattern
(a) 32,000 (c) 16,000
(b) 18,480 (d) 12,500

245. Calculate the spacing of plants raised in quincunx planting if there are 3200 plants/ha in plantation
(a) 2.5 m (b) 4 m (c) 2 m (d) 5 m

246. Restocking blanks in an artificially regenerated area with fresh sowing or planting is known as
(a) Casuality replacement
(b) Beating up
(c) Dibling
(d) Both (a) and (b)

247. A crop of trees or shrubs grown to foster the growth of another and more important tree crop in early stages is known as
(a) Cover crop
(b) Nurse crop
(c) Under planting
(d) Enrichment planting

248. *Gmelina* and *Tephrosia* are raised as a nurse crop for
(a) Sal
(b) *Ricinus communis*
(c) *Dipterocarpus turbinatus*
(d) All of the above

249. *Cajanus* and *Ricinus communis* are introduced as cover crop in
(a) Teak plantation
(b) Sal taungya plantation
(c) *Dipterocarpus turbinatus* plantation
(d) None of the above

250. A subsidary crop of low plants introduced in a plantation to afford soil cover between or below the main crop is known as
(a) Enrichment planting.
(b) Cover crop
(c) Nurse crop
(d) Beating up

251. Examples of cover crops is
(a) *Indigofera tinctoria*
(b) *Leucaena glauca*
(c) *Tephrosia candida*
(d) All of the above

252. Planting of valuable species under the main crop is known as

(a) Enrichment planting.
(b) Underplanting.
(c) Both (a) and (b)
(d) None of these

253. A register containing details of all the work carried out in plantation area recorded from time to time is known as
(a) Plantation journal
(b) Field register
(c) Both (a) and (b)
(d) None of the above

254. Contour trenches are usually made on slopes upto
(a) 10% (c) 40%
(b) 20% (d) 8%

255. Size of contour trench is
(a) 1 m long & 15 cm deep
(b) 3 m long & 30 cm deep
(c) 5 m long & 1 m deep
(d) None of the above

256. Which of the following is the indicative of poor Kallar soil?
(a) *Salvadora* species
(b) *Tamarix* species
(c) Both (a) and (b)
(d) None of these

257. Which of the following is the indicative of soils free from salts
(a) *Prosopis spicigera*
(b) *Prosopis juliflora*
(c) *Capparis* species
(d) Both (a) and (c)

258. The desert afforestation work comprises of
(a) Sand dune stabilization
(b) Creation of windbreak & shelterbelt
(c) Development of fuel and fodder blocks
(d) All of the above.

259. Species suitable for afforestation of denuded hill slopes in temperate areas is
(a) Chir pine
(b) Blue pine
(c) *Robinia pseudocasia*
(d) *Acacia catechu*

260. Species suitable for planting canal banks is
(a) *Eycalyptus* species
(b) *Ailanthus* species
(c) *Acacia catechu*
(d) All of the above

261. Operation carried out for the benefit of forest crop at any stage of life between seedling and mature stage is known as
(a) Tending
(b) Thinning
(c) Improvement felling
(d) Pruning

262. Regeneration felling and ground operations are not included in
(a) Silvicultural treatment.
(b) Cultural operation
(c) Tending operation
(d) None of the above

263. Cleanings in plantation are done to
(a) Improve light conditions
(b) Reduce root competition
(c) Both (a) and (b)
(d) None of these

264. Felling made in an immature stand for the purpose of improving

the growth and form of the tree without permanently breaking the canopy of the tree
(a) Tending
(b) Felling
(c) Improvement felling
(d) Cleaning

265. Thinning's are done in crop to
(a) Maintain hygienic conditions
(b) Increase net yield of timber
(c) Shortening the rotation age
(d) All of the above

266. Standard tree classification adopted in Indian forestry divided into .... classes.
(a) 3 (b) 5 (c) 4 (d) 7

267. In tree classification, the diseased trees are denoted by
(a) k (b) s (c) D (d) M

268. In Sissoo plantation, the alternate trees in alternate rows are removed is
(a) Ordinary thinning
(b) Mechanical thinning
(c) Free thinning
(d) Maximum thinning

269. The grover formula for mechanical thinning for deodar is
(a) D = d (c) D = 1.5(d+4)
(b) D = 2d (d) None of these

270. Ordinary thinning is also known as
(a) Low thinning
(b) German thinning
(c) Thinning from below
(d) All of the above

271. Sclerophyllus forest are commonly found in
(a) Temperate region
(b) Tropical region
(c) Mediterranean region
(d) None of the above

272. Thinning involving removal of inferior individuals of a crop starting from the suppressed class than dominated and lastly dominants are known as
(a) Low thinning
(b) German thinning
(c) Thinning from below
(d) French thinning

273. Which of the following thinning is seldom carried out for comparative research purpose?
(a) Moderate thinning
(b) Light thinning
(c) Heavy crown thinning
(d) Light crown thinning

274. The dead, dying, diseased, and suppressed trees are removed in
(a) Heavy thinning
(b) Very heavy thinning
(c) Moderate thinning
(d) Light thinning

275. .... grade is ordinarily used for yield table computation.
(a) Heavy thinning
(b) Light crown thinning
(c) Light thinning
(d) Moderate thinning

276. Generally ordinary thinning is applied to
(a) Shade demander
(b) Shade bearer
(c) Light demander
(d) All of the above

277. The Crown thinning is also known as
(a) French thinning

(b) Thinning from above
(c) High thinning
(d) All of the above

278. Thinning carried out in moderately shade tolerant species is
(a) Free thinning
(b) Crown thinning
(c) Maximum thinning
(d) Ordinary thinning

279. Free thinning was devised by
(a) Heck (c) Gehrhardt
(b) Craib (d) Trevor

280. The method of thinning in which attention in concentrated on evenly spaced selected stems is known as
(a) Elite thinning
(b) Advance thinning
(c) Free thinning
(d) Both (a) and (c)

281. The maximum thinning has been developed by
(a) Heck
(b) Criab
(c) Gehrhardt
(d) All of the above

282. Thinning done in anticipation of suppression is known as
(a) Criab thinning
(b) Free thinning
(c) Maximum thinning
(d) Ordinary thinning

283. Stand density index was proposed by
(a) Troup (c) Reineke
(b) Schimper (d) Brandis

284. The planned interval which lapses between successive thinning's is known as
(a) Rotation
(b) Physiological cycle
(c) Thinning cycle
(d) None of the above

285. Thinning carried out in irregular crop is
(a) Ordinary thinning
(b) Crown thinning
(c) Maximum thinning
(d) Selection thinning

286. In India bud pruning is applied in
(a) *Salix* species
(b) *Populus* species
(c) Both (a) and (b)
(d) None of these

287. Bric planting is done in
(a) Desert
(b) Lateritic soils
(c) Hills
(d) Saline-alkaline soils

288. Which of the following is correctly matched?
(a) D = d, Sagreiya's formula
(b) D = 1.5(d+4), Howard formula
(c) D – 2d, Glover formula
(d) D = 1 ½ d, Warren formula

289. Cauliflory is common in
(a) Cacao (c) *Ficus palmata*
(b) *Artocarpus* (d) Both (a) and (b)

290. Tending operation done in sapling crop involving removal or topping of inferior growth of the same age including the individuals of the favoured species when they are interfering with the better grown individuals is termed as
(a) Cleaning (c) Pruning
(b) Thinning (d) Weeding

291. A plant which avoids calcareous soil is
(a) Calciphile (c) Both (a) and (b)
(b) Calciphuge (d) None of these

292. Tending operation comprises of
(a) Cleaning (c) Thinning
(b) Weeding (d) All of these

293. U.P. forest department based on experiences of ravine afforestation work recommends not planting *Acacia nilotica* in ravine valley due to
(a) Thorny species
(b) Light demanding species
(c) Frost tender species
(d) All of the above

294. Maximum forest area in India comes under
(a) Tropical wet evergreen forest
(b) Tropical dry deciduous forest
(c) Tropical dry evergreen forest
(d) None of the above

295. A forest, for a given site consists of ideal growing stock, age class distribution is called
(a) Normal forest
(b) Abnormal forest
(c) High forest
(d) Regular forest

296. The differences of rotation age less than 25% in case, where stand is not harvested for 100 years is termed as
(a) Unevenaged forest
(b) Evenaged forest
(c) Normal forest
(d) Abnormal forest

297. The family of teak is
(a) Dipterocarpaceae
(b) Tiliaceae
(c) Lamiaceae
(d) None of the above

298. In sub-tropical forest, which of the following is a light demander species
(a) *Pinus wallichiana*
(b) *Quercus incana*
(c) *Pinus roxburghii*
(d) Both (a) and (c)

299. Fruiting and flowering on stem is known as
(a) Stem flowering
(b) Vivipary
(c) Cauliflory
(d) None of the above

300. In degraded hilly lands trees are planted
(a) Along contour bunds
(b) In between the contours
(c) Along bench terraces
(d) All of the above

301. The complex of physical and biological factors which determine the vegetation or forest of an area is termed as
(a) Site (c) Habitat
(b) Ecosystem (d) Edaphic

302. The factors of locality are also referred to
(a) Site factors
(b) Habitat factors
(c) Both (a) and (b)
(d) None of the above

303. Short day induces dormancy in
(a) Junipers (c) Salix
(b) Populus (d) Deodar

304. Root coiling is common in seedling raised in
(a) Root trainer

(b) Polythene bags
(c) Seed bed
(d) All of the above

305. Twisted fibre in Chir pine is due to
(a) Genetical (c) Borer
(b) Physiological (d) Environment

306. Study of succession does not helps in
(a) Classification of vegetation into forest types
(b) Artificial regeneration
(c) Classification of successional stages
(d) None of the above

307. Most limiting factor for understory is
(a) Moisture (c) Nutrients
(b) Minerals (d) Light

308. Oak seeds are dispersed by
(a) Wind (c) Gravity.
(b) Animals (d) Birds

309. Study of different phases of plant growth is
(a) Phenological spectrum
(b) Phenognamy
(c) Phenograph
(d) None of the above

310. Major constituent required for plant energy or metabolism is
(a) Carbon (c) Hydrogen
(b) Oxygen (d) Nitrogen

311. The seedlings in Nursery mainly dieback due to deficiency of
(a) Magnesium (c) Zinc
(b) Boron (d) Copper

312. In Temperate species in which the leaves change colour to yellow, red, purple due to
(a) Chlorosis
(b) Yellow pigments
(c) Chlorophyll
(d) Carotenoids

313. Fodder tree specie in which cut and carry system is followed is
(a) *Tectona grandis*
(b) *Leucaena leucocephala*
(c) *Albizzia procera*
(d) All of the above

314. Root nodules present in *Sesbania grandiflora* is
(a) *Rhizobium*
(b) *Frankia*
(c) *Cyanobacteria*
(d) All of the above

315. $11A/C_1$ denotes the
(a) Mangrove forest
(b) Shola forest
(c) Thorn forest
(d) Southern montane wet temperate forest

316. $W = A \times D/P \times N \times 100$, is the formula for calculating z
(a) Weight of fruit
(b) Weight of seed
(c) Quantity of seed
(d) None of the above

317. Pinaceae is the largest family of the order
(a) Cyeadales (c) Coniferales
(b) Malvales (d) Monaliales

318. Father of Dendrology is
(a) Aristotle
(b) Dietrich Brandis
(c) Birbal Sahni
(d) Gregor Mendel

319. Deodar is most commonly found in elevation ranging from

(a) 400- 800 m (c) 1000-1500 m
(b) 800-1500 m (d) 1800-2600 m

320. The first to propose a system of classification based on the habit of plants is
(a) Theophrastus (c) Hutchinson
(b) Joseph Hooker (d) Bentham.

321. Among the Artificial system of classification, the best known is
(a) Hooker's system
(b) Linnaean system
(c) Bentham's system
(d) Dong system

322. Two sub-division of angio-sperms are
(a) Leguminosae and Malvaceae
(b) Evergreen and deciduous
(c) Dicotyledons and monocotyledons
(d) Ectoderm and endoderm

323. Which of the following are examples of monocotyledons family?
(a) Liliaceae and Poaceae
(b) Rutaceae and Malvaceae
(c) Leguminosae and Papilionaceae
(d) None of the above

324. The forest type found at 0-10 m altitude in India is called
(a) Tropical moist deciduous type
(b) Alpine and cold desert type
(c) Lower temperate montane type
(d) Marine and littoral type

325. Mangrove and Jute are found in which type of forest
(a) Estuarine type
(b) Marine and littoral type
(c) Sub-tropical montane type
(d) Temperate type

326. Gunny bags is made from
(a) Cotton fibres
(b) Coconut fibres
(c) Jute fibres
(d) Stem fibres

327. Mechanical thinning can be carried out at those places where
(a) Uniformly space and uniform growth
(b) Different space and different growth
(c) Different space and uniform growth
(d) Uniform space and different growth

328. A fire resistant bamboo species is/are
(a) *Melocanna bambosoides*
(b) *Dendrocalamus hamiltonii*
(c) Both (a) and (b)
(d) None of the above

329. Tress develop with more than one apical meristem are termed as
(a) Monoaxial
(b) Polyaxial
(c) Orthotropic
(d) All of the above

330. The optimum number of trees per unit area is inversely proportional to the square of the height of the tree is
(a) N/H^2 relationship
(b) N/D relationship
(c) H/N^2 relationship
(d) None of the relationship

331. Root pruning is beneficial for
(a) *Eucalyptus* (c) Both (a) and (b)
(b) *Albizzia* (d) None of these

332. In early stages of plant life
(a) Shoot growth = Root growth
(b) Shoot growth < Root growth.
(c) Shoot growth > Root growth.
(d) All of the above

Answer Keys									
1. (c)	**2.** (a)	**3.** (b)	**4.** (c)	**5.** (d)	**6.** (a)	**7.** (c)	**8.** (a)	**9.** (b)	**10.** (d)
11. (a)	**12.** (b)	**13.** (d)	**14.** (d)	**15.** (b)	**16.** (d)	**17.** (a)	**18.** (c)	**19.** (a)	**20.** (d)
21. (a)	**22.** (b)	**23.** (d)	**24.** (a)	**25.** (c)	**26.** (d)	**27.** (c)	**28.** (b)	**29.** (d)	**30.** (d)
31. (a)	**32.** (d)	**33.** (b)	**34.** (c)	**35.** (a)	**36.** (b)	**37.** (c)	**38.** (a)	**39.** (d)	**40.** (b)
41. (a)	**42.** (c)	**43.** (d)	**44.** (d)	**45.** (c)	**46.** (d)	**47.** (d)	**48.** (d)	**49.** (a)	**50.** (c)
51. (d)	**52.** (b)	**53.** (b)	**54.** (a)	**55.** (c)	**56.** (b)	**57.** (b)	**58.** (c)	**59.** (d)	**60.** (a)
61. (c)	**62.** (d)	**63.** (b)	**64.** (a)	**65.** (a)	**66.** (d)	**67.** (c)	**68.** (b)	**69.** (a)	**70.** (c)
71. (b)	**72.** (d)	**73.** (a)	**74.** (d)	**75.** (c)	**76.** (a)	**77.** (c)	**78.** (c)	**79.** (a)	**80.** (b)
81. (a)	**82.** (d)	**83.** (b)	**84.** (a)	**85.** (c)	**86.** (a)	**87.** (b)	**88.** (b)	**89.** (c)	**90.** (a)
91. (d)	**92.** (b)	**93.** (d)	**94.** (c)	**95.** (c)	**96.** (b)	**97.** (c)	**98.** (a)	**99.** (d)	**100.** (b)
101. (b)	**102.** (d)	**103.** (a)	**104.** (c)	**105.** (c)	**106.** (b)	**107.** (a)	**108.** (a)	**109.** (c)	**110.** (b)
111. (a)	**112.** (d)	**113.** (b)	**114.** (c)	**115.** (a)	**116.** (d)	**117.** (b)	**118.** (a)	**119.** (d)	**120.** (a)
121. (d)	**122.** (c)	**123.** (b)	**124.** (c)	**125.** (d)	**126.** (d)	**127.** (a)	**128.** (b)	**129.** (c)	**130.** (d)
131. (c)	**132.** (a)	**133.** (b)	**134.** (d)	**135.** (d)	**136.** (c)	**137.** (d)	**138.** (c)	**139.** (a)	**140.** (c)
141. (a)	**142.** (b)	**143.** (d)	**144.** (c)	**145.** (a)	**146.** (d)	**147.** (c)	**148.** (b)	**149.** (c)	**150.** (b)
151. (a)	**152.** (a)	**153.** (c)	**154.** (d)	**155.** (b)	**156.** (a)	**157.** (b)	**158.** (b)	**159.** (d)	**160.** (a)
161. (c)	**162.** (b)	**163.** (a)	**164.** (b	**165.** (c)	**166.** (d)	**167.** (b)	**168.** (a)	**169.** (c)	**170.** (d)
171. (d)	**172.** (d)	**173.** (d)	**174.** (a)	**175.** (d)	**176.** (b)	**177.** (b)	**178.** (d)	**179.** (d)	**180.** (a)
181. (d)	**182.** (c)	**183.** (d)	**184.** (b)	**185.** (c)	**186.** (a)	**187.** (c)	**188.** (c)	**189.** (a)	**190.** (b)
191. (d)	**192.** (b)	**193.** (c)	**194.** (a)	**195.** (b)	**196.** (d)	**197.** (c)	**198.** (a)	**199.** (b)	**200.** (d)
201. (a)	**202.** (a)	**203.** (b)	**204.** (c)	**205.** (b)	**206.** (a)	**207.** (d)	**208.** (c)	**209.** (d)	**210.** (c)
211. (a)	**212.** (d)	**213.** (b)	**214.** (b)	**215.** (a)	**216.** (c)	**217.** (b)	**218.** (d)	**219.** (d)	**220.** (c)
221. (d)	**222.** (b)	**223.** (d)	**224.** (a)	**225.** (b)	**226.** (a)	**227.** (d)	**228.** (c)	**229.** (b)	**230.** (a)
231. (d)	**232.** (b)	**233.** (b)	**234.** (c)	**235.** (d)	**236.** (d)	**237.** (b)	**238.** (a)	**239.** (c)	**240.** (d)
241. (d)	**242.** (c)	**243.** (c)	**244.** (b)	**245.** (a)	**246.** (d)	**247.** (b)	**248.** (c)	**249.** (d)	**250.** (b)
251. (d)	**252.** (c)	**253.** (a)	**254.** (b)	**255.** (b)	**256.** (c)	**257.** (d)	**258.** (d)	**259.** (b)	**260.** (d)
261. (a)	**262.** (c)	**263.** (c)	**264.** (b)	**265.** (d)	**266.** (b)	**267.** (a)	**268.** (b)	**269.** (a)	**270.** (d)
271. (c)	**272.** (b)	**273.** (b)	**274.** (d)	**275.** (a)	**276.** (c)	**277.** (d)	**278.** (b)	**279.** (d)	**280.** (d)
281. (c)	**282.** (a)	**283.** (c)	**284.** (c)	**285.** (d)	**286.** (c)	**287.** (a)	**288.** (d)	**289.** (d)	**290.** (a)
291. (b)	**292.** (c)	**293.** (c)	**294.** (b)	**295.** (s)	**296.** (b)	**297.** (c)	**298.** (d)	**299.** (c)	**300.** (d)
301. (a)	**302.** (c)	**303.** (b)	**304.** (b)	**305.** (a)	**306.** (c)	**307.** (d)	**308.** (c)	**309.** (a)	**310.** (d)
311. (b)	**312.** (d)	**313.** (b)	**314.** (a)	**315.** (d)	**316.** (c)	**317.** (c)	**318.** (c)	**319.** (d)	**320.** (a)
321. (b)	**322.** (c)	**323.** (a)	**324.** (d)	**325.** (a)	**326.** (c)	**327.** (a)	**328.** (c)	**329.** (b)	**330.** (a)
331. (b)	**332.** (b)								

4

Silviculture Systems

1. The regeneration, harvesting and tending operations are important aspect for
(a) Forest management
(b) Silvicultural system
(c) Thinning
(d) Pruning

2. Silvicultural systems in India are based on
(a) Mode of regeneration
(b) Pattern of felling
(c) Both (a) and (b)
(d) None of the above

3. System of concentrated regeneration depend upon
(a) Clear felling system
(b) Shelterwood strip system
(c) Two storeyed high forest system
(d) Both (a) and (b)

4. Felling made in view to assist regeneration under shelterwood system is
(a) Seeding felling
(b) Regeneration felling
(c) Selection felling
(d) None of the above

5. Removal of the last sheltered after regeneration has been affected under shelterwood system is
(a) Final felling
(b) Seeding felling
(c) Secondary felling
(d) Selective felling

6. Opening the canopy of mature stand to provide condition for securing regeneration from seed of tree retained for that purpose is
(a) Final felling
(b) Secondary felling
(c) Seeding felling
(d) None of the above

7. Felling carried out between seeding felling and final felling is called as
(a) Regeneration felling
(b) Selective felling
(c) Section felling
(d) None of the above

8. Interval between seeding felling and final felling is called
(a) Regeneration interval
(b) Rotation
(c) Felling Cycle
(d) None of the above

9. Most artificial Silvicultural system is
(a) Shelterwood system
(b) Clear strip system
(c) Clearfelling system
(d) Group system

10. Change from one Silviculture system to another is termed as
(a) Conversion period
(b) Conversion
(c) Both (a) and (b)
(d) None of the above

11. Conversion takes place within
(a) 20–30 years (c) 50–60 years
(b) 150–200 years (d) 100–120 years

12. When equal or equiproductive areas of matured crop are successively clearfelled is called
(a) Clearstrip system
(b) Clearfelling system
(c) Alternate strip system
(d) All of the above

13. Clear felling followed by natural regeneration can be achieved by
(a) Seed stored in area
(b) Seed received from outside
(c) Advance growth
(d) All of the above

14. Clear felling followed by natural regeneration through seed received from outside is
(a) Babul seed inundated with Indus river
(b) *Morus indica* seed by birds and water
(c) *Avicennia* in India
(d) All of the above

15. Clear felling was first Introduced by
(a) Troup
(b) Gayer
(c) Heinrich von cotta
(d) Eberhard

16. Examples of Clearfelling followed by artificial regeneration by Taungya are
(a) Sukana technique for Sal
(b) Allapalli technique for Teak
(c) Nilambore technique for Teak
(d) Both (a) and (c)

17. *Cupressus* and *Cryptomeria* plantations of West Bengal are example of
(a) Clear felling system followed by natural regeneration
(b) Clearfelling followed by artificial regeneration
(c) Uniform system
(d) Group system

18. Example of Clearfelling followed by natural regeneration from seed stored in the area is/are
(a) *Acacia mearnsii* in Tamil Nadu
(b) *Pinus pinaster* in Landes, France
(c) Both (a) and (b)
(d) None of the above

19. Examples of the Clearfelling from natural regeneration from advance growth
(a) South Raipur technique for Sal
(b) Saranda technique for Teak
(c) Both (a) and (b)
(d) None of the above

20. Clearfelling and Clear Strip system is suitable for
(a) Shade demanding species
(b) Shade bearing species
(c) Light demanding species
(d) None of the above

21. Silvicultural system in which clearfelling is done in the form of strip which progress successively in one direction usually against the wind direction is called
(a) Progressive strip system
(b) Alternate strip system
(c) Clearfelling
(d) All of the above

22. Example of the Clear Strip system is
(a) *Acacia mearnsii*
(b) *Pinus pinaster*

(c) *Pinus keysia*
(d) *Pinus roxburghii*

23. According to Troupe width of the clearfelled strip in Alternate strip system is
(a) 36–54 m (c) 15–20 m
(b) 12–20 m (d) 36–80 m

24. System of successive regeneration felling is known as
(a) Clearfelling system
(b) Selection system
(c) Shelterwood system
(d) Improvement felling

25. Alternate Strip system was successful in case of
(a) Sal
(b) Fir
(c) *Acacia mearnsii*
(d) All of the above

26. Uniform Shelterwood system is also known as
(a) Shelterwood Compartment system
(b) Compartment system
(c) Both (a) and (b)
(d) None of the above

27. In India, generally ……… number of seed bearers are retained per hectare
(a) 12-15 (c) 20-25
(b) 45-50 (d) 8-10

28. Which of the following is not a regeneration felling
(a) Seeding felling
(b) Seedling felling
(c) Secondary felling
(d) Final felling

29. Large number of seed bearers are retained on the
(a) N-NE aspect (c) N-NW aspect
(b) S-SW aspect (d) S-SE aspect

30. Type of crop obtained in clear felling system is
(a) Unevenaged
(b) Evenaged
(c) Irregular
(d) All of the above

31. Introduction of fast-growing exotics is possible in
(a) Uniform system
(b) Shelterwood strip system
(c) Group system
(d) Clear felling system

32. Silvicultural system in which Clearfelling is done in the form of strips alternate with unfelled strips is known as
(a) Uniform Strip system
(b) Alternate Strip system
(c) Clear Strip system
(d) Group and Strip system

33. In Deodar/Spruce………number of seed bearers are retained per hectare
(a) 20–25 (c) 45–50
(b) 75–80 (d) 12–18

34. In Silver fir and Spruce forest, the number of secondary felling is/are
(a) One (c) Three
(b) Two (d) Four

35. A part of the forest set aside to be regenerated is termed as
(a) Regeneration interval
(b) Regeneration period
(c) Periodic block
(d) None of the above

36. Area under the fixed periodic block is worked according to
(a) FS *R/P (c) FS/P*R
(b) FS *P/R (d) None of these

37. Type of the periodic block in which the area of block and the period is not fixed
(a) Single periodic block
(b) Floating periodic block
(c) Permanent periodic block
(d) Both (a) and (b)

38. Floating periodic block was first introduced in
(a) France (c) America
(b) Germany (d) Both (a) and (b)

39. Floating periodic block is also known as
(a) Quartier bleu method
(b) Single periodic Block method
(c) Quartier blanc
(d) Both (a) and (b)

40. Felling's made under high forest system usually towards the end of rotation is termed as
(a) Regeneration felling
(b) Final felling
(c) Preparatory felling
(d) Selection felling

41. Uniform system for Deodar and Kail was started by ....in 1912
(a) Trevor
(b) Heinrich von cotta
(c) Eberhard
(d) Gayer

42. The Group system was developed by
(a) Trevor (c) Troupe
(b) Eberhard (d) Karl Gayer

43. Group system is also known as
(a) Bavarian Femelschlag
(b) Swiss Femelschlag
(c) Baden Femelschlag
(d) Both (b) and (c)

44. The Group system was first tried for
(a) Deodar (UP) (c) Deodar (HP)
(b) Sal (Assam) (d) Acacia (HP)

45. Which system is applied in moist Deodar forest in H.P. and U.P.?
(a) Indian Irregular Shelterwood system
(b) Uniform Shelterwood system
(c) Selection system
(d) Group system

46. The modification of the Uniform system is/are
(a) Shelterwood Strip system
(b) Strip and Group system
(c) Indian Irregular Shelterwood system
(d) Both (a) and (c)

47. Wagner Blender's Saumschlag is the modification of
(a) Shelterwood Strip system
(b) Uniform system
(c) Group system
(d) None of the above

48. Silvicultural system in which regeneration fellings are carried out in narrow strips extending in east-west direction and advancing from north to south is
(a) Wedge system.
(b) Strip and Group system
(c) Wagner Blender's Saumschlag
(d) None of the above

49. The width of strip in Wagner's Blender Saumschlag is usually......... the height of tree
(a) Triple (c) Double
(b) Half (d) Equal

50. The Wedge system is developed by
(a) Gayer (c) Eberhard
(b) Trevor (d) Troupe

51. The most widely distributed Bamboo in India is
(a) *Bambusa arundinacea*
(b) *Dendocalamus strictus*
(c) *Bambusa vulgaris*
(d) None of above

52. Bamboo is generally worked on felling cycle of
(a) 1 year (c) 2 years
(b) 3 years (d) 4 years

53. The Irregular shelterwood system is practised in
(a) Germany (c) Canada
(b) Europe (d) India

54. The regeneration period of Bavarian Femelschlag is usually
(a) 50–60 years (c) 5–10 years
(b) 60–70 years (d) 20–30 years

55. The Swiss and Baden Femelschlag the regeneration period is
(a) 50–60 years (c) 20–30 years
(b) 30–40 years (d) 10–20 years

56. The crop produced in Irregular Shelterwood system is
(a) Unevenaged (c) Evenaged
(b) Irregular (d) Both (a) and (c)

57. In natural selection Sal is best achieved under
(a) Clearfelling system
(b) Shelterwood system
(c) Coppice system
(d) None of the above

58. In ....system generally large quantity of advance growth, trees upto 40 cm retained as future crop
(a) Irregular Shelterwood system
(b) Coppice with reserve
(c) Indian Irregular Shelterwood system
(d) Coppice with standards

59. Andaman's canopy lifting shelterwood system is also known as
(a) Indian Irregular Shelterwood system
(b) Irregular Shelterwood system
(c) Group system
(d) Selection system

60. The alternate strip system was experimentally applied to
(a) *Acacia mearnsii* (c) Sal
(b) *Cedrus deodara* (d) Fir

61. Clear felling is done in the form of strips
(a) To regenerate area naturally from seed from adjoining area
(a) To protect young crop against wind, snow, etc...
(a) Both (a) and (b)
(d) None of the above

62. Important considerations for the application of Clear felling system is/are
(a) Species and composition of crop
(b) Potential productivity of site
(c) Factors of locality
(d) All of the above

63. The number of secondary fellings depends upon
(a) Species
(b) Progress of its regeneration
(c) Light requirement
(d) Both (a) and (b)

64. Factors affecting length of regeneration period in uniform system
(a) Light requirement

(b) Fire
(c) Incidence of grazing and browsing
(d) All of the above

65. The time lapse between the successive main fellings on same area is
(a) Seedling felling
(b) Final felling
(c) Regeneration period
(d) None of the above

66. Felling cycle in India is usually of
(a) 5 years (c) 10 years
(b) 8 years (d) 11 years

67. Silvicultural system in which felling and regeneration is distributed over the whole of the area and resultant crop is Unevenaged
(a) Selection system
(b) Group selection system
(c) Strip and group system
(d) Improvement felling

68. The tropical evergreen and semi-evergreen forest in Assam, Kerala, Karnataka, Maharashtra and Tamil Nadu are worked under
(a) Selection system
(b) Group selection system
(c) High forest with reserve system
(d) Two storied high forest system

69. Pick the odd one out
(a) Improvement felling
(b) High forest with reserve system
(c) Two storied high forest system
(d) None of the above

70. In two storied high forest system, the crop in each storey is..............
(a) Unevenaged (c) Regular
(b) Evenaged (d) Irregular

71. In two Storied High forest system, lower storey is obtained by
(a) Natural regeneration by seed stored in the area
(b) Natural regeneration by seed received from outside
(c) Under planting
(d) Both (b) and (c)

72. Chir pine and Bamboo growing in dry deciduous forest are worked under
(a) Two storeyed high forest system
(b) High forest with reserve system
(c) Group selection system
(d) None of the above

73. The main aim of the High forest with Reserve system is to get
(a) Small size timber
(b) Large size timber
(c) Both (a) and (b)
(d) None of the above

74. Improvement felling is generally applied to
(a) Evenaged forest
(b) Un-evenaged forest
(c) Mixed Un-evenaged forest
(d) None of the above

75. The crop arising from the Adventitious bud of the stump of felled tree is known as
(a) Coppice system
(b) Selection system
(c) Accessary system
(d) Clearfelling system

76. Coppicing season is from
(a) September–October
(b) June–July
(c) October–November
(d) November–March

77. The *Eucalyptus* in Nilgiris is worked under
(a) Shelterwood coppice system
(b) Simple coppice system
(c) Coppice with standards
(d) Coppice with reserve system

78. In which of the following system the main objective is to produce large sized timber in addition to poles, fuel, and small sized timber
(a) Simple Coppice system
(b) Shelterwood Coppice system
(c) Coppice of two rotation system
(d) Coppice with reserve system

79. In Frost prone areas the most suitable system is
(a) Coppice of two rotation system
(b) Coppice with reserve system
(c) Coppice with standard
(d) Shelterwood coppice system

80. In *Eucalyptus* species the usual stump height is
(a) 15–20 cm
(b) 5–10 cm
(c) 10–15 cm
(d) 20–25 cm

81. The most acceptable system in areas having more demand for small sized timber and fuelwood, the most acceptable system is
(a) Selection system
(b) Pollard system
(c) Simple coppice system.
(d) Coppice with standard system

82. In Coppice with standard system, the standards are retained for
(a) Enrichment of coppice
(b) Supply of large size timber
(c) Protection against frost
(d) All of the above

83. The Coppice with standard system applicable to
(a) *Anogeissus pendula*
(b) Jamun belts
(c) Sal forest
(d) All of the above

84. Which is the best system for biodiversity conservation
(a) Shelterwood system
(b) Selection system
(c) Clear felling system
(d) Coppice system

85. The Sal and other mixed tree species of moist deciduous forest are managed under
(a) Selection system
(b) Indian Irregular Shelterwood system
(c) Uniform system
(d) Clear felling system

86. The regeneration period and rotation age of *Pinus roxburghii* under uniform shelterwood system is
(a) 25 and 100 years
(b) 30 and 100 years
(c) 30 and 120 years
(d) None of the above

87. Planting of valuable species in degraded forest is known as
(a) Enrichment planting
(b) Improvement felling
(c) Planting out
(d) Afforestation

88. Selection system is best suited for
(a) Light demander
(b) Shade bearer
(c) Shade demander
(d) All of the above

89. The Coppice with reserve system was first introduced by
(a) Troupe (c) RN Dutta
(b) KP Sagreiya (d) Gayer

90. In Coppice with Reserve system, the reservation is done by
(a) Species
(b) Area
(c) Trees
(d) All of the above

91. The Coppice with reserve system is applicable to
(a) Evergreen forest
(b) Moist Deciduous forest
(c) Dry Deciduous forest
(d) Moist Evergreen forest

92. *Acacia catechu* in H.P. work under
(a) Coppice selection system.
(b) Selection system
(c) Coppice with reserve system
(d) Clearfelling

93. *Hardwickia binata* in Andhra Pradesh worked under
(a) Simple Coppice system
(b) Shelterwood system
(c) Coppice with reserve system
(d) Pollard system

94. Dauerwald is a...........term
(a) Latin (c) French
(b) German (d) American

95. Dauerwald concept was introduced by
(a) Alfred Moller (c) RN Dutta
(b) Trevor (d) Brandis

96. Removal of sound and mature trees above a fixed diameter/girth limit is known as
(a) Selective felling
(b) Final felling
(c) Preparatory felling
(d) Selection felling

97. The steep hill slopes and broken rugged terrain are worked under
(a) Coppice Selection system
(b) Selection system
(c) Improvement felling
(d) Shelterwood system

98. The Ideal example of the Dauerwald system is
(a) Shelterwood system
(b) Improvement felling
(c) Group system
(d) Selection system

99. .... was the first forester to raise voice against Clearfelling
(a) Gurnaud
(b) Biolley
(c) JB Rain tree
(d) All of the above

100. Bamboo is worked under
(a) Selection system
(b) Coppice system
(c) Culm selection system
(d) Uniform system

Answer Keys									
1. (b)	**2.** (a)	**3.** (d)	**4.** (b)	**5.** (a)	**6.** (c)	**7.** (d)	**8.** (a)	**9.** (c)	**10.** (b)
11. (d)	**12.** (b)	**13.** (d)	**14.** (d)	**15.** (c)	**16.** (d)	**17.** (a)	**18.** (c)	**19.** (a)	**20.** (c)
21. (a)	**22.** (c)	**23.** (a)	**24.** (c)	**25.** (c)	**26.** (c)	**27.** (a)	**28.** (b)	**29.** (b)	**30.** (b)
31. (d)	**32.** (b)	**33.** (c)	**34.** (b)	**35.** (c)	**36.** (b)	**37.** (d)	**38.** (a)	**39.** (d)	**40.** (c)
41. (a)	**42.** (d)	**43.** (a)	**44.** (c)	**45.** (d)	**46.** (d)	**47.** (a)	**48.** (c)	**49.** (b)	**50.** (c)
51. (b)	**52.** (d)	**53.** (b)	**54.** (d)	**55.** (a)	**56.** (d)	**57.** (b)	**58.** (c)	**59.** (a)	**60.** (c)
61. (c)	**62.** (d)	**63.** (d)	**64.** (d)	**65.** (b)	**66.** (c)	**67.** (a)	**68.** (b)	**69.** (d)	**70.** (b)
71. (d)	**72.** (a)	**73.** (b)	**74.** (c)	**75.** (a)	**76.** (d)	**77.** (b)	**78.** (c)	**79.** (d)	**80.** (a)
81. (c)	**82.** (d)	**83.** (d)	**84.** (b)	**85.** (a)	**86.** (c)	**87.** (a)	**88.** (b)	**89.** (c)	**90.** (d)
91. (c)	**92.** (a)	**93.** (d)	**94.** (b)	**95.** (a)	**96.** (d)	**97.** (b)	**98.** (d)	**99.** (a)	**100.** (c)

5

Forest Management Law and Policy

1. The practical application of the scientific, technical, and economic principles of forestry is termed as
(a) Forest regulation
(b) Forest management
(c) Forest valuation
(d) Forest production

2. Forest and wildlife were included in the concurrent list during the year
(a) 1976 (c) 1975
(b) 1980 (d) 1898

3. Forest was included in the concurrent list (42nd constitutional amendment) during .... schedule.
(a) Third (c) Fourth
(b) Sixth (d) Seventh

4. The total number of forest policies in India upto 2019 are
(a) Two (c) Four
(b) Three (d) Five

5. Who promulgated the Forest (Conservation) Ordinance, 1980
(a) President
(b) Prime minister
(c) Forest minister
(d) None of the above

6. National Commission on Agriculture was constituted in the year
(a) 1980 (c) 1976
(b) 1876 (d) 1975

7. Working plans are invariably based on the principle of
(a) Progressive yield
(b) Sustained yield
(c) Intermediate yield
(d) None of the above

8. Social forestry includes
(a) Farm forestry
(b) Extension forestry
(c) Recreation forestry
(d) All of the above

9. According to NFP-1952, the Indian forest has been classified into
(a) Four types (c) Three types
(b) Five types (d) Two types

10. Smallest permanent working plan unit in India is
(a) Block
(b) Compartment
(c) Sub-compartment
(d) None of the above

11. The main territorial division of the forest bounded by natural features is
(a) Coupe
(b) Compartment
(c) Sub-Compartment
(d) Block

12. At Central level, all the Government owned forest are administered by

(a) Chief Conservator of Forest
(b) Deputy Conservator
(c) Inspector General of Forest
(d) None of the above

13. The smallest functional territorial unit which forms the base of the Indian forest administration is
(a) Beat
(b) Range
(c) Block
(d) All of the above

14. In the state forest are managed by
(a) PCCF (c) ACCF
(b) CCF (d) Both (a) and (c)

15 ....is the instrument used for the forest management
(a) Working circle
(b) Felling series
(c) Working plan
(d) None of the above

16 On the basis of silviculture management forests are classified as
(a) Working circle
(b) Cutting sections
(c) Coupe
(d) All of the above

17. The periodic blocks which retain their territorial identity at each working plan revision is/are termed as
(a) Fixed
(b) Floating
(c) Permanent
(d) Both (a) and (b)

18. The time that elapses between successive main felling on the same area is termed as
(a) Felling cycle
(b) Felling series
(c) Regeneration cycle
(d) None of the above

19. Working circle is further divided into
(a) Working plan
(b) Cutting section
(c) Felling series
(d) Period block

20. Felling series is further divided into
(a) Coupe
(b) Cutting section
(c) Compartment
(d) Felling series

21. .... is the backbone of forest management.
(a) Progressive yield
(b) Sustained yield
(c) Intermediate yield
(d) Maximum sustained yield

22. An age class with one year interval is termed as
(a) Age-gradation (c) Both (a) and (b)
(b) Age-interval (d) Evenage

23. .... is a very important unit in the management and administration of forest.
(a) Beat (c) Range
(b) Section (d) Division

24 The material that a forest can yield annually in perpetuity is termed as
(a) Progressive yield
(b) Sustained yield
(c) Both (a) and (b)
(d) None of the above

25. Sustained yield may be
(a) Annual (c) Monthly
(b) Periodic (d) Both (a) and (b)

26. The sustained yield principle is applicable to............... Forestry.
(a) Production (c) Social
(b) Protection (d) Both (a) and (b)

27. The "concept of the progressive yield" advocated by German forester
(a) Brandis (c) Huber
(b) Von mantle (d) Hartig

28 The "concept of progressive yield" envisages at
(a) Continuous yield
(b) Raising productivity of soil and crop
(c) Raising productivity of soil
(d) Raising productivity of crop

29. The principle of progressive yield is
(a) Dynamic (c) Static
(b) Increasing (d) Decreasing

30. The aim of sixth Silviculture conference in early stages is/are
(a) Sustained yield
(b) Progressive yield
(c) Increasing yield
(d) All of the above

31. The period between the formation and final felling of forest crop is termed as
(a) Rotation period
(b) Production period
(c) Both (a) and (b)
(d) None of the above

32. Rotation that coincides with the natural lease of life of species on a given site.
(a) Silvicultural rotation
(b) Physical rotation
(c) Technical rotation
(d) None of the above

33. Physical rotation is applicable to
(a) Protection forests
(b) Park lands
(c) Roadside avenues
(d) All of the above

34. Silvicultural rotation may be useful in forests managed for
(a) Aesthetic and recreation purpose
(b) Production
(c) Protection
(d) All of the above

35. Which type of rotation is adopted in industrial firms?
(a) Rotation of maximum volume production
(b) Financial rotation
(c) Technical rotation
(d) Physical rotation

36. .... aims at producing the maximum material of specified dimension/quality for specific purpose such as railway sleepers, paper-wood, match-wood.
(a) Financial rotation
(b) Physical rotation
(c) Silvicultural rotation
(d) None of the above

37. The rotation that yields the maximum quantity of material is/are
(a) Economic rotation
(b) Rotation of maximum volume production
(c) Silvicultural rotation
(d) All of the above

38. For the production of firewood, paper and pulp, the best rotation is
(a) Rotation of maximum volume production
(b) Physical rotation
(c) Silvicultural rotation
(d) Economic rotation

39. The rotation of maximum volume production is usually .... than financial rotation.
(a) Longer

(b) Smaller
(c) Larger
(d) None of the above

40. Which type of rotation yields the largest volume per unit area per annum?
(a) Economic rotation
(b) Rotation of maximum volume production
(c) Silvicultural rotation
(d) Financial rotation

41 Common practice in forestry is to adopt a combination of
(a) Financial rotation and Rotation of highest income
(b) Technical rotation and Silvicultural rotation
(c) Maximum volume production and rotation of highest income
(d) Financial rotation and Rotation of maximum volume production

42. Which of the following is calculated without interest and irrespective of time when items of income or expenditure occurs?
(a) Rotation of highest income
(b) Financial rotation
(c) Economic rotation
(d) Both (b) and (c)

43. The rotation which is most profitable is
(a) Technical rotation
(b) Rotation of highest income
(c) Economic rotation
(d) Rotation of maximum volume production

44. From national point of view which rotation is important?
(a) Technical rotation
(b) Rotation of highest income/ revenue
(c) Financial rotation
(d) All of the above

45. Which rotation yields the highest net return on the invested capital?
(a) Financial rotation
(b) Rotation of highest income
(c) Rotation of maximum volume production
(d) All of the above

46. Faustmann's formula is used to calculate
(a) Sustained yield
(b) Progressive yield
(c) Soil expectation value
(d) Land expectation value

47. Which of the rotation concerned with money return?
(a) Financial rotation
(b) Rotation of maximum volume production
(c) Rotation of revenue
(d) Both (b) and (c)

48. Soil expectation value is
(a) X/0.0p
(b) Faustmann's formula
(c) Both (a) and (b)
(d) None of the above

49. Length of the rotation depends upon
(a) Economic considerations
(b) Growth rate
(c) Silvicultural Characters of species
(d) All of the above

50. The rotation age of the Chir pine worked under chakrata division
(a) 100 years

(b) 90 years
(c) 120 years
(d) 60 years

51. The rotation age of the *Eucalyptus globulus* worked under simple coppice in Nilgiris
(a) 12 years (c) 10 years
(b) 8 years (d) 20 years

52. The change from one silvicultural system to another is termed as
(a) Conversion period
(b) Conversion
(c) Production
(d) All of the above

53. The conversion period is usually .... than/to rotation
(a) Equal
(b) Less
(c) More
(d) All of the above

54. Which of the following forms the part of "trinity of norms" in forestry
(a) Normal increment
(b) Normal growing stock
(c) Normal series of age gradations
(d) All of the above

55. "Trinity of norms" was given by
(a) Hartig (c) Brandis
(b) Osmastan (d) Watson

56. The concept of "Normal forest" was introduced by
(a) German foresters
(b) French foresters
(c) American foresters
(d) None of the above

57. Normal forest is a conception of forest management based on the principle of
(a) Progressive yield
(b) Intermediate yield
(c) Sustained yield
(d) None of the above

58. The Normal forest is also called as
(a) Regulated forest
(b) Fully regulated forest
(c) Managed forest
(d) All of the above

59. Which of the following is a nearest approach to the theoretical normality?
(a) Virgin forest
(b) Plantations
(c) Natural forest
(d) Both (a) and (c)

60 Normal forest is created by
(a) Progressive scientific treatment
(b) Artificial regeneration
(c) Natural regeneration
(d) All of the above

61. A forest is called abnormal forest if
(a) Overstocked
(b) Understocked
(c) Increment is subnormal
(d) All of the above

62. The kind of the abnormality arising due to reduction in rotation age
(a) Under-stocked
(b) Over-stocked
(c) Poor stand density
(d) All of the above

63. A forest is said to be understocked due to
(a) Extension in rotation age
(b) Poor density
(c) Reduction in rotation age
(d) Both (a) and (b)

64. Worst form of abnormality
(a) Normal increment volume in an abnormal forest

(b) Increment is sub normal
(c) Normal GS but abnormal distribution of age classes
(d) Under-stocked.

65. Law used to estimate the proportionate distribution of trees in selection forest
(a) De Liocourt's law
(b) Mayer's formula
(c) Faustamann formula
(d) None of the above

66. According to Ninth Silvicultural conference at F.R.I., Dehradun held in 1956. The De Liocourt law is also applicable to
(a) Tropical wet evergreen forest
(b) Southern moist deciduous forest
(c) Hill sal forest
(d) All of the above

67. In general, the De Liocourt law holds good for
(a) Tropical forest
(b) Temperate forest
(c) Both (a) and (b)
(d) None of the above

68. In the De Liocourt's law a, aq^{-1}, aq^{-2} aq^{-3}...........aq^{n-1}. The "q" refers to
(a) Constant
(b) Decreasing number of trees per hectare
(c) Coefficient of reduction
(d) None of the above

69. The linear graph between the log number of stems against their mid-diameters indicate
(a) Balanced crop
(b) Unbalanced crop
(c) Mature crop
(d) Immature crop

70. The increase in the growth that takes place in a particular year is termed as
(a) CAI
(b) MAI
(c) PAI
(d) All of the above

71. The increments are usually taken during the interval of
(a) 2–4 years (c) 5–10 years
(b) 9–10 years (d) 1–2 years

72. The average annual increment for a short period is
(a) MAI (c) CAI
(b) PAI (d) TAI

73. The increase in girth, diameter, basal area, height of the individual tree or crop during a given period is
(a) Yield (c) CAI
(b) Growing stock (d) Increment

74. The total increments up to a given age divided by that age is
(a) MAI (c) CAI
(b) Final MAI (d) PAI

75. The M.A.I. for the entire rotation period is
(a) Current annual increment
(b) Mean annual increment
(c) Final mean annual increment
(d) Periodic annual increment

76. The M.A.I. and C.A.I. coincides .... times in the life of a crop.
(a) 1
(b) 3
(c) 2
(d) None of the above

77. Which of the following is correct?
(a) To begin with CAI keeps below MAI

(b) The CAI attains the maximum before MAI
(c) The MAI is falling when CAI is rising
(d) While the MAI is more than CAI, MAI is rising

78. The maximum average volume per unit area per is obtained when
(a) MAI > CAI
(b) MAI < CAI
(c) CAI = PAI
(d) CAI = MAI

79. The age at which the M.A.I. and C.A.I. curve intersect is known as
(a) Culminating point
(b) Rotation of maximum volume production
(c) Both (a) and (b)
(d) None of the above

80. Generally in seedling and sapling stage C.A.I. is
(a) Rapidly increasing
(b) Declining
(c) Small
(d) Large

81. C.A.I. can be
(a) Positive
(b) Negative
(c) Zero
(d) All of the above

82. M.A.I. can never be
(a) Negative (c) Positive
(b) Zero (d) Both (a) and (b)

83 The increment of a crop is expressed in the form of
(a) CAI (c) Percentage
(b) MAI (d) Both (a) and (b)

84. For an individual tree increment can be found using stem analysis by the means of
(a) Pressler increment borer
(b) Wedge prism
(c) Both (a) and (b)
(d) None of the above

85. Which of the following formula is applicable to ring porus trees?
(a) Schneider's formula
(b) Pressler's formula
(c) Both (a) and (b)
(d) None of the above

86. Which of the following is used to calculate the increment percent using number of rings
(a) Pressler's formula
(b) Schneider's formula
(c) Both (a) and (b)
(d) None of the above

87. A fast growing species yields a minimum of
(a) 15 m^3/ha/year (c) 25 m^3/ha/year
(b) 20 m^3/ha/year (d) 10 m^3/ha/year

88. In younger plantations, the height increment should not be less than
(a) 10 cm/annum (c) 60 cm/annum
(b) 50 cm/annum (d) 80 cm/annum

89. Fast growing tropical pine species is/are
(a) *Pinus caribeae*
(b) *Pinus radiata*
(c) *Pinus patula*
(d) All of the above

90. Method used for determination of yield based on successive inventories
(a) Biolley's methode du controle
(b) Hartig method
(c) Brandis diameter class method
(d) Von mantle's formula

91. The term capital in forest management usually refers to
(a) Growing stock (c) Land
(b) Normal forest (d) Forest soil

92. Sum of all the trees growing in a forest is called as
(a) Yield (c) Rotation
(b) Growing stock (d) Increment

93. According to the De Liocourt law percentage reduction in stem number from one diameter class to next
(a) Constant (c) Increases
(b) Decreases (d) Both (b) and (c)

94. The actual growing stock can be enumerated most accurately by
(a) Complete enumeration
(b) Sample plot measurement
(c) Sample enumeration
(d) All of the above

95. In maximum volume production which option is correct?
(a) CAI and MAI cuts together
(b) MAI becomes maximum
(c) CAI and MAI becomes equal
(d) All of the above

96 What does 'C' denote in the equation for calculating 'growing stock' GS = I × r × C
(a) Swiss constant
(b) Flury's constant
(c) Reducing factor
(d) None of the above

97. In selection forest, Munger evolved the formula for measuring growing stock based on
(a) CAI (c) MAI
(b) PAI (d) Final MAI

98. The percent ratio of normal yield to the normal growing stock is called
(a) Massion's ratio
(b) Exploitation percent
(c) Utilisation percent.
(d) All of the above.

99. The number of stems that can be removed annually/periodically is called
(a) Growing stock
(b) Yield
(c) Final yield
(d) Sustained yield

100. Intermediate yield is
(a) Yield harvested from a natural forest
(b) Yield obtained from forest between two harvesting seasons
(c) Yield harvested from thinning
(d) None of the above

101. The material that a forest can yield annually/periodically in perpetuity is
(a) Normal yield (c) Yield
(b) Total yield (d) Sustained yield

102. Intermediate yield is also known as
(a) Main yield
(b) Subsidiary yield
(c) Total yield
(d) Final yield

103. The yield from the normal forest is called as
(a) Normal yield
(b) Sustained yield
(c) Final yield
(d) Total yield

104. Term generally applied to the determination of the yield and the prescribed means of realising it is called

(a) Yield capacity
(b) Yield determination
(c) Yield regulation
(d) None of the above

105. The forest and wildlife were brought under concurrent list as per
(a) 40th amendment
(b) 44th amendment
(c) 48th amendment
(d) 42nd amendment

106. The period during which a change from one Silvicultural system to another is called
(a) Conversion
(b) Conversion period
(c) Production period
(d) All of the above

107. Indian method used to estimate yield based on volume and increment using number of trees is
(a) Smythies' safe-guarding formula
(b) Brandis diameter class
(c) Smythies' modification
(d) Volume unit method

108. Working plan is generally prepared for the period of
(a) 5–6 years (c) 10 years
(b) 8–10 years (d) 5 years

109. What is the unit of working plan in India?
(a) Working Circle
(b) Forest Division
(c) Felling series
(d) Block

110. The main objective of yield regulation is to achieve .... from the forest.
(a) Progressive yield
(b) Sustained yield
(c) Intermittent yield
(d) None of the above

111. The W.P.O. in India is
(a) Deputy conservator of forest
(b) Conservator of forest
(c) Assistant conservator of forest
(d) None of the above

112. The constitution of Working plan is generally decided by
(a) Conservator of forest
(b) Deputy conservator of forest
(c) Conservator of forest
(d) Deputy ranger

113. The constitution of Working plan is approved by
(a) Assistant conservator of forest
(b) Chief conservator of forest
(c) Deputy conservator of forest
(d) None of the above

114. Which of the following working plan map depicts the distribution of forest types, main forest species, blanks etc...
(a) Stock map
(b) Reference map
(c) Enumeration map
(d) Regeneration survey map

115. "Enumeration map" are prepared in the scale of
(a) 1:5000
(b) 1:50,000
(c) 1:15,000
(d) 1:10,000

116. Working plan map are prepared in the scale of
(a) 1:10,000
(b) 1:60,000
(c) 1:40,000
(d) 1:5000

117. The stock maps are generally prepared by
(a) WP Rangers (c) Both (a) and (b)
(b) ACF's (d) None of these

118. The crop composition >75% of teak may be classified as
(a) Pure Teak (c) Mixed Teak
(b) Mainly Teak (d) Even-age Teak

119. In forest, the standard colour given to excellent (81–100%) regeneration
(a) Blue (c) Yellow
(b) Red (d) Green

120. In working plan, Himalayan moist temperate forest is depicted by
(a) Orange (c) Yellow
(b) Green (d) Red

121. In working plan, sub-tropical pine forest is depicted by
(a) Orange (c) Pink
(b) Blue (d) Violet

122. Basis of the yield regulation in Von mantle's formula is
(a) Area and volume
(b) Increment and volume
(c) Volume
(d) None of the above

123. Von mantles formula of yield regulation is
(a) 8V/3r (c) 9V/4r
(b) 2V/(r-x) (d) 2GS/r

124. Which of the following is known as the formula of glorious simplicity
(a) French method
(b) Von mantles formula
(c) Masson's formula
(d) None of the above

125. In Von mantles formula, yield during the rotation is .... the growing stock.
(a) Twice (c) Equal
(b) Half (d) None of these

126. Massion formula for yield is
(a) r/200 (c) 100/r
(b) 200/r (d) 200/2r

127. The exploitation percentage is calculated using
(a) French method
(b) Howard method
(c) Masson's formula
(d) Simmon's formula

128. In Howard's modification, crop is measured down to diameter corresponding to rotation age
(a) 1/3 (c) 1/4
(b) 2/3 (d) 1/2

129. Howard's formula used for calculating yield is
(a) 8V/3r (c) 9V/4r
(b) 2V/r-x (d) 2GS/r

130. Simmons modifications formula is
(a) 8V/3r
(b) 2V/r-x
(c) $\frac{(2n^2 \times V)}{r(n^2 - 1)}$
(d) 9V/4r

131. Smythe's modification formula is
(a) 8V/3r
(b) 2V/r-x
(c) $\frac{(2n^2 \times V)}{r(n^2 - 1)}$
(d) 9V/4r

132. Burma modification formula is
(a) 8V/3r
(b) $\frac{(2n^2 \times V)}{r(n^2 - 1)}$
(c) 2V/r-x
(d) 9V/4r

133. A forest area organised for particular object and managed under one Silvicultural system where one set of working plan prescriptions is known as
(a) Working plan
(b) Working circle
(c) Working series
(d) Forest management

134. The base of Napier logarithms is
(a) 2.71828 (c) 2.95432
(b) 2.1562 (d) 3.99562

135. Growing stock is calculated by
(a) Planimeter
(b) By counting the square
(c) Both (a) and (b)
(d) None of the above

136. Yield table provides data at an interval of
(a) 10 year (c) 5 year
(b) 15 year (d) 2 year

137. The increment determination in regular forest is done by
(a) Area method
(b) MAI method
(c) Yield table
(d) All of the above

138. Andrew's formula is used for calculating increment in
(a) Regular forest
(b) Even-age forest
(c) Irregular forest
(d) Both (a) and (c)

139. The forest capital is
(a) Sustained yield
(b) Progressive yield
(c) Growing stock
(d) None of the above

140. The actual growing stock of the forest is calculated using by
(a) Complete enumeration
(b) Partial enumeration
(c) Sample plot measurement
(d) All of the above

141. The concept of sustainability was first introduced in
(a) 1987 (c) 1976
(b) 1874 (d) 1850

142. The permanent allotment method was evolved by
(a) Hartig (c) Simmon
(b) Cotta (d) Hufnagl

143. The permanent allotment method was modified by
(a) Hufnagl (c) Klipstein
(b) Howard (d) Osmaston

144. If growing stock is 300, then the exploitation percentage will be
(a) 2/r% (b) 20/r% (c) 1/r% (d) 3/r%

145. Karl calculated the annual yield from
(a) MAI (c) PAI
(b) CAI (d) Final MAI

146. The "Smythies' safeguard formula" was evolved for
(a) Sal forest of UP
(b) Sal forest of HP
(c) Teak forest of UP
(d) None of the above

147. Working plan officer is
(a) RFO (b) ACF
(c) DFO
(d) None of these

148. The wildlife protection act was formulated in the year
(a) 1927 (c) 1872
(b) 1972 (d) 1756

149. The wildlife protection act consist of
(a) 7 chapters and 66 sections
(b) 6 chapters and 66 sections
(c) 7 chapters and 60 sections
(d) 6 chapter and 60 sections

150. First Indian forest act was drafted in the year
(a) 1927 (c) 1980
(b) 1872 (d) 1976

151. The last forest policy was formulated in
(a) 2004 (c) 1952
(b) 1988 (d) 1980

152. The first Inspector general of forest was
(a) Brandis (c) Karl gayer
(b) Troup (d) Chaturvedi

153. The National commission on agriculture submitted its report during
(a) 1954 (c) 1978
(b) 1870 (d) 1976

154. The first National Forest Policy was formulated and based on the recommendation of
(a) Dr Voelker (c) Brandis
(b) Dr Brandis (d) Troup

155. First Forest Policy of India was enacted in
(a) 1927 (c) 1980
(b) 1988 (d) 1952

156. The headquarters of Inspector General of Forest is
(a) Banglore (c) Mumbai
(b) Delhi (d) None of these

157. The first Forest Policy was enacted during
(a) 1952 (c) 1894
(b) 1884 (d) 1957

158. Conservation of ecological fragile ecosystem and preservation of biological diversity in terms of flora and fauna was emphasized in
(a) Forest policy, 1894
(b) Forest Policy, 1952
(c) Forest policy, 1988
(d) None of the above

159. Ban on deforestation for the purpose of agriculture is as per
(a) Forest policy, 1894
(b) Forest Policy, 1952
(c) Forest policy, 1988
(d) None of the above

160. Second National Forest policy of India was issued in
(a) 1989 (c) 1956
(b) 1988 (d) 1952

161. NFP, 1988 prescribed .... area under forest in plains.
(a) 33% (c) 30%
(b) 66% (d) 60%

162. Which NFP encourages the efficient utilization of forest produce and maximizing the substitution of wood?
(a) 1988 NFP (c) 1952 NFP
(b) 1894 NFP (d) All of these

163. Joint forest planning and management scheme has been passed in NFP during

(a) 1962 (b) 1894
(c) 1980
(d) None of the above

164. According to NFP 1952, the minimum forest area for environment stability should be
(a) 3% (c) 33%
(b) 60% (d) 44%

165. Forest laws are
(a) General laws
(b) Constitutional laws
(c) National laws
(d) Special laws

166. Forest Conservation act was enacted in the year
(a) 1980 (c) 1972
(b) 1982 (d) 1940

167. The project tiger was initiated as per advice of
(a) Rajeev Gandhi (c) Indira Gandhi
(b) Feroz Gandhi (d) None of these

168. The project tiger was launched in
(a) 1972 (c) 1973
(b) 1980 (d) All of these

169. Wildlife protection Act, 1972 has
(a) 7 schedules
(b) 5 schedules
(c) 8 schedules
(d) None of the above

170. Central Zoo Authority and recognisation of zoo's comes under which chapter of Wildlife Protection Act
(a) IV A (c) IV B
(b) IV C (d) IV D

171. The Indian Board of Wildlife (IBWL) was set up in
(a) 2002 (c) 1951
(b) 1952 (d) 1980

172. Wildlife conservation strategy (2002) was announced during 21st meeting of
(a) WWF (c) IBWL
(b) BNHS (d) WII

173. Conservation of ecologically fragile ecosystem and preservation of biological diversity in terms of flora and fauna was emphasized in
(a) NFP-1988 (b) NFP-1952
(c) NFP-1894
(d) All of the above

174. The Declaration of National park by state government is covered under which section of Wildlife (Protection) Act, 1972?
(a) Section 18 (c) Section 35
(b) Section 38 (d) Section 20

175. The declaration of National park or Wildlife sanctuary by central government is covered under which section of Wildlife (Protection) Act, 1972?
(a) Section 38 (c) Section 35
(b) Section 18 (d) Section 20

176. The declaration of Wildlife sanctuary by state government is covered under which section of Wildlife (Protection) Act, 1972?
(a) Section 38 (c) Section 35
(b) Section 21 (d) Section 18

177. The Protected forest are classified under ....chapter of Indian Forest Act, 1927
(a) II (b) III (c) IV (d) V

178. Sub chapter 6.2 of Indian Forest Act (1927), deals with
(a) Protected forest
(b) Penal fees and procedures
(c) Duty on timber and other products
(d) Reserve forest

Answer Keys									
1. (b)	**2.** (a)	**3.** (d)	**4.** (b)	**5.** (a)	**6.** (c)	**7.** (b)	**8.** (d)	**9.** (a)	**10.** (b)
11. (d)	**12.** (c)	**13.** (a)	**14.** (b)	**15.** (c)	**16.** (d)	**17.** (c)	**18.** (a)	**19.** (c)	**20.** (b)
21. (d)	**22.** (a)	**23.** (c)	**24.** (b)	**25.** (d)	**26.** (a)	**27.** (d)	**28.** (b)	**29.** (a)	**30.** (c)
31. (c)	**32.** (b)	**33.** (d)	**34.** (a)	**35.** (c)	**36.** (d)	**37.** (b)	**38.** (a)	**39.** (c)	**40.** (b)
41. (d)	**42.** (a)	**43.** (c)	**44.** (b)	**45.** (a)	**46.** (c)	**47.** (d)	**48.** (c)	**49.** (d)	**50.** (a)
51. (c)	**52.** (b)	**53.** (d)	**54.** (d)	**55.** (b)	**56.** (a)	**57.** (c)	**58.** (b)	**59.** (b)	**60.** (a)
61. (d)	**62.** (b)	**63.** (d)	**64.** (c)	**65.** (a)	**66.** (d)	**67.** (b)	**68.** (c)	**69.** (a)	**70.** (a)
71. (c)	**72.** (b)	**73.** (d)	**74.** (a)	**75.** (c)	**76.** (c)	**77.** (b)	**78.** (d)	**79.** (c)	**80.** (c)
81. (d)	**82.** (d)	**83.** (d)	**84.** (a)	**85.** (a)	**86.** (b)	**87.** (b)	**88.** (c)	**89.** (d)	**90.** (a)
91. (a)	**92.** (b)	**93.** (a)	**94.** (c)	**95.** (d)	**96.** (b)	**97.** (a)	**98.** (c)	**99.** (b)	**100.** (c)
101. (d)	**102.** (b)	**103.** (a)	**104.** (c)	**105.** (d)	**106.** (b)	**107.** (a)	**108.** (c)	**109.** (b)	**110.** (b)
111. (a)	**112.** (c)	**113.** (b)	**114.** (a)	**115.** (c)	**116.** (b)	**117.** (d)	**118.** (a)	**119.** (d)	**120.** (b)
121. (a)	**122.** (c)	**123.** (d)	**124.** (b)	**125.** (a)	**126.** (b)	**127.** (c)	**128.** (d)	**129.** (a)	**130.** (c)
131. (b)	**132.** (d)	**133.** (b)	**134.** (a)	**135.** (c)	**136.** (c)	**137.** (d)	**138.** (d)	**139.** (c)	**140.** (d)
141. (a)	**142.** (b)	**143.** (c)	**144.** (d)	**145.** (b)	**146.** (a)	**147.** (c)	**148.** (b)	**149.** (a)	**150.** (a)
151. (b)	**152.** (a)	**153.** (d)	**154.** (a)	**155.** (d)	**156.** (b)	**157.** (c)	**158.** (c)	**159.** (b)	**160.** (b)
161. (a)	**162.** (a)	**163.** (d)	**164.** (c)	**165.** (d)	**166.** (a)	**167.** (c)	**168.** (c)	**169.** (d)	**170.** (a)
171. (b)	**172.** (c)	**173.** (a)	**174.** (c)	**175.** (a)	**176.** (d)	**177.** (c)	**178.** (d)		

6

Forest Mensuration

1. The term "Forest mensuration" is derived from word
(a) German (c) Latin
(b) Burmese (d) French

2. How many types of "Pramana" are their
(a) Five (c) Three
(b) Four (d) six

3. Bhaskaracharya's famous book "Lilavati" contains
(a) Tables on measures and weight
(b) Chapter on mensuration
(c) Tables regarding linear measurement
(d) None of the above

4. "Patiganita" book is written by
(a) Anantpala
(b) Sripati
(c) Lalla
(d) All of the above

5. Forest mensuration deals with the measurement of
(a) Diameter
(b) Age
(c) Volume
(d) All of the above

6. British system of measurement was enacted in
(a) 1962
(b) 1856
(c) 1956
(d) 1895

7. French system of measurement was introduced in
(a) 1962 (c) 1960
(b) 1959 (d) 1976

8. Examples of the active sensors is/are
(a) Radar
(b) Photographic system
(c) Lidar
(d) Both (a) and (c)

9. One Acre =..............yards
(a) 1760 yards (c) 8436 yards
(b) 4632 yards (d) None of these

10 Distance can be measured directly by
(a) Relaskop (c) Biltmore stick
(b) Scale (d) Both (a) and (b)

11. The breast height measured in India is
(a) 1.37 m (c) 1.30 m
(b) 4.5 feet (d) Both (a) and (b)

12. Wedge prism is used for
(a) Diameter measurement
(b) Girth measurement
(c) Horizontal point sampling
(d) All of the above

13. .... is considered as the index of fertility
(a) Biomass
(b) Volume
(c) Height
(d) None of these

14. The total height of the tree can be determined using
(a) Crown length and Crown height
(b) Stump height
(c) Both (a) and (b)
(d) None of the above

15. "Varahmihira" refers in which book of mensuration?
(a) Aryabhatika
(b) Bhramgupta sidhanta
(c) Patiganita
(d) None of the above

16. The oldest reference regarding measuring is given in
(a) Rigveda (c) Yajurveda
(b) Atharvaveda (d) Samaveda

17. The Bhaskaracharya's famous book "Lilavati" was translated in Persian by
(a) Arya Bhatt (c) Lalla
(b) Faizi (d) Sripati

18. Thickness of the bark is measured by using
(a) Presseler increment borer
(b) Swedish bark gauge
(c) Wedge prism
(d) None of the above

19. In hill, the measurement is taken towards
(a) Up hill side
(b) Down hill side
(c) Both (a) and (b)
(d) None of these

20. If forking occurs above breast height than it is considered as
(a) Two trees
(b) Single tree
(c) Both (a) and (b)
(d) None of the above

21. Sector fork is used for measuring
(a) Diameter (c) Height
(b) Volume (d) Form

22. Height of tree can be measured by
(a) Barr
(b) Stroud dendrometer
(c) Relaskop
(d) All of the above

23. The choice of instrument for diameter and girth measurement depends upon
(a) Degree of accuracy required
(b) Condition of the logs
(c) Standing/felled tree
(d) All of the above

24. Metal callipers generally used in India are made up of
(a) Aluminium alloy and brass
(b) Aluminium alloy
(c) Aluminium
(d) None of the above

25. In India, callipers are available in the length of
(a) 50 cm (c) 35 cm
(b) 75 m (d) 100 cm

26. ....callipers are graduated to give girth of a tree/log
(a) Flurry calliper
(b) Girth calliper
(c) Fromme calliper
(d) All of the above

27. Tree measuring tapes usually having length of
(a) 10 m (c) 20m
(b) 25 feet (d) 5 m

28. The relationship between diameter and girth of a tree is given by
(a) $1/\pi$
(b) 0.3182

(c) 7/22
(d) All of the above

29. Instruments used for measuring upper stem diameter is/are
(a) Dendrometers
(b) Finnish parabolic calliper
(c) Both (a) and (b)
(d) None of the above

30. Instrument used for measuring diameter at any point of stem
(a) Wheeler penta prism calliper
(b) Finnish parabolic calliper
(c) Spiegel relaskop
(d) All of these

31. Bole height is also known as
(a) Commercial bole height
(b) Length of clear main stem
(c) Both (a) and (b)
(d) None of the above

32. Height can be measured by using
(a) Relaskop
(b) Single pole method
(c) Christian's hypsometer
(d) All of the above

33. Instrument not based on trigonometric principles is
(a) Clinometer
(b) Brandis hypsometer
(c) Blume leiss hypsometer
(d) Smythies hypsometer

34. Instrument based on the principle of similar triangle is/are
(a) Modified Smythies hypsometer
(b) Brandis hypsometer
(c) Both (a) and (b)
(d) None of the above

35. Relaskop was developed by
(b) Hirata (c) Shiffel
(b) Huber (d) Bitterlich

36. Basal area of the crop is measured by using
(a) Ravi multimeter
(b) Relaskop
(c) Wedge prism
(d) Conimeter

37. The form of the tree at the base resembles to
(a) Cone
(a) Paraboloid
(c) Neloid
(d) Circle

38. The form of the tree at the top resembles to
(a) Cylindrical (c) Neloid
(b) Paraboloid (d) Cone

39. The form of the tree in middle resembles to
(a) Cylindrical
(b) Neloid
(c) Cone
(d) Paraboloid

40. "Girder theory" was proposed by
(a) Metzger (c) Brandis
(b) Hirata (d) Lorey

41. Normal form factor is also known as
(a) Absolute form factor
(b) Breast height form factor
(c) True form factor
(d) Artificial form factor

42. Form factor is defined as
(a) Ratio of volume of tree to volume of cylinder having same height and area of cross-section
(b) Ratio of volume of a tree to product of basal area and height
(c) Both (a) and (b)
(d) None of the above

43. In absolute form factor volume refers to
(a) Whole tree above ground level
(b) Part of tree above the point of measurement
(c) Both (a) and (b)
(d) None of the above

44. In India, ……….. is commonly used
(a) Artificial form factor
(b) Absolute form factor
(c) Normal form factor
(b) True form factor

45. The percentage error that can result if the plane of the calliper is not placed at right angle to the axis of the tree is calculated by
(a) 100 (secθ – 1)
(b) D (secθ +1)
(c) 100 (sinθ – 1)
(d) None of these

46. Form quotient was redefined by
(a) A Schiffel (c) Bitterlich
(c) Brandis (d) Tor Jonson

47. Form quotient was postulated by
(a) Tor Jonson (c) Lorey
(b) A Schiffel (d) Hirata

48. Form quotient is defined as ratio between
(a) Mid diameter and dbh
(b) Diameter and dbh
(c) Both (a) and (b)
(d) None of the above

49. Diameter taper tables are also known as
(a) Form class taper tables
(b) Ordinary taper tables
(c) Both (a) and (b)
(d) None of the above

50. Tree form equation is/are given by
(a) Smalian's formula
(b) Huber's formula
(c) Hojer's formula
(d) All of the above

51. Bole surface area is given by
(a) $S = g*l$
(b) $S = \frac{(g_1 + g_2)}{2} \times l$
(c) Both (a) and (b)
(d) None of the above

52. The volume of the frustum of a solid when logs are stacked is given by
(a) Newton's formula
(b) Prismoidal formula
(c) Hojer's formula
(d) None of the above

53. When logs are stacked than volume is calculated by using
(a) Newton's formula
(b) Huber's formula
(c) Smalian's formula
(d) None of the above

54. Smalian's formula always
(a) Under estimates the volume
(b) Gives positive error
(c) Over estimates the volume
(d) Both (b) and (c)

55. "Girder theory" was first proposed by
(a) Schwendener (c) Metzger
(b) Heck (d) Mayer

56. Huber's formula always
(a) Under estimates the volume
(b) Gives negative error
(c) Both (a) and (b)
(d) None of the above

57. "Hoppus formula" is used to calculate
(a) Volume of logs
(b) Basal area of crop
(c) Height
(d) Volume of stand

58. Quater girth formula is also known as "Hoppus rule" in
(a) Germany
(b) Britain
(c) France
(d) All of the above

59. Quarter girth formula is used to calculate
(a) Bole surface area
(b) Volume of the firewood
(c) Volume of the log
(d) None of the above

60. Quarter girth formula gives ...... percentage of true volume
(a) 78.5 (c) 63.6
(b) 21.5 (d) 36.4

61. In case of sawn timber, Quater girth formula gives ...... percentage of true volume
(a) 78.5 (c) 63.6
(b) 21.5 (d) 36.4

62. Timber calculators are used to calculate
(a) Volume of timber
(b) Volume of firewood
(c) Both (a) and (b)
(d) None of the above

63. The solid volume of firewood is measured by using
(a) Xylometric method
(b) Specific gravity
(c) Both (a) and (b)
(d) None of the above

64. In western countries, the volume of the stacked wood is expressed in terms of
(a) Kg (c) m^2
(b) Cord (d) Cm^2

65. The size of the "CHATTIES" in Dehradun is
(a) 7.2 m × 1.8 m × 1.5 m
(b) 1.8 m × 1 m × 1 m
(c) 1.8 m × 1.8 m × 1 m
(d) All of the above

66. Volume of the standing tree can be directly measured by using
(a) Spiegel Relaskop
(b) Wheeler penta prism
(c) Both (a) and (b)
(d) None of the above

67. Volume tables are truly applied to
(a) Individual trees
(b) Group of trees
(c) Both (a) and (b)
(d) None of these

68. Volume of a tree depends upon
(a) Straightness (c) Height
(b) Form (d) Both (b) and (c)

69. In the restricted area .... variable's is/are used for preparation of volume tables
(a) Three (c) Two
(b) One (d) Four

70. Volume tables are used in extensive area only
(a) Regional volume tables
(b) General volume tables
(c) Standard volume table
(d) None of the above

71. Volume tables generally not prepared in India is/ae
(a) Three variable volume table

(b) Form class volume tables
(c) Both (a) and (b)
(d) None of the above

72. Local volume tables are generally derived from
(a) Regional volume table
(b) General volume table
(c) Standard volume table
(d) Form class volume table

73. Volume table based on three variables is known as
(a) General volume table
(b) Local volume table
(c) Regional volume table
(b) Form class volume table

74. Which of the following does not give volume of round wood?
(a) Commercial volume table
(b) Sawn outturn volume table
(c) Both (a) and (b)
(d) None of the above

75. Volume tables can be prepared through
(a) Graphical method
(b) Least square fit method
(c) Alignment chart method
(d) All of the above

76. In graphical method ....trees are satisfactory for whole range of diameter and height for the preparation of volume table
(a) 1000 (c) 500
(b) 100 (d) 2000

77. The accuracy of the volume tables can be checked by using
(a) Height-diameter check
(b) Average deviation check
(c) Aggregate check
(d) All of the above

78. Aggregate check difference should not exceed ....in the volume table
(a) 5% (c) 3%
(b) 1% (d) 10%

79. In height-diameter check the difference should not exceed .... in the volume table
(a) 3% (c) 5%
(b) 15% (d) 10%

80. In relative check the difference should not exceed .... in the volume table
(a) 3% (b) 4%
(b) 10% (d) 15%

81. In case of exceptional importance which check is applied
(a) Aggregate check
(b) Relative check
(c) Average deviation check
(b) All of the above

82. Volume tables applicable in mixed forest is/are
(a) Tariff table
(b) Tree quality volume table
(c) Both (a) and (b)
(d) None of the above

83. The portion of the stem or log which is unmerchantable is termed as
(b) Diseased (c) Burr
(b) Cull (d) Croocked

84. Weight of wood is affected by
(a) Bark and foreign material
(b) Moisture content
(c) Specific gravity
(b) Both (a) and (b)

85. Which of the following is true?
(a) Specific gravity decreases from top to base of a stem

(b) Specific gravity increases from pith to cambium of a stem
(c) Both (a) and (b)
(d) None of the above

86. Moisture content in air dried wood is
(a) 0% (c) 12%
(b) 30% (d) 20%

87. Density of Bark as compared to wood is
(a) Lower (c) Equal
(b) Higher (d) Both (a) and (b)

88. The age of the standing tree can be determined by
(a) General appearance
(b) By taking three periodic measurements
(c) Stump analysis
(d) Both (a) and (b)

89. Average annual increment for a short period is known as
(a) CAI (c) MAI
(b) PAI (d) Final MAI

90. Total increment is defined as
(a) Increment that a tree or crop put on from the origin upto an age
(b) Mean volume of tree or crop at desired age
(c) Both (a) and (b)
(d) None of the above

91. The increment that a tree puts on in a year is known as
(a) Total increment
(b) Periodic annual increment
(c) Mean annual increment
(d) None of the above

92. Final MAI is
(a) Calculated for a portion
(b) Calculated at rotation age
(c) Both (a) and (b)
(d) None of the above

93. CAI can be
(a) Positive
(b) Negative
(c) Zero
(d) All of the above

94. The point at which MAI and CAI meet is known as
(a) Age of maximum production
(b) Rotation age
(c) Both (a) and (b)
(d) None of the above

95. Increment percentage can be determined by using
(a) Schneider's formula
(b) Compound interest formula
(c) Presseler formula
(b) All of the above

96. Diameter corresponding to mean basal area of 250 biggest diameter/ha is termed as
(a) Mean dia (c) Crop dia
(b) Top dia (d) None of these

97. The height corresponding to mean diameter of 250 biggest diameter/ha is termed as
(a) Top height (c) Both (a) and (b)
(b) Crop height (d) None of these

98. Site Quality can be assessed by using
(a) Crop height (c) Top height
(b) Height (d) Both (a) and (c)

99. Crop height can be determined using
(a) Gumpel's formula
(b) Smalian's formula
(c) Lorey's formula
(d) Andre's formula

100. The mean age of the evenaged crop can be determined by using
(a) Smalian's and Heyer's formula
(b) Gumpel's formula
(c) Andre's formula
(d) All of the above

101. Temporary sample plots are laid for
(a) Preparing yield table
(b) Enumeration survey
(c) Both (a) and (b)
(d) None of the above

102. For the preparation of yield tables
(a) Permanent sample plots are laid
(b) Temporary sample plots are laid
(c) Both (a) and (b)
(d) None of the above

103. Sample plots are laid in
(a) Even age forest
(b) Pure forest
(c) Well stocked forest
(d) All of the above

104. Optimum number of plots for each thinning regime while preparing of yield table are
(a) 500 (c) 600
(b) 1000 (d) 100

105. Buffer strip should be at least
(a) 5 m wide
(b) 15 m wide
(c) 15 cm wide
(d) 25 m wide

106. In India, sample plots are generally thinned to
(a) C and D grade Ordinary thinning
(b) A and B grade Ordinary thinning
(c) Light crown thinning
(d) None of the above

107. Situation map is laid on scale of
(a) 1:250 (c) 1:25000
(b) 1:2500 (d) 1:5000

108. Plot chart is laid on scale of
(a) 1:500 (c) 1:5000
(b) 1:250 (d) 1:2500

109. Volume of the sample plot can be calculated by using
(a) Hartig method
(b) Urich method
(c) Draudt's method
(d) All of the above

110. Hartig method is used for
(a) Permanent sample plot
(b) Temporary sample plot
(c) Both (a) and (b)
(d) None of the above

111. Block's method is used for
(a) Permanent sample plot
(b) Groups formed from highest to lowest diameter
(c) Volume of sample plot
(d) All of the above

112. Who suggested the volume curve method?
(a) Schwappach (c) Huber
(b) Speidel (d) Mayer

113. Prusian institute method is suggested by
(a) Meyer (c) Speidel
(b) A Schiffel (d) None of these

114. Forest inventory is defined by
(a) Loetsch and Haller
(b) Hirata
(b) Bitterlich
(d) None of the above

115. The term "Cruise" is commonly used for forest inventory in
(a) Germany

(b) North America
(c) India
(d) France

116. Optimum number of stages for forest survey is/are
(a) One (c) Three
(b) Six (d) Four

117. Sampling based on subjective judgement of observer is/are
(a) Systemic
(b) Multistage
(c) Sampling with varying probability
(d) All of the above

118. Bamboo enumeration is done by using
(a) Multistage sampling
(b) Systemic sampling
(c) Partial sampling
(d) Multiphase sampling

119. Shape of the sampling units is/are
(a) Cluster
(b) Circular
(c) Topographic units
(d) All of the above

120. The concept of the Point sampling was given by
(a) Hirata (c) Gyde Land
(b) Bitter Lich (d) Mayer

121. Point sampling is also known as
(a) Poly aerial plot sampling
(b) Angle count crusing
(c) Pointless cruising
(d) All of the above

122. ....is measure of value of an angle expressed in "Sine"
(a) Diopter
(b) Plot radius factor
(c) Tree factors
(d) BAF

123. One diopter is equal to ....
(a) 0.57°
(b) 34'36"
(c) Sin 57°
(d) All of the above

124. Callibration distance factor is ......times plot radius factor
(a) 1000 (c) 200
(b) 100 (d) 10

125. Instrument used in Horizontal sampling is/are
(a) Simple angle gauge
(b) Relaskop
(c) Wedge prism
(d) All of the above

126. Who developed vertical point sampling?
(a) Bitter Lich (c) Hirata
(b) Meyer (d) Brandis

127. Vertical sampling is used to determine
(a) Top height
(b) Mean diameter
(c) Mean stand height
(d) Volume of stand

128. Who coined the term "Stand"
(a) Smalian (c) Brandis
(c) Meyer (d) Huber

129. Growth of stand is a function of
(a) Stand density (c) Site index
(b) Site quality (d) Both (a) and (c)

130. If density of the crown lies between 0.75 and 1 than it is called
(a) Dense (c) Thin
(b) Closed (d) Open

131. The relationship between tree height and age is known as
(a) Site quality (c) Site index
(b) Top height (d) Site vigour

132. Aerial remote sensing is done at a height of
(a) 5000–15,000 m
(b) 5000–15000 km
(c) 150–350 km
(d) 160–320 m

133. Aerial photograph is interpreted by using
(a) Planimeter (c) Relaskop
(b) Stereoscope (d) None of these

134. False colour photographs are sensitive to
(a) UV region
(b) Visible region
(c) IR region
(d) None of the above

135. Large scale photograph has scale of
(a) 1:40000 to 1:70000
(b) 1:20000 to 1:40000
(c) 1:5000 to 1:10000
(d) 1:5000 to 1:20000

136. If optical axis of camera is more than 30° is termed as
(a) Vertical photograph
(b) Oblique photograph
(c) Both (a) and (b)
(d) None of the above

137. Space remote sensing is done at
(a) 160–320 m
(b) 500–1000 km
(c) 160–320 km
(d) None of these

138. First "ERTS" was launched in
(a) 1972 (c) 1973
(b) 1968 (d) 1978

139. Ideal time for Aerial photograph is
(a) 10–11am and 1–2 pm
(b) 9–11 am and 1–3 pm
(c) 10–11 am and 2–3 pm
(b) 9–11 am and 1–2 pm

140. The diameter measurement using calliper involves
(a) Systematic error
(b) Negative error
(c) Positive error
(d) Both (b) and (c)

141. Upper stem diameter can be measured using
(a) Dendrometers
(b) Callipers
(c) Relaskop
(d) All of the above

142. The errors in case of tape measurement is/are
(a) Positive error
(b) Systematic error
(c) Negative error
(d) Both (a) and (b)

143. The usual diameter classes used in India are
(a) 1 inch, 2 inch, 4 inch
(b) 2 cm, 5 cm, 10 cm
(c) 5 cm, 10 cm, 15 cm
(d) Both (a) and (b)

144. The trees attaining maturity at 50 cm dbh falls in diameter class of
(a) 5 cm (c) 20 cm
(b) 10 cm (d) 1 inch

145. The usual girth classes followed in India are
(a) 5 cm, 15 cm, 30 cm
(b) 5 cm, 10 cm, 15 cm
(c) 5 cm, 10 cm, 20 cm
(d) 10 cm, 15 cm, 30 cm

146. The maximum spread of the crown along its widest diameter is termed as
(a) Crown length (c) Crown width
(b) Crown spread (d) Both (b) and (c)

147. Sources of errors in Height measurement are
(a) Personal errors
(b) Errors due to lean of trees
(c) Errors due to observation
(d) All of the above

148. Height classes generally used in India are
(a) 5 m, 10 m, 15 m
(b) 1 m, 3 m, 5 m
(c) 1 m, 2 m, 5 m
(d) 5 m, 10 m, 3 m

149. The trees attaining the height between 15-25 m at maturity falls in height class of
(a) 3 m (b) 1 m (c) 5 m (d) 2 m

150. The horizontal projection of tree crown on ground is termed as
(a) Crown length (c) Crown height
(b) Crown width (d) None of these

151. Volume of felled tree consists of
(a) Volume of stem wood
(b) Volume of branch wood
(c) Volume of root wood
(d) All of the above

152. The usual length of the log is
(a) 3 m (c) 5 m
(b) 35 m (d) 4 m

153. The quarter girth formula is used to calculate
(a) Volume of standing tree
(b) Volume of firewood
(c) Both (a) and (b)
(d) None of the above

154. Timber calculators are used to calculate volume of
(a) Round timber
(b) Sawn timber
(c) Both (a) and (b)
(d) None of these

155. Which of the following volume table not prepared in India?
(a) Form class volume table
(b) Tariff table
(c) Both (a) and (b)
(d) None of the above

156. Stump analysis is done to find
(a) Age-height relations
(b) Age-volume relations
(c) Age-diameter relations
(d) All of the above

157. Stem analysis is done to find
(a) Age-height relations
(b) Age-volume relations
(c) Age-diameter relations
(d) All of the above

158. Crop diameter is based on the
(a) Mean diameter
(b) Mean basal area
(d) Both (a) and (b)
(d) None of the above

159. The sample plot size varies from
(a) 0.2-0.5 ha
(b) 1-2 ha
(c) 0.1-1 ha
(d) 0.1-0.2 ha

160. In square, spacing canopy closure is
(a) 0.7854
(b) $2\pi/8$
(c) $\pi/4$
(d) All of the above

161. In equilateral triangular, spacing canopy closure is
(a) 0.9068 (c) Both (a) and (b)
(b) $\pi/2\sqrt{3}$ (d) None of these

162. The fractional quality classes running within quality classes are
(a) 0.0-2.0 (c) 0.0-0.5
(b) 0.0-0.2 (c) 0.0-2.0

163. Yield tables are applicable to
(a) Unevenaged forest
(b) Mixed forest
(c) Evenaged forest
(d) None of the above

164. The focal length of wide-angle camera is
(a) 12 cm (c) 15 cm
(b) 20 cm (d) 17 cm

165. The instrument used in vertical point sampling is
(a) Relaskope (c) Wedge prism
(b) Conimeter (d) None of these

166. In vertical point sampling, instrument substends a critical angle of
(a) 30° (c) 40°
(b) 50° (d) 45°

167. Horizontal sampling is used to determine
(a) Basal area per hectare
(b) Number of stems per hectare
(c) Volume per hectare
(b) All of the above

168. Bark gauge is used to measure
(a) Bark thickness (c) Bark volume
(b) Bark length (d) None of these

169. Form of the tree can be found from
(a) Volume tables (c) Yield tables
(b) Taper tables (d) None of these

170. One cubic metre of timber in metric system is equal .... in FPS
(a) 35.00 cft (c) 35.314 cft
(b) 40 cft (d) 30 cft

171. In England, U.K. dbh is measured at
(a) 1.30 m (c) 1.37 m
(b) 1.34 m (d) 1.39 m

172. The Sun synchronous satellite is located at an altitude between
(a) >36,000 Km
(b) 3,000–36,000 Km
(c) 5,000–10,000 Km
(d) 300–1,000 Km

173. Stand density is considered overstocked when
(a) < 1 (c) = 1
(b) > 1 (d) None of these

174. The sample plot with minimum perimeter for a given area is
(a) Circular (c) Square
(b) Rectangular (d) Triangular

175. Form height is the product of form factor and
(a) Bole height
(b) Diameter
(c) Basal area
(d) Tree height

176. Spectral signature of green vegetation in a satellite image is
(a) Green (c) Blue
(b) Red (d) Yellow

177. 1 hectare is equal to .... m^2.
(a) 1000 (c) 100000
(b) 100 (d) 10000

178. For determination of increment percent which of the following instrument is based on counting number of growth rings

(a) Schneiders formula
(b) Presslers formula
(c) Compound interest formula
(d) None of the above

179. The proportion of error involved in use of calliper and tape is
(a) 20:1 (c) 21:2
(b) 21:1 (d) 10:1

180. The relative completeness of a canopy is termed as
(a) Stand density
(b) Canopy density
(c) Canopy cover
(d) All of the above

181. Which of the following component is not considered for biomass?
(a) Wood (c) Leaves
(b) Bark (d) Root

182. FAO recommended to measure diameter at
(a) 1.37 m (c) 4.5 ft
(b) 1.30 m (d) 4 ft

183. Basic assumption on which the tree measuring instruments are devised
(a) Both base & tip of the tree are visible
(b) Tree is vertical
(c) Both (a) and (b)
(d) None of the above

184. Which of the following is correctly matched?
(a) Avamana- measure of height
(b) Tulamana- measure of diameter
(c) Kalapramana- measure of time
(d) Mana- measure of weight

185. In metric system girth of a tree is always taken in
(a) Centimetres to the nearest millimetre
(b) Metres to the nearest centimetres
(c) Foot and to the nearest inches
(d) None of the above

186. Basal area of a tree refers to
(a) Cross sectional area at stem height
(b) Cross sectional area at breast height
(c) Cross sectional area at any height
(d) All of the above

187. While measuring the height from Abney's level, how far a person should stand from tree
(a) 10-15 m (c) 15-20 m
(b) 20-30 m (d) 30-40 m

188. Sampling intensity (percent) of sub-tropical pine forest is
(a) 2-10% (c) 10%
(b) 2-5% (d) 5%

189. Biltmore stick is used to measure
(a) Height (c) Volume
(b) Basal area (d) Diameter

190. Distance can be measured directly by
(a) Wedge prism (c) Scale
(b) Relaskope (d) All of these

191. ....is the most important measurement in the management of forest.
(a) Volume (c) Height
(b) Diameter (d) Form

192. While measuring the perimeter of an ellipse the value of e^2 is
(a) >1 (c) = 1
(b) <1 (d) Both (a) and (b)

193. Chako concluded that the cross section of stem cannot be considered as

(a) Circle (c) Triangular
(b) Ellipse (d) Both (a) and (b)

194. The wooden scale comes in the sizes of
(a) 30 cm (c) 60 cm
(b) 90 cm (d) Both (a) and (c)

195. Callipers exceeding .... are not used for measuring tree diameter
(a) 120 cm (c) 100 cm
(b) 130 cm (d) 60 cm

196. ....when measures the angle of inclination as percentage of horizontal distance is known as hypsometer.
(a) Abney level (c) Clinometer
(b) Dumpy level (d) Both (a) and (b)

197. Girder theory deals with the
(a) Tree height (c) Tree volume
(b) Tree form (d) Tree diameter

198. The top and bottom of a tree as measured with the help of Hega-altimeter are 130 and 15 percent respectively on a level ground at a distance of 10 m than the height of tree is
(a) 16 m (c) 14.5 m
(b) 20 m (d) 12 m

199. The formula for converting gob into gub is
(a) $g = g' - 2\pi h$ (c) $g = d - 2\pi t$
(b) $g = g' - 2\pi t$ (d) None of these

200. Smalian's formula for calculating volume of log is
(a) $S \times l$
(b) $\left(\frac{S_1 + S_2}{2}\right) \times l$
(c) $S_m \times l$
(d) $\left(\frac{S_1 + 4S_m + S_2}{6}\right) \times l$

201. Volume of stem from ground level down to 20 cm diameter over bark is
(a) Commercial volume
(b) Standard outturn
(c) Standard stem small wood timber
(d) Standard stem timber

202. Random sampling method is
(a) Sequential sampling
(b) Selective sampling
(c) Stratified sampling
(d) Systematic sampling

203. Volume of stem wood in round between 20 cm and 5 cm diameter over bark is called
(a) Standard stem small wood
(b) Commercial wood
(c) Standard wood
(d) Sawn outturn wood

204. In which of the following volume of the stump is excluded?
(a) Commercial volume
(b) Sawn outturn volume
(c) Standard volume
(a) Both (a) and (b)

205. General volume table is based on
(a) Diameter only
(b) Height only
(c) Diameter and Height
(d) Diameter, Height, Form

206. Which formula is used to calculate the mean age of evenaged crop using area occupied by each age class?
(a) Andre's formula
(b) Heyer's formula
(c) Gumpel's formula
(d) None of the above

207. Which formula is used to calculate the mean age of Un-evenaged crop using number of trees
(a) Andre's formula
(b) Gumpel's formula
(c) Smalian's formula
(d) All of the above

208. The factor which indicate the distance at which a given angle gauge will exactly lower a one metre wide rectangular target is called
(a) BAF
(b) Plot radius factor
(c) Callibration radius factor
(d) Basal area

209. The ratio between the Basal area of the tree and area of the circular plot of the tree is
(a) $K^2/4$
(b) $K^2/8$
(c) $4K^2/5$
(d) $K/5$

210. The number of trees tallied in horizontal sampling multiplied by a constant factor gives
(a) Basal area
(b) Basal area per hectare
(c) Number of stems per hectare
(b) Diameter per hectare

211. The process of sampling in which a series of sampling plots are selected randomly or systematically distributed over the whole area, when trees around these points are viewed through any angle gauge at breast height than the tree having an angle greater than critical angle are counted is called
(a) Vertical point sampling
(b) Systematic sampling
(c) Horizontal point sampling
(d) Variable Point crusing

212. No slope correction is required in Wedge prism upto an angle of
(a) 18°
(b) 25°
(c) 40°
(d) 30°

213. A slope angle of 25% causes an error of about …….. in counting of tally trees by Wedge prism.
(a) 10%
(b) 30%
(c) 20%
(d) 40%

214. The Basal area factor of a Wedge prism is calculated on the basis of the
(a) Acute angle of prism
(b) Critical angle of prism
(c) Obtuse angle of prism
(d) None of the above

215. A tree whose image do not touch the tree stem in horizontal point sampling is called
(a) Full tallied
(b) Half tallied
(c) Non tallied
(d) All of the above

216. A tree whose image just touch the tree stem in horizontal point sampling is called
(a) Full tallied
(b) Non tallied
(c) Half tallied
(d) None of the above

217. CVP index is used to calculate
(a) Soil productivity
(b) Site quality
(c) Soil fertility
(d) Site index

218. Wedge prism works on principle of
(a) Bending of light rays passing through the eyes of a person

(b) Turning of light rays passing through prism
(c) Bending of light rays passing through the trees
(d) Bending of light rays passing through the prism

219. The trees are tallied using wedge prism if the image
(a) Do not overlap directly viewed tree
(b) Overlap the directly viewed tree
(c) Both (a) and (b)
(d) None of the above

220. Map projection of earth surface is/are
(a) Orthogonal
(b) Central
(c) Perpendicular
(d) Both (a) and (b)

221. Aerial photograph and map give...............dimensional view respectively.
(a) 2D and 3D
(b) 3D and 2D
(c) Both (a) and (b)
(d) None of these

222. Actual image of the object and the ground is seen in
(a) Aerial photograph
(b) Plane table survey
(c) Map
(d) None of the above

223. The area covered by the vertical photograph is
(a) Square
(b) Circle
(c) Triangular
(d) Rectangular

224. The area covered by the oblique photograph is
(a) Rectangular
(b) Trapezoidal
(c) Square
(d) Triangular

225. The most commonly used filter for remote sensing in forestry is
(a) Yellow
(b) Red
(c) Green
(d) Blue

226. Which filter is used to delineate water bodies?
(a) Blue
(b) White
(c) Black
(d) Red

227. The red filter absorbs
(a) Energy of lower wavelengths of visible spectrum
(b) Energy of higher wavelengths of visible spectrum
(d) Energy of lower wavelengths of infrared spectrum
(d) Energy of higher wavelengths of infrared spectrum

228. The standards of weight and measurement act came into force in
(a) 1946
(b) 1663
(c) 1956
(d) 1976

229. The most commonly used instrument for measurement of crown diameter is
(a) Micro meter wedge
(b) Stereoscope
(c) Planimeter
(d) Tape

230. If the Sampling size is increasing and goes on changing, the consequence would be
(a) Biased
(b) Inconsistent
(c) Unbiased
(d) Consistent

231. In natural forest enumeration, sampling intensity is
(a) 0.01% (c) 1%
(b) 2% (d) 5%

232. The top and bottom of a tree as measured with the help of Hega altimeter are 130 and 15 percent respectively on a level ground at a distance of 10 m than the height of tree is
(a) 16 m (c) 20 m
(b) 14.5 m (d) 12 m

233. CVP index is used to calculate
(a) Soil productivity
(b) Site quality
(c) Soil fertility
(d) None of the above

Answer Keys									
1. (c)	**2.** (b)	**3.** (a)	**4.** (d)	**5.** (d)	**6.** (c)	**7.** (a)	**8.** (d)	**9.** (d)	**10.** (d)
11. (d)	**12.** (d)	**13.** (c)	**14.** (c)	**15.** (b)	**16.** (a)	**17.** (b)	**18.** (b)	**19.** (a)	**20.** (b)
21. (a)	**22.** (d)	**23.** (d)	**24.** (b)	**25.** (a)	**26.** (b)	**27.** (d)	**28.** (d)	**29.** (c)	**30.** (a)
31. (b)	**32.** (d)	**33.** (d)	**34.** (a)	**35.** (d)	**36.** (b)	**37.** (c)	**38.** (c)	**39.** (d)	**40.** (a)
41. (c)	**42.** (c)	**43.** (b)	**44.** (a)	**45.** (a)	**46.** (d)	**47.** (b)	**48.** (a)	**49.** (b)	**50.** (c)
51. (c)	**52.** (d)	**53.** (c)	**54.** (d)	**55.** (a)	**56.** (c)	**57.** (a)	**58.** (b)	**59.** (c)	**60.** (a)
61. (c)	**62.** (a)	**63.** (c)	**64.** (b)	**65.** (b)	**66.** (d)	**67.** (c)	**68.** (d)	**69.** (b)	**70.** (d)
71. (a)	**72.** (b)	**73.** (d)	**74.** (b)	**75.** (d)	**76.** (a)	**77.** (d)	**78.** (b)	**79.** (c)	**80.** (a)
81. (c)	**82.** (b)	**83.** (b)	**84.** (d)	**85.** (b)	**86.** (c)	**87.** (a)	**88.** (d)	**89.** (b)	**90.** (a)
91. (d)	**92.** (b)	**93.** (d)	**94.** (c)	**95.** (d)	**96.** (b)	**97.** (a)	**98.** (c)	**99.** (c)	**100.** (d)
101. (b)	**102.** (a)	**103.** (d)	**104.** (c)	**105.** (b)	**106.** (a)	**107.** (c)	**108.** (b)	**109.** (d)	**110.** (b)
111. (d)	**112.** (b)	**113.** (d)	**114.** (a)	**115.** (c)	**116.** (c)	**117.** (a)	**118.** (d)	**119.** (d)	**120.** (c)
121. (d)	**122.** (a)	**123.** (d)	**124.** (b)	**125.** (d)	**126.** (c)	**127.** (c)	**128.** (b)	**129.** (d)	**130.** (a)
131. (c)	**132.** (a)	**133.** (b)	**134.** (c)	**135.** (d)	**136.** (b)	**137.** (c)	**138.** (a)	**139.** (d)	**140.** (d)
141. (d)	**142.** (d)	**143.** (d)	**144.** (b)	**145.** (a)	**146.** (c)	**147.** (d)	**148.** (b)	**149.** (a)	**150.** (d)
151. (d)	**152.** (a)	**153.** (d)	**154.** (c)	**155.** (c)	**156.** (c)	**157.** (d)	**158.** (b)	**159.** (d)	**160.** (d)
161. (c)	**162.** (a)	**163.** (c)	**164.** (c)	**165.** (b)	**166.** (d)	**167.** (d)	**168.** (a)	**169.** (b)	**170.** (c)
171. (a)	**172.** (d)	**173.** (b)	**174.** (a)	**175.** (d)	**176.** (b)	**177.** (d)	**178.** (a)	**179.** (b)	**180.** (b)
181. (d)	**182.** (b)	**183.** (c)	**184.** (c)	**185.** (a)	**186.** (b)	**187.** (c)	**188.** (d)	**189.** (d)	**190.** (d)
191. (a)	**192.** (b)	**193.** (d)	**194.** (d)	**195.** (a)	**196.** (c)	**197.** (b)	**198.** (c)	**199.** (b)	**200.** (b)
201. (d)	**202.** (c)	**203.** (a)	**204.** (d)	**205.** (c)	**206.** (c)	**207.** (a)	**208.** (c)	**209.** (a)	**210.** (b)
211. (c)	**212.** (a)	**213.** (a)	**214.** (b)	**215.** (c)	**216.** (c)	**217.** (d)	**218.** (d)	**219.** (b)	**220.** (a)
221. (b)	**222.** (c)	**223.** (d)	**224.** (b)	**225.** (a)	**226.** (d)	**227.** (b)	**228.** (c)	**229.** (a)	**230.** (c)
231. (a)	**232.** (b)	**233.** (d)							

7

Tree Improvement and Genetics

1. Tree improvement means
(a) Specifically, geared programme to achieve specific goals
(b) Practical application of forest genetics
(c) Application of principles of tree breeding and Silviculture in plantation forestry
(d) None of the above

2. The applied use of forest genetics to actually improve the quality of the trees is
(a) Forest Tree Improvement
(b) Forest Tree Breeding
(c) Forest Genetics
(d) None of the above

3. The activities specifically geared up to achieve the specific goals of improved trees is
(a) Forest genetics
(b) Forest Tree Improvement
(c) Forest Tree Breeding
(d) All of the above

4. Tree improvement is the combination of
(a) Forest genetics and breeding
(b) Forest tree breeding and Silviculture
(c) Both (a) and (b)
(d) None of the above

5. The physical appearance of tree is termed as
(a) Phenotype
(b) Genotype
(c) Both (a) and (b)
(d) None of the above

6. The performance of a tree depends upon
(a) Genotype
(b) Environment
(c) Genotype × Environment
(d) All of the above

7. Progeny test is done to know the genetic worth of
(a) Parents
(b) Population
(c) Offspring
(d) All of the above

8. If female parent is known and male parent is unknown in a tree species called as
(a) Half sib
(b) Full sib
(c) Both (a) and (b)
(d) None of these

9. If both male and female parent are known in a tree species called as
(a) Full sib
(b) Half sib
(c) Both (a) and (b)
(d) None of these

10. The age at which the stand of the tree is to be harvested is termed as
(a) Economic age
(b) Age of maximum production
(c) Rotation age
(d) All of the above

11. Meiosis is also known as
(a) Equational division
(b) Reductional division
(c) Both (a) and (b)
(d) None of the above

12. Which of the following is important for the "Tree improvement programme"?
(a) Additive variance
(b) Non- Additive variance
(c) Both (a) and (b)
(d) None of the above

13. Hereditary variation due to interaction among gene loci is known as
(a) Dominance
(b) Linkage
(c) Epistasis
(d) All of the above

14. Which of the following maintain high degree of genetic variation?
(a) Out cross breeding
(b) Inbreeding
(c) Line breeding
(d) None of the above

15. Polyploidy is common in
(a) Gymnosperms
(b) Angiosperm
(c) Both (a) and (b)
(d) None of the above

16. The chromosome number of *Sequoia sempervirens* is
(a) 88 (b) 22 (c) 66 (d) 44

17. The forces that increase the variability in natural stand is/are
(a) Mutation
(b) Natural selection
(c) Gene flow
(d) Both (a) and (b)

18. The forces that decrease the variability in natural stand is/are
(a) Gene drift
(b) Natural selection
(c) Gene flow
(d) Both (a) and (c)

19. The sudden heritable change in the DNA sequence is called as
(a) Mutation (c) Genetic drift
(b) Translation (d) Transcription

20. The chance of fluctuation in allele frequency within a population is termed as
(a) Gene flow
(b) Mutation
(c) Genetic drift
(d) None of the above

21. Gene flow at species level take place through a process called as
(a) Introgression
(b) Interaction
(c) Regression
(d) Both (a) and (c)

22. The chemical most commonly used for inducing polyploidy is
(a) Colchicine
(b) UV-rays
(c) Acriflavin
(d) All of the above

23. What is the minimum number of codons required for amino acids containing five amino acids (ignore start and stop codons)?
(a) 10 (c) 5
(b) 20 (d) 15

24. Triploids are normally
(a) Sterile (c) Fertile
(b) Male sterile (d) Female sterile

25. The donor plant from which plant propagated vegetatively is
(a) Explant (c) Clone
(b) Ramet (d) Ortet

26. A very commonly used surface sterilant for explants in plants in plant tissue culture is
(a) NaCl (c) $KMnO_4$
(b) $HgCl_2$ (d) $CaCl_2$

27. NBPGR is located at
(a) Lucknow (c) New Delhi
(b) Chennai (d) Hyderabad

28. The ability of a cell to develop a complete organism is called as
(a) Apomixis (c) Parthenocarpy
(b) Totipotency (d) Growth

29. A form of multitrait selection that combines information on all traits of interest into single index is
(a) Selection index
(b) Tandem selection
(c) Independent culling
(d) Mass selection

30. The stand established with special stock from a balanced mixture of seeds from at least 60 good parents and gradually culled is
(a) CSO (c) SPA
(b) SSO (d) ESSO

31. The repetition of cross where the sexual function of parents is reversed
(a) Back cross
(b) Reciprocal cross
(c) Test cross
(d) None of the above

32. The pattern of pollination set up between individuals is/are
(a) Mating design
(b) Experimental design
(c) Field design
(d) All of the above

33. The reduced ability of some hybrids to produce viable gametes is
(a) Heterosis
(b) Hybrid sterility
(c) Hybrid vigour
(d) Both (a) and (b)

34. Which of the following is a start codon?
(a) GUA (c) GUG
(b) UAG (d) UAA

35. The nitrogenous base which is not found in DNA
(a) Adenine (c) Cytosine
(b) Thymine (d) Uracil

36. The process by which protein is synthesized from RNA known as
(a) Replication
(b) Translation
(c) Transcription
(d) None of the above

37. The process of RNA synthesis from DNA is known as
(a) Transcription
(b) Replication
(c) Reverse transcription
(d) All of the above

38. SCA can be
(a) Positive
(b) Negative
(c) Both (a) and (b)
(d) None of the above

39. Termination of the shedding of pollen of a plant or flower prior to receptivity on the same plant or flower is termed as
(a) Protogyny
(b) Pritoandry

(c) Dichogamy
(d) All of the above

40. Reproduction of an unfertilised egg is known as
(a) Parthenogenesis
(b) Parthenocarpy
(c) Both (a) and (b)
(d) None of the above

41. The female cells produced by sexual cell divisions and give rise to embryo sac is called
(a) Microspore (c) Macrospore
(b) Megaspore (d) Both (b) and (c)

42. One or more set of similar alleles are called as
(a) Heterozygous
(b) Homogeneous
(c) Homozygous
(d) Heterogeneous

43. One or more set of unlike alleles are called as
(a) Heterozygous
(b) Homogeneous
(c) Homozygous
(d) Heterogeneous

44. The proportion of an allele in a population compared to other alleles of that gene is called
(a) Allelic frequency
(b) Gene frequency
(c) Gene pool
(d) None of the above

45. The maturation of male and female organs on the same plant at separate organs is
(a) Autogamy
(b) Dichogamy
(c) Allogamy
(d) All of the above

46. In Complete pedigree design
(a) Both parents are known
(b) One parent known
(c) Both (a) and (b)
(d) None of the above

47. Plant composed of two or more genetically different tissues growing separately but adjacent to each other in one plant is
(a) Burr (c) Variegation
(b) Callus growth (d) Chimera

48. Crossing an offspring (F_1) to recessive parents is known as
(a) Test cross
(b) Back cross
(c) Reciprocal cross
(d) None of the above

49. Crossing of an offspring to either one of its parents is known as
(a) Test cross
(b) Reciprocal cross
(c) Back cross
(d) None of the above

50. The phenotypic differences between males and females of same species is called
(a) Polymorphism
(b) Dimorphism
(c) Both (a) and (b)
(d) None of the above

51. Broad genetic base is found in
(a) Clonal seed orchard
(b) Extensive seedling seed orchard
(c) Seed production areas
(d) Seedling seed orchard

52. PCR is used for
(a) Reverse transcribing RNA to DNA
(b) Digesting DNA

(c) Copying Plasmids
(d) All of the above

53. The seed orchard has
(a) Production area
(b) Buffer area
(c) Both (a) and (b)
(d) None of the above

54. The original geographic area from which seeds or another propagule material is obtained
(a) Seed source
(b) Provenance
(c) Geographic source
(d) None of the above

55. The "Coff harbour provenance" is popularly known as
(a) Seed source
(b) Provenance
(c) Geographic source
(d) All of the above

56. The tree grown outside its natural range is called
(a) Indigenous
(b) Adapted
(c) Exotic
(d) Both (a) and (c)

57. A gradient in measurable characteristic is termed as
(a) Ecotype
(b) Land race
(c) Cline
(d) All of the above

58. The population of an individual that has become adapted to specific environment in which it has been planted is termed as
(a) Ecotype
(b) Land race
(c) Cline
(d) All of the above

59. The average performance of a progeny when matted to several other individuals in a population is termed as
(a) GCA
(b) SCA
(c) Both (a) and (b)
(d) All of the above

60. The average performance of a progeny of a cross when matted to one of its parents is known as
(a) GCA
(b) SCA
(c) Both (a) and (b)
(d) None of the above

61. Breeding value of an individual is
(a) ½ GCA
(b) GCA
(c) 2 × GCA
(d) All of the above

62. The ratio of the total genetic variation in population to phenotypic variation
(a) Narrow sense heritability
(b) Genetic gain
(c) Broad sense heritability
(d) None of the above

63. The Narrow sense heritability is
(a) Additive variance/Phenotypic variance
(b) Non-additive variance and Additive variance/Phenotypic variance
(c) Non-additive variance/Phenotypic variance
(d) None of the above

64. The range of heritability is
(a) 0 to 1
(b) $-\infty$ to 0
(c) $-\infty$ to $+\infty$
(d) 0 to $+\infty$

65. If the value of heritability is zero, it exhibits
(a) Additive variance
(b) Non-additive variance
(c) Both (a) and (b)
(d) None of the above

66. Genetic gain is
(a) h/S (b) $h/\sqrt{S}$ (c) hS^2 (d) h^2S

67. Which of the following involves selection of individual solely on the basis of phenotypes?
(a) Progeny testing
(b) Tandem selection
(c) Mass selection
(d) Family selection

68. Which of the following works best in low heritability?
(a) Family plus with in family selection
(b) Family selection
(c) Tandem selection
(d) Independent culling

69. Which of the following has slowest rate of inbreeding?
(a) Sibling selection
(b) Tandem selection
(c) Within family selection
(d) None of the above

70. Which of the following involves selection of one trait at a time?
(a) Selection Index
(b) Independent culling
(c) Recurrent selection
(d) Tandem selection

71. Which of the following involves many cycles of selection and breeding?
(a) Selection Index
(b) Independent culling
(c) Recurrent selection
(d) Tandem selection

72. Which type of selection is mostly used in forest tree improvement?
(a) Selection Index
(b) Independent culling
(c) Recurrent selection
(d) Tandem selection

73. The winner from the selection programme is
(a) Elite tree (c) Candidate tree
(b) Plus tree (d) Check tree

74. The plus tree means
(a) Genetically superior
(b) Phenotypically superior but not progeny tested
(c) Both (a) and (b)
(d) None of the above

75. The minimum area required for SPA is
(a) 1 ha (c) 10 ha
(b) 20 ha (d) 4 ha

76. The reliable regression curve for height or volume can be made with
(a) 10 trees
(b) 100 trees
(c) 50 trees
(d) None of the above

77. The minimum pollen dilution zone for pines is
(a) 400-500 ft
(b) 50-100 ft
(c) 100-200 ft
(d) 600-800 ft

78. Nested design is also known as
(a) Hierarchical parent design
(b) North Carolina design
(c) Both (a) and (b)
(d) None of the above

79. Individual propagules originating from ortet are
(a) Clone
(b) Ramet
(c) Both (a) and (b)
(d) None of the above

80. Which of the following does not take place in evolution of plants?
(a) Crossing over
(b) Mutation
(c) Linkage
(d) None of the above

81. Different cross combinations of F_1 generation parent type remaining same from generation to generation is
(a) Genetic gain
(b) Specific combining ability
(c) General combining ability
(d) Self incompatibility

82. Tree stands maintained for Breeding and Conservation purpose is called as
(a) Breeding orchard
(b) Special orchard
(c) Seedling seed orchard
(d) Seed production areas

83. The immediate production of genetically superior trees is met through
(a) Clonal seed orchard
(b) Seed orchards
(c) Seedling seed orchard
(d) Seed production areas

84. The author of "Introduction to Forest Tree Genetics" is
(a) Jonathan Wright
(b) Zobel
(c) PK Khosla
(d) None of the above

85. Enzyme involved in the formation of RNA from DNA (transcription)
(a) Topoisomerase
(b) DNA ligase
(c) RNA polymerase
(d) DNA polymerase

86. Identify the important method of tree breeding
(a) Introduction
(b) Selection
(c) Hybridization
(d) All of the above

87. Identify the most important component of general breeding strategies
(a) Identification of the superior phenotypes
(a) Establishment of clonal seedling seed orchard
(c) Carrying out clonal mating
(d) All of the above

88. Race is due to
(a) Geographic variation
(b) Environment variation
(c) Genetic variation
(d) All of the above

89. Emasculation is
(a) Removing of male portion
(b) Removing of female portion
(c) Removing of all reproductive portion
(d) None of the above

90. Father of "Tissue culture" is
(a) Haber land
(b) Tensely
(c) Funk
(d) None of the above

91. Germ plasm theory was proposed by
(a) Wiesman
(b) Darwin
(c) Lamarck
(d) Mendel

92. ....is a geographical area where maximum genetic variation is found for a particular forest tree species.
(a) Provenance
(b) Land race
(c) Centre of origin
(d) Geographical source

93. Geographic variability is tested through
(a) Multilocational trials
(b) Provenance trials
(c) Progeny testing
(d) None of the above

94. Study of pollens is
(a) Palynology (c) Dendrology
(b) Phenology (d) Paleology

95. The phase of mitosis in which chromosomes line up along the equatorial plane of the cell is called
(a) Prophase (c) Telophase
(b) Metaphase (d) Anaphase

96. The 9:7 ratio in F_2 generation depicts
(a) Dominance
(b) Co-dominance
(c) Epistasis
(d) None of the above

97. Which of the following is a physical mutagen?
(a) Alfa rays
(b) X- rays
(c) UV- rays
(d) All of the above

98. Which of the following is responsible for carrying amino acids to the site of protein synthesis?
(a) t-RNA
(b) r-RNA
(c) m-RNA
(d) All of the above

99. The structural change in the chromosome due to orientation in reverse direction is termed as
(a) Duplication (c) Translocation
(b) Inversion (d) Deletion

100. Commonly used surface sterilant in tissue culture
(a) $KMnO_4$ (c) $HgCl_2$
(b) $CaCl_2$ (d) NaCl

Answer Keys

1. (c)	**2.** (a)	**3.** (c)	**4.** (b)	**5.** (a)	**6.** (d)	**7.** (c)	**8.** (a)	**9.** (a)	**10.** (c)
11. (b)	**12.** (a)	**13.** (c)	**14.** (a)	**15.** (b)	**16.** (c)	**17.** (d)	**18.** (d)	**19.** (a)	**20.** (c)
21. (a)	**22.** (a)	**23.** (d)	**24.** (a)	**25.** (d)	**26.** (b)	**27.** (c)	**28.** (b)	**29.** (a)	**30.** (d)
31. (b)	**32.** (a)	**33.** (b)	**34.** (c)	**35.** (d)	**36.** (b)	**37.** (a)	**38.** (c)	**39.** (b)	**40.** (a)
41. (d)	**42.** (c)	**43.** (a)	**44.** (b)	**45.** (b)	**46.** (a)	**47.** (d)	**48.** (a)	**49.** (c)	**50.** (b)
51. (d)	**52.** (a)	**53.** (c)	**54.** (b)	**55.** (a)	**56.** (c)	**57.** (c)	**58.** (b)	**59.** (a)	**60.** (b)
61. (c)	**62.** (c)	**63.** (a)	**64.** (a)	**65.** (b)	**66.** (d)	**67.** (c)	**68.** (A)	**69.** (c)	**70.** (d)
71. (c)	**72.** (c)	**73.** (a)	**74.** (b)	**75.** (d)	**76.** (c)	**77.** (a)	**78.** (a)	**79.** (b)	**80.** (c)
81. (b)	**82.** (a)	**83.** (b)	**84.** (a)	**85.** (c)	**86.** (d)	**87.** (a)	**88.** (b)	**89.** (a)	**90.** (a)
91. (d)	**92.** (c)	**93.** (b)	**94.** (a)	**95.** (b)	**96.** (c)	**97.** (d)	**98.** (a)	**99.** (b)	**100.** (c)

8

Forest Products and Utilization

1. Drugs are mainly composed of
(a) Glucosides
(b) Organic compounds
(c) Alkaloids
(d) All of the above

2. Which of the following is not true regarding drugs?
(a) Product of living cells
(b) Composed of inorganic compounds
(c) Waste product of metabolism of plant
(d) Founds in seeds, flowers, roots

3. Concentrated active constituents of plants are commonly found in
(a) Seed
(b) Roots
(c) Both (a) and (b)
(d) None of the above

4. Active principles in smaller quantity are found in which part of the plant
(a) Bark
(b) Leaves
(c) Flowers
(d) All of the above

5. Drugs in maximum can contration can be obtained from
(a) Fruits, Bark, Stem
(b) Flower, Root, Leaves
(c) Both (a) and (b)
(d) None of the above

6. Which of the following is stimulant?
(a) Strychnine (c) Cocaine
(b) Chloroform (d) Chloral

7. Drugs from underground parts is obtained from
(a) *Abroma angusta*
(b) *Picrorhiza kurrooa*
(c) *Rheum modi*
(d) All of the above

8. The commercial source of liquorice is
(a) *Urgenia indica*
(b) *Cephaelis ipecacuanha*
(c) *Hemidesmus indicus*
(d) None of the above

9. The principle content in liqourice varies from
(a) 2–14% (c) 5–10%
(b) 15–20% (d) 20–30%

10. which of the following is a substitute for "sarsaparilla" obtained from *Smilax medica*?
(a) *Glycyrhiza glabra*
(b) *Nardostachys jatamansi*
(c) *Hemidesmus indicus*
(d) *Alstonia scholaris*

11. The scientific name of "Indian rubarb" is
(a) *Sassuria lappa*
(b) *Rheum emodi*

(c) *Rheum officinale*
(d) *Ephedra gerardiana*

12. The active principle of Kuth is
(a) Essential oil
(b) Glucoside
(c) Alkaloid
(d) All of the above

13. The "true squill" is obtained from
(a) *Urginea scilla*
(b) *Rheum emodi*
(c) *Gaultheria fragrantissima*
(d) None of the above

14. Which of the following yields "dita" bark of commerce?
(a) *Cinchona ledgeriana*
(b) *Cassia fistula*
(c) *Alstonia scholaris*
(d) None of the above

15. The quinine content obtained from *Cinchona hybrida* is
(a) 14%
(b) 20-30%
(c) 5-10%
(d) None of the above

16. Drugs obtained from wood in
(a) *Bixa* species
(b) *Cannabis sativa*
(c) *Ephedra* species
(d) All of the above

17. .... is generally used as a substitute for "aldrenaline"
(a) Ephedrine (c) Quinine
(b) Liquorice (d) Cannabinol

18. The drug "stramonium" is obtained from dried leaves and flowering tops of
(a) *Datura innoxia*
(b) *Datura metal*
(c) *Datura stramonium*
(d) All of the above

19. The wintergreen oil is commercially obtained from
(a) *Gaultheria fragrantissima*
(b) *Cedrus deodara*
(c) *Aquilaria agallocha*
(d) None of the above

20. The "Henbane" drug is commercially obtained from
(a) Leaves of *Hyoscyamus niger*
(b) Leaves and flowers of *Hyoscyamus niger*
(b) Flowers of *Hyoscymus niger*
(c) Seeds of *Hyoscymus niger*

21. The immature leaves and flowering tops of "*Artemisia* species" yields
(a) Strychnine
(b) Glucosides
(c) Santonin
(d) Both (b) and (c)

22. The "Cannabinol" obtained from hemp is
(a) Resin
(b) Gum
(c) Both (a) and (b)
(d) None of the above

23. Commercially known as "purging cassia"
(a) *Cassia occidentalis*
(b) *Cassia fistula*
(c) *Cassia senna*
(d) All of the above

24. The commercial source of "strychnine" is
(a) *Cannabis sativa*
(b) *Berberis aristata*
(c) *Hyocymus niger*
(d) *Strychnos nux-vomica*

25. The scientific name of "Indian liquorice" is
(a) *Glycyrrhiza glabra*
(b) *Abrus precatorius*
(c) *Saussurea lappa*
(d) None of the above

26. Essential oils are
(a) By product of carbohydrate and and fat
(b) Compound of glycerine and fatty acids
(c) Both (a) and (b)
(d) None of the above

27. The method of extraction of essential oil depends upon
(a) Quantity
(b) Quality
(c) Stability of compound
(d) All of the above

28. In advance stage steam distillation is done at
(a) Atmospheric pressure
(b) Atmospheric pressure involving partial vacuum
(c) Below atmospheric pressure
(d) None of the above

29. The steam pressure higher than atmospheric pressure is employed for
(a) *Vetiver zizanoides*
(b) *Santalum album*
(c) Both (a) and (b)
(d) None of the above

30. Which of the following is not a grass oil?
(a) Palmarosa oil
(b) Citronella oil
(c) Khus oil
(d) Wintergreen oil

31. Sandalwood oil is obtained from
(a) Heartwood (c) Sapwood
(b) Root (d) Leaf

32. Which of the following yields Palmarosa oil?
(a) *Cymbopogon martinii Var. sofia*
(b) *Cymbopogon martinii Var. motia*
(c) *Cymbopogon nardus*
(d) *Cymbopogon flexuosus*

33. Root oil is obtained from
(a) *Sassurea lappa*
(b) Screw pine
(c) *Acorus calamus*
(d) *Abelmoschus moschatus*

34. The chief constituent of mint oil is
(a) Citral
(b) Carvone
(c) Menthol
(d) All of the above

35. Which of the following is/are not soluble in water?
(a) Fixed oil (c) Essential oil
(b) Resins (d) Both (a) and (b)

36. Tung oil is extracted from
(a) *Aleurites fordii*
(b) *Actinodaphne hookeri*
(c) *Calophyllum inophyllum*
(d) *Diploknema butresia*

37. A valuable fat known as "Phulwara butter" extracted from
(a) *Garcinia indica*
(b) *Diploknema butresia*
(c) *Madhuca indica*
(d) *Vateria indica*

38. The "piney tallow" is extracted from Vateria indica
(a) Seed (c) Stem
(b) Flower (d) Root

39. A valuable edible fat known as "Kokam butter" is yield from
(a) *Vateria indica*
(b) *Madhuca indica*
(c) *Azadirachta indica*
(d) *Garcinia indica*

40. Which oil is used as a substitute for biodiesel?
(a) *Pongamia pinnata*
(b) *Jatropha curcus*
(c) Both (a) and (b)
(d) None of the above

41. Indian butter tree is
(a) *Garcinia indica*
(b) *Azadirachta indica*
(c) *Vateria indica.*
(d) *Pongamia pinnata*

42. Waxes are found on the
(a) Endodermis of leaves and fruits
(b) Epidermis of leaves and fruits
(c) Whole part of leaves and fruits
(d) All of the above

43. The famous Carnuva wax is obtained from
(a) *Copernicia cerifera*
(b) *Sapium sebiferum*
(c) *Vateria indica*
(d) *Terminalia chebula*

44. Fruit wax is obtained from
(a) *Pongamia pinnata*
(b) Quebracho
(c) *Melia azedarach*
(d) *Acacia nilotica*

45. The formation of essential oil in *Aquilaria agallocha* is due to the activity of
(a) Insect
(b) Fungi
(c) Both (a) and (b)
(d) None of the above

46. The soft fibres are obtained from
(a) Stem
(b) Leaves
(c) Both (a) and (b)
(d) None of the above

47. The commercial surface fibre obtained from
(a) Linen
(b) Cotton
(c) Jute
(d) Both (a) and (b)

48. Paper making fibres contain
(a) Parenchyma cells
(b) Sclerenchyma cells
(c) Collenchyma cells
(d) All of the above

49. The method of extracting fibres from one another is technically known as
(a) Beating
(b) Scratching
(c) Scraping
(d) Retting

50. The well-known "Kittul" fibre obtained from
(a) *Caryota urens*
(b) *Sterculia urens*
(c) *Grewia optiva*
(d) *Sterculia vilosa*

51. Silk cotton tree is
(a) *Bombax ceiba*
(b) *Calotropis gigantea*
(c) *Coccus nucifera*
(d) *Ceiba pentandra*

52. Coconut coir is extracted from
(a) Pericarp
(b) Mesocarp
(c) Endocarp
(d) All of the above

53. Example of tree that produces valuable stem fibre for rope making
(a) *Adenanthera pavonina*

(b) *Sterculia vilosa*
(c) *Bombax ceiba*
(d) *Ceiba pentandra*

54. Which of the following are fodder grasses?
(a) *Cynadon dactylon*
(b) *Apluda mutica*
(c) *Heteropogon contortus*
(d) All of the above

55. The canes generally attain maturity at the age of
(a) 10 years
(b) 30-40 years
(c) 5 years
(d) None of the above

56. Grasses used for matting are
(a) *Sacchrum munja*
(b) *Eulaliopsis binata*
(c) *Desmostachya bipinnata*
(d) Both (a) and (c)

57. Which of the following are thatching grasses?
(a) *Impreta cylindrica*
(b) *Saccharum spontaneum*
(c) Both (a) and (b)
(d) None of the above

58. Thorny bamboo is
(a) *Bambusa arundinacea*
(b) *Arundinaria falcata*
(c) *Dendrocalamus strictus*
(d) *Bambusa vulgaris*

59. Bamboo is
(a) Found in humid and tropical region
(b) Deciduous or evergreen
(c) Grass
(d) All of the above

60. Which of the following bamboo species is found in high altitude forests?
(a) *Dendrocalamus strictus*
(b) *Arundinaria spathiflora*
(c) *Melocanna* species
(d) *Calamus* species

61. Which of the following bamboo species is found in fur, spruce, deodar forest in Western Himalayas?
(a) *Arundinaria falcata*
(b) *Bambusa balcooa*
(c) *Arundinaria spathiflora*
(d) *Bambusa tulda*

62. ….,.. is found in Oak-Deodar forest.
(a) *Arundinaria falcata*
(b) *Bambusa balcooa*
(c) *Melocanna* species
(d) None of the above

63. Shrubby bamboo species having more than 100 culms is
(a) *Calamus* species
(b) *Bambusa tulda*
(c) *Arundinaria falcata*
(d) *Dendrocalamus hamiltonii*

64. The species employed for making Kraft paper is
(a) *Melocanna bambusoides*
(b) *Dendrocalamus longispathus*
(c) *Bambusa balcooa*
(d) None of the above

65. Insect resistant bamboo species is/are
(a) *Bambusa balcooa*
(b) *Bambusa tulda*
(c) Both (a) and (b)
(d) None of the above

66. Thin walled bamboo species is
(a) *Cephalostachyum pergracile*
(b) *Melocanna bambusoides*
(c) *Dendrocalamus hamiltonii*
(d) None of the above

67. The scientific name of tarai/muli bamboo is
(a) *Bambusa vulgaris*
(b) *Dendrocalamus strictus*
(c) *Arundinaria falcata*
(d) *Melocanna bambusoides*

68. The commercial name of canes is
(a) Male bamboo
(b) Dammer
(c) Rattans
(d) None of the above

69. Canes are generally found in
(a) Evergreen forest
(b) Dry deciduous forest
(c) Temperate forest
(d) All of the above

70. Tans and dyes are simple chemical compounds of
(a) C,H,O
(b) C,H,O along with N
(c) C,H,O along with K
(d) None of the above

71. In *Quebracho*, the ratio of tannin to non-tannin content is
(a) 10:1
(b) 1:10
(c) 100:1
(d) None of the above

72. Wood tan is obtained from
(a) *Cassia fistula*
(b) *Acacia senegal*
(c) *Quebracho colorado*
(d) *Calophyllum inophyllum*

73. Astringent liquid is obtained from the bark of
(a) *Acacia mollisima*
(b) *Emblica officinalis*
(c) *Mallotus philippensis*
(d) None of the above

74. Fruit tan is obtained from
(a) *Terminalia arjuna*
(b) *Terminalia chebula*
(c) *Emblica officinalis*
(d) All of the above

75. Pyrogallol class of tan is obtained from
(a) *Caesalpinia echinata*
(b) *Terminalia* species
(c) *Acacia concinna*
(d) None of the above

76. Leaf tan is obtained from
(a) *Cassia fistula*
(b) *Anogeissus latifolia*
(c) *Artocarpus* species
(d) All of the above

77. Dyeing art was revolutionised in 1856 during the accidental invention of
(a) Brazilian dye
(b) Cutch dye
(c) Aniline dye
(d) None of the above

78. Brazilian dye is obtained from
(a) *Caesalpinia sappan*
(b) *Alnus* species
(c) *Mallotus philippensis*
(d) *Quebracho colorado*

79. Santalin dye is obtained from
(a) *Pterocarpus marsupium*
(b) *Pterocarpus santalinus*
(c) *Santalum album*
(d) Both (b) and (c)

80. The well-known fruit dye "kamela" is obtained from
(a) *Mallotus philippensis*
(b) *Bixa orellana*
(c) *Lawsonia inermis*
(d) None of the above

81. The principle colouring agent in Kamela dye is
(a) Indicans
(b) Rottlerin
(c) Both (a) and (b)
(d) None of the above

82. Henna dye is obtained from
(a) Root
(b) Flower
(c) Stem
(d) Leaf

83. The king of dye stuff is
(a) *Acacia catechu*
(b) *Caesalpinia sappan*
(c) *Indigofera tinctoria*
(d) None of the above

84. The red colour dye is obtained from
(a) *Toona ciliata*
(b) *Rubia cordifolia*
(c) *Caesalpinia sappan*
(d) All of the above

85. Annato dye is obtained from
(a) *Bixa orellana* flower and fruit
(b) *Bixa orellana* bark
(c) *Bixa orellana* root
(d) All of the above

86. The root dye is obtained from
(a) *Rubia cordifolia*
(b) *Berberis aristata*
(c) *Punica granatum*
(d) All of the above

87. Which of the following species is most suitable for matchwood industry?
(a) *Albizzia* species
(b) *Calophyllum* species
(c) *Ailanthus triphysa*
(d) *Dalbergia sissoo*

88. *Diospyros melanoxylon* is an important source of beedi leaf belongs to family
(a) Diospyraceae
(b) Ebenaceae
(c) Euphorbeaceae
(d) Melanoxylaceae

89. Wavy grain is observed in
(a) Maple
(b) Teak
(c) Sissoo
(d) Red sanders

90. Best quality cricket bats are made from
(a) Female salix
(b) Male poplar
(c) Male salix
(d) Female poplar

91. Which of the following wood emit crackle and sparkle on burning?
(a) *Pinus roxburghii*
(b) *Diospyros tomentosa*
(c) Both (a) and (b)
(d) None of the above

92. Species used for making aircraft
(a) Fur
(b) Spruce
(c) Sitka spruce
(d) All of the above

93. The excellent material for the aeroplane propeller blades is
(a) Impreg
(b) Compreg
(c) Staypack
(d) Staywood

94. Which of the following species are used for making agricultural implements?
(a) Axelwood
(b) *Messua ferrea*
(c) Babul
(d) All of the above

95. Species are used for making civets
(a) *Swintonia floribunda*
(b) *Anacardium occidentale*
(c) Both (a) and (b)
(d) None of the above

96. Species popularly used for making mathematical instruments
(a) *Juglans* species
(b) *Asculus* species
(c) Teak
(d) All of the above

97. The guitar is made from
(a) White dhup
(b) Mahogany
(c) Teak
(d) Spruce

98. Harmonium is made from
(a) Toona
(b) Teak
(c) Rosewood
(d) Walnut

99. The braille paper is manufactured from
(a) *Eucalyptus terticornis*
(b) *Eucalyptus hybrid*
(c) *Eucalyptus globulus*
(d) None of the above

100. The root fibre is obtained from
(a) *Grewia optiva*
(b) *Hardwickia binnata*
(c) Cotton
(d) *Butea monosperma*

101. The principle constituent of cutch is
(a) Catechu
(b) Catechin
(c) Catechu tannic acid
(d) Both (a) and (b)

102. Chilgoza is obtained from
(a) *Pinus gerardiana*
(b) *Pinus wallichiana*
(c) *Pinus radiata*
(d) *Pinus roxburghii*

Answer Keys

1. (d)	**2.** (b)	**3.** (c)	**4.** (d)	**5.** (b)	**6.** (a)	**7.** (d)	**8.** (d)	**9.** (a)	**10.** (c)
11. (b)	**12.** (d)	**13.** (a)	**14.** (c)	**15.** (d)	**16.** (c)	**17.** (a)	**18.** (c)	**19.** (a)	**20.** (b)
21. (c)	**22.** (a)	**23.** (b)	**24.** (d)	**25.** (b)	**26.** (a)	**27.** (d)	**28.** (b)	**29.** (c)	**30.** (d)
31. (a)	**32.** (b)	**33.** (a)	**34.** (c)	**35.** (d)	**36.** (a)	**37.** (b)	**38.** (a)	**39.** (d)	**40.** (c)
41. (a)	**42.** (b)	**43.** (a)	**44.** (c)	**45.** (c)	**46.** (a)	**47.** (b)	**48.** (b)	**49.** (d)	**50.** (a)
51. (d)	**52.** (b)	**53.** (b)	**54.** (d)	**55.** (c)	**56.** (d)	**57.** (c)	**58.** (a)	**59.** (d)	**60.** (b)
61. (c)	**62.** (a)	**63.** (c)	**64.** (b)	**65.** (c)	**66.** (a)	**67.** (d)	**68.** (c)	**69.** (a)	**70.** (b)
71. (a)	**72.** (c)	**73.** (a)	**74.** (d)	**75.** (b)	**76.** (b)	**77.** (c)	**78.** (a)	**79.** (b)	**80.** (a)
81. (b)	**82.** (d)	**83.** (c)	**84.** (d)	**85.** (a)	**86.** (d)	**87.** (c)	**88.** (b)	**89.** (d)	**90.** (a)
91. (c)	**92.** (d)	**93.** (b)	**94.** (d)	**95.** (c)	**96.** (d)	**97.** (a)	**98.** (b)	**99.** (c)	**100.** (d)
101. (c)	**102.** (a)								

9

Wood Science and Logging

1. Square shape pith is generally observed in
(a) Walnut (c) Teak
(b) Sal (d) Spruce

2. In some hardwood species, heartwood does not show clear distinctive features
(a) Mango, Bombax
(b) Sal, Mango, Teak
(c) Sissoo, Sandal, Bombax
(d) None of the above

3. .... is rich in starch.
(a) Heartwood
(b) Hardwood
(c) Sapwood
(d) All of the above

4. From anatomical point of view ,the sapwood and heartwood are
(a) Different
(b) Same
(c) Both (a) and (b)
(d) None of the above

5. In................. regions, growth ring is a reliable index of age.
(a) Temperate
(b) Tropical
(c) Both (a) and (b)
(d) None of the above

6. Generally the distinct annual rings are visible in
(a) Sal, Morus, Mango, Teak
(b) Pinus, Toona, Jamun
(c) Teak, Deodar, Mango
(d) Deodar, Toona, Morus, Teak

7. If 8 concentric rings or more per centimetre of radius is the indicative of
(a) Fast growth
(b) Slow growth
(c) Medium growth
(d) All of the above

8. If the growth rings is 5-25 mm per centimetre of radius is the indicative of
(a) Slow growth
(b) Fast growth
(c) Both (a) and (b)
(d) None of the above

9. Latewood is also known as
(a) Spring wood (c) Autumn wood
(b) Summer wood (d) Winter wood

10. The general alignment and direction of cell is termed as
(a) Texture (c) Grain
(b) Fibre (d) None of these

11. ...... refers to the size of the cells.
(a) Texture
(b) Grain
(c) Both (a) and (b)
(d) None of the above

12. Sal is the example of
(a) Straight grain
(b) Wavy grain
(c) Spiral grain

(d) Interlocked grain

13. Fine textured wood is
(a) *Terminalia myriocarpa*
(b) *Morus* species
(c) *Tectona grandis*
(d) None of the above

14. The occurrence of Pores/Vessels is a common feature of
(a) Softwoods
(b) Hardwoods
(c) Both (a) and (b)
(d) None of the above

15. Example of ring porus wood in India
(a) Sal (c) Haldu
(b) Teak (d) Fir

16. Example of diffused porus wood is
(a) Mulberry (c) Toon
(b) Teak (d) Sal

17. The vessels are distinctly visible to naked eye in
(a) Semul
(b) Kokko
(c) Both (a) and (b)
(d) None of the above

18. .... is the common feature of conifers.
(a) Trachieds (c) Fibres
(b) Vessels (d) Both (a) and (b)

19. A piece of wood having no vessels belong to
(a) Teak (c) Deodar
(b) Mango (d) Neem

20. Irregular patches of soft tissues formed as a result of injury in cambium due to insect attack is
(a) Riple marks (c) Rays
(b) Shake (d) Pith flecks

21. Quarter sawing is generally done in
(a) Teak (c) Neem
(b) Oaks (d) Deodar

22. .... are the plates of horizontally oriented parenchyma cells which run in radial direction from pith towards the bark.
(a) Rays (c) Ripple marks
(b) Vessels (d) Pith flecks

23. Silver grain effect is commonly seen in quarter sawned timber like
(a) Oak
(b) *Corallia* species
(c) Both (a) and (b)
(d) None of the above

24. In hardwoods, fine inconspicuous rays are found in
(a) Ebony (c) Sal
(b) Teak (d) Oak

25. Ripple marks are visible in
(a) Satin wood
(b) *Pterocarpus marsupium*
(c) *Holoptelea integrifolia*
(d) All of the above

26. Predominately solitary and scattered canals filled with white resinous deposit found in
(a) Chirpine (c) Fur
(b) Vellapine (d) Deodar

27. Resin canals are found in
(a) Fir (c) Chir pine
(b) Cypress (d) Sal

28. Intercellular canals are arranged in long tangential bands present in
(a) Sal
(b) Hopea
(c) Deodar
(d) All of the above

29. Radial or Horizontal canals is commonly found in
(a) *Parashia insignis*
(b) *Boswellia serrata*
(c) *Lannea coromandelica*
(d) All of the above

30. In hardwoods, mechanical support is provided by
(a) Trachieds (c) Fibre
(b) Vessels (d) Heartwood

31. In non-porus wood conduction and mechanical support is provided by
(a) Vessels
(b) Fibres
(c) Tracheids
(d) All of the above

32. In porus wood, the conduction of sap is performed by
(a) Vessels
(b) Fibres
(c) Tracheids
(d) All of the above

33. .... carries the sap from the roots to the leaves in a tree
(a) Phloem
(b) Xylem
(c) Cambium
(d) All of the above

34. In trees .... carries manufactured food material from leaves down to branches
(a) Xylem
(b) Phloem
(c) Heartwood
(d) Sapwood

35. .... gives the rigidity and strength to the tree.
(a) Sapwood (c) Bark
(b) Cambium (d) Heartwood

36. Annual rings in tree are composed of
(a) Spring wood
(b) Summer wood
(c) Both (a) and (b)
(d) None of the above

37. On the basis of specific gravity, the very light wood is
(a) *Cryptomeria japonica*
(b) *Tectona grandis*
(c) *Ailanthus grandis*
(a) *Acacia nilotica*

38. Speed of the sound in wood is least in
(a) Radial direction
(b) Tangential direction
(c) Longitudinal direction
(d) All of the above

39. Tyloses inclusions are generally found in
(a) Hardwood
(b) Sapwood
(c) Both (a) and (b)
(d) None of the above

40. The cross banded construction of the wood is called as
(a) Laminated wood
(b) Laminae wood
(c) Core boards
(d) Plywood

41. Inner bark is composed of
(a) Secondary xylem
(b) Cork
(c) Secondary phloem
(d) Phellogen

42. Major components of the wood is/are
(a) Cellulose
(b) Hemi-cellulose

(c) Lignin
(d) All of the above

43. Cellulose constitutes ....% of wood

(a) 10–25% (c) 50–60%
(b) 45–50% (d) 5–10%

44. Hemi-cellulose is generally extracted by using

(a) Water
(b) Dilute alkali
(c) Dilute acid
(d) All of the above

45. The lignin content of wood varies from

(a) 20-35%
(b) 10-15%
(c) 1-2%
(d) None of the above

46. Extractives are found in

(a) Cell wall
(b) Lumina of cell
(c) Inside cell
(d) All of the above

47. Inorganic extractives is/are

(a) Free Silica
(b) Organic acids
(c) Minerals
(d) All of the above

48. In certain species like Mlanje cedar or *Araucaria* species contain essential oil

(a) 1–2% (c) > 8%
(b) 5–6% (d) > 12%

49. In sandalwood essential oil content is

(a) < 1%
(b) 8%
(c) 2–3%
(d) None of the above

50. In softwood the resin content varies from

(a) 0.8–25% (c) 5–10%
(b) 2–3% (d) 20–25%

51. Oleo resin is obtained from

(a) Sapwood
(b) Sapwood of living tree
(c) Aged stump
(d) All of the above

52. Wood rosin is extracted using

(a) Gasoline
(b) Benzene
(c) Both (a) and (b)
(d) None of the above

53. The water held by the capillary forces in the cell cavities is termed as

(a) Free water
(b) Bound water
(c) Adsorbed water
(d) Hygroscopic water

54. The water held in cell walls due to hydrogen bonding is termed as

(a) Bound water
(b) Hygroscopic water
(c) Adsorbed water
(d) All of the above

55. Which of the following is true for fibre saturation point?

(a) Cell cavity free of "free water"
(b) Cell wall saturated with "bound water"
(c) Both (a) and (b)
(d) None of the above

56. Any feature that lowers the technical quality, commercial value or grade of timber is known as

(a) Blemish
(b) Defect

(c) Both (a) and (b)
(d) None of the above

57. Pick the odd one out
(a) Knots
(b) Reaction wood
(c) Cross grain
(d) Warping

58. A knot in which the fibres are intergrown with surrounding wood is
(a) Dead knot (c) Live knot
(b) Black knot (d) Enclosed knot

59. A knot in which the fibres are not completely intergrown with surrounding wood is
(a) Dead knot
(b) Black knot
(c) Encased knot
(d) All of the above

60. The separation of fibres along the grain is termed as
(a) Cross-grain
(b) Shakes
(c) Warping
(d) None of the above

61. Chir pine, Qak are prone to
(a) Cup shake
(b) Ring shake
(c) Heart shake
(d) Star shake

62. The abnormal wood generally found on the underside of branches and leaning tree is
(a) Reaction wood
(b) Tension wood
(c) Compression wood
(d) All of the above

63. Common example of the woody climbers
(a) *Butea monosperma*
(b) *Bauhinia vahlii*
(c) *Acacia pennata*
(d) All of the above

64. In warping, distortion in the longitudinal plane is called as
(a) Spring (c) Cupping
(b) Bowing (d) Twist

65. Wood boring marine animals belong to the classes?
(a) Molluscs
(b) Crustaceans
(c) Both (a) and (b)
(d) None of the above

66. Important Molluscs genera that attacks the timber in sea water is/are
(a) Teredo
(b) Bankia
(c) Martesia
(d) All of the above

67. Important Crustacean genera that attack timber in sea water is/are
(a) Limnoria
(b) Cherula
(c) Sphaeroma
(d) All of the above

68. The Crustaceans and Molluscs that attack the timber in sea water are popularly known as
(a) Shipworms
(b) Timber feeders
(c) Insects
(d) Both (b) and (c)

69. In white rot wood is subjected to ......which imparts white colour to wood
(a) Oxidation
(b) Hydrolysis
(c) Both (a) and (b)
(d) None of the above

70. In wood, decay and stain is caused by
(a) Fungi (c) Bacteria
(b) Actinomycetes (d) Virus

71. Fungi causing stain in wood belongs to
(a) Fungi imperfecti
(b) *Ceratocystis* species
(c) Both (a) and (b)
(d) None of the above

72. Wood rotting fungi belongs to
(a) Hymenomycetes
(b) Ascomycetes
(c) Fungi imperfecti
(d) All of the above

73. Defects arising due to conversion and wood working is known as
(a) Waney
(b) Collapse
(c) Warping
(d) Case hardening

74. Seasoning is the process of
(a) Cooling of wood
(b) Harvesting of wood
(c) Drying of wood
(d) None of the above

75. Which of the following cannot be easily killed by girdling?
(a) *Adina cordifolia*
(b) *Lagerstromia tomentosa*
(c) *Homalium tomentosa*
(d) All of the above

76. Examples of non-refractory wood
(a) Sal
(b) Deodar
(c) Teak
(d) All of the above

77. Woods that can be rapidly seasoned free from defect even in open air and Sun is known as
(a) Highly refractive wood
(b) Non refractive wood
(c) Moderately refractive wood
(d) None of the above

78. The most common method of stacking timber in air seasoning is
(a) Horizontal (c) Vertical
(b) Box piling (d) None of these

79. Most common method of stacking in railway sleepers is
(a) Close crib method
(b) Open crib method
(c) One in nine method
(d) All of the above

80. Vertical stacking is mainly used for
(a) Non refractory wood
(b) Refractory wood
(c) Moderately refractory wood
(d) None of the above

81. .... is most widely used for preventing end splitting in logs, sleepers and squares.
(a) Paraffin wax
(b) Bituminous paint
(c) Rosin
(d) Thick coal tar

82. .... is used at higher temperature in kiln seasoning to prevent end splitting.
(a) Glass oil paint
(b) Rosin and lamp black
(c) Paraffin wax
(d) Both (a) and (b)

83. The most suitable time for conversion of highly refractive wood for air seasoning is towards
(a) End of summers

(b) End of rainy season
(c) Hot and dry season
(d) None of the above

84. In Kiln seasoning which of the following factors are controlled
(a) Temperature
(b) Relative humidity
(c) Air circulation
(d) All of the above

85. The durable wood has service life of
(a) > 10 years (c) 5–10 years
(b) > 15 years (d) < 5 years

86. The average life of properly treated wooden railway sleepers is
(a) 10–15 years (c) 20–25 years
(b) > 5 years (d) > 20 years

87. The chief characteristics of wood preservatives are
(a) High toxicity and permanency
(b) High penetrability
(c) High stability
(d) All of the above

88. Coal tar, wood tar, water gas tar creosotes are .... wood preservatives.
(a) Water soluble
(b) Oil type
(c) Organic solvent
(a) Water soluble- leachable type

89. Proprietary preservative contains high boiling creosote as a base in
(a) Solignum
(b) Creosant
(c) Both (a) and (b)
(d) None of the above

90. .... is popularly known as ASCU.
(a) Arsenic-copper-chromate
(b) Arsenic pentoxide
(c) Zinc-meta-arsenite
(d) None of the above

91. Zinc chloride is
(a) Oil type preservative
(b) Water soluble-fixed type
(c) Organic solvent type
(d) Water soluble- leachable type

92. The elements of ASCU, Arsenic Penta Oxide, Copper sulphate and potassium or sodium dichromate are in the proportion of
(a) 1:4:3 (c) 1:2:3
(b) 1:3:4 (d) 1:4:2

93. When the wood is sublimated in $HgCl_2$, the process is known as
(a) Steeping (c) Kyanising
(b) Powellizing (d) Dipping

94. The diffusion process in which the timber is immersed in sugar solution containing arsenic trioxide is known as
(a) Powellizing
(b) Kyanising
(c) Osmose process
(d) All of the above

95. Boucherie method is used to treat
(a) Dry bamboo and timber with bark
(b) Green bamboo and timber with bark
(c) Green bamboo and timber without bark
(d) None of the above

96. The amount of the preservative retained in the wood is expressed in terms of
(a) Kg/m^3
(b) Kg/m^2

(c) Both (a) and (b)
(d) None of the above

97. Which of the following process is used for the timbers of low absorptive properties and readily leachable preservatives?
(a) Lowry process
(b) Rueping process
(c) Bethal process
(d) All of the above

98. Maximum penetration of the preservative with minimum net absorption found in
(a) Bethal process
(b) Full cell process
(c) Empty cell process
(d) None of the above

99. Rueping process is used to treat
(a) Bamboo
(b) Mixed species
(c) Timber containing sapwood and heartwood
(d) Both (a) and (b)

100. Green timbers are specially treated using
(a) Boulten process
(b) Steam cum vaccum process
(c) Both (a) and (b)
(d) None of the above

101. Incisions are useful in treatment of
(a) Refractory timber
(b) Non-refractory timber
(c) Both (a) and (b)
(d) None of the above

102. The built up, bonded products consisting of wholly wood or in combination with metals is termed as
(a) Improved wood
(b) Composite wood
(c) Compressed wood
(d) All of the above

103. The outer faces of the plywood are known as
(a) Face
(b) Core
(c) Back
(d) Both (a) and (b)

104. The simplest plywood construction is of
(a) Three plies
(b) Two plies
(c) Five plies
(d) None of these

105. The veneers are produced using
(a) Rotatory peeling
(b) Slicing
(c) Both (a) and (b)
(d) None of the above

106. The time between the spreading of glue and application of full pressure is termed as
(a) Assembly time
(b) Application time
(c) Lubrication time
(d) All of the above

107. The built-up product of wood layers laid with their grains parallel is termed as
(a) Coreboard
(b) Plywood
(c) Laminated wood
(d) None of the above

108. The maximum length of log for rotatory peeling is
(a) 200-220 cm
(b) 100-120 cm
(c) 255-280 cm
(d) None of the above

109. The built-up board having core of light material on both sides with

relatively thin material having high strength properties
(a) Fibre board
(b) Core board
(c) Laminated wood
(a) Sandwich board

110. Hard boards are a type of
(a) Fibre board
(b) Sandwich board
(c) Particle board
(d) Core board

111. The wood impregnated with synthetic material and compressed to to improve strength and durability is
(a) Composite wood
(b) Compressed wood
(c) Modified wood
(d) None of the above

112. The heating of wood in absence of air to produce wood alcohol is
(a) Wood scarification
(b) Destructive distillation
(c) Wood gasification
(d) None of the above

113. During burning of wood which cell component degrades first
(a) Cellulose
(b) Hemi-cellulose
(c) Lignin
(a) Pectin

114. Lightest wood in the world is
(a) Champaka (c) Rubber
(b) Erythrina (d) Balsa

115. Major characteristic of raw material in paper industry is
(a) Tracheids
(b) Fibre length
(c) Vessels
(d) Both (a) and (c)

116. Axes are generally used for
(a) Trimming (c) Splitting
(b) Grubbing (d) All of these

117. The weight of the ordinary axe for felling poles and saplings is
(a) 3–4 kg (c) 1.5–2 kg
(b) 0.7 kg (d) 1 kg

118. ....action of the axe is inversely proportional both to the sharpness of edge and thickness of blade.
(a) Severing (c) Crushing
(b) Shearing (d) All of these

119. The standard length of axe handle is
(a) 90 cm (c) 120 cm
(b) 75 cm (d) 150 cm

120. An obliquely driven axe involves
(a) Shearing action
(b) Crushing action
(c) Severing action
(d) All of the above

121. While splitting logs with axe .... action takes place.
(a) Severing (c) Crushing
(b) Shearing (d) Both (a) and (b)

122. ....is used for felling bamboos
(a) Axe (c) Bill hooks
(b) Cant hook (d) Cleaver

123. ....is used for cross cutting, ripping logs into planks, scantlings
(a) Saw (c) Cleaver
(b) Axe (c) Both (a) and (b)

124. The most commonly used saw is
(a) M-tooth saw
(b) Raker tooth saw

(c) V-tooth saw
(d) Peg tooth saw

125. The saws are made up of
(a) Iron (c) Steel
(b) Aluminium (d) Both (a) and (b)

126. The entire opening between the two adjacent teeth is known as
(a) Space (c) Gauge
(b) Kerf (d) Gullet

127. The width of the cut made by the saw is known as
(a) Pitch (c) Set
(b) Kerf (d) Pitch

128. The extend to which saw teeth bend, is termed as
(a) Set (c) Kerf
(b) Pitch (d) Face

129. While sawing more sawdust is produced by
(a) Softwoods
(b) Hardwoods
(c) Both (a) and (b)
(d) None of the above

130. The most commonly type of saw used in Himalayan forests is
(a) Delhi saw (c) Peg tooth saw
(b) Frame saw (d) Pit saw

131. The technique of proper levelling of the saw teeth is known as
(a) Splicing
(b) Jointing
(c) Setting
(d) None of the above

132. ...... is used for rolling, stopping, turning logs
(a) Cant hook and Pickaroons
(b) Bill hooks
(c) Cleaver
(a) Wedges

133. Saw used by single man for cutting upto 20 cm diameter is
(a) Cross cut saws (c) Delhi saw
(b) Ripping saw (d) Bow saw

134. In India, axe handle is
(a) Oval (c) Rectangular
(b) Round (d) Both (a) and (b)

135. European and American axes usually have .... shaped handle.
(a) Round (c) Oval
(b) Circular (d) Rectangular

136. The saw having triangular teeth is known as
(a) Peg tooth saw
(b) Raker tooth saw
(c) M- tooth saw
(d) Lance tooth saw

137. The distance between the two adjacent teeth is known as
(a) Face (c) Gauge
(b) Space (d) Pitch

138. The thickness of the saw blade is known as
(a) Space (c) Gullet
(b) Kerf (d) Gauge

139. In axe, the blade should be
(a) Concave
(b) Convex
(c) Both (a) and (b)
(d) None of the above

140. The edges of the axe blade should be
(a) Curved (c) Straight
(b) Circular (d) Square

141. ....is used for preventing the splitting of the butt end of the trees, from splitting during the felling.
(a) Cable puller

(b) Pickaroons
(c) Stem tightener
(d) None of the above

142. In Himalayas, at elevation more than 1800 m felling is carried out
(a) From October- March
(b) From March onwards
(c) In Summers
(d) In all seasons

143. In plains, sub montane tract of India felling is carried out during
(a) March-April
(b) May-June
(c) October-March
(d) None of the above

144. From the seasoning point of view felling should not be carried out in
(a) Summers (c) Winters
(b) Rainy season (d) Both (a) and (c)

145. On hilly ground the tree should be felled in
(a) Downhill direction
(b) Uphill direction
(c) Any direction
(d) None of the above

146. On contours and undulating ground tree should be felled in
(a) Downhill direction
(b) Uphill direction
(c) Both (a) and (b)
(d) None of the above

147. In hilly areas felling should proceed from
(a) Down of the slope
(b) Middle of the slope
(c) Any place
(a) Top of the slope

148. In felling alone with axe, an under cut is given to the width of
(a) 2/3 of diameter
(b) 2/5 of diameter
(c) 2/7 of diameter
(a) 1/5 of diameter

149. Felling by roots is commonly done in
(a) *Santalum album*
(b) *Acacia catechu*
(c) Both (a) and (b)
(d) None of the above

150. .... are used to prevent the jamming of the saw and for assisting the felling of the tree.
(a) Pickaroons (c) Cleaver
(b) Wedges (d) Axe

151. .... cut is given in the direction of fall
(a) Under cut (c) Felling cut
(b) Saw cut (d) Both (a) and (c)

152. The rounded logs split with the axe is known as
(a) Scantlings (c) Hakeries
(b) Poles (d) Both (a) and (c)

153. ....are square or rectangular generally under 15 cm in section.
(a) Beams (c) Scantlings
(b) Hakeries (d) Waney baulks

154. Wooden slabs upto 5 cm thickness with varying lengths and width is
(a) Planks (c) Scantlings
(b) Beams (d) Waney baulks

155. The roughly sawn timber is known as
(a) Timber (c) Log
(b) Lumber (d) Hakeries

156. Cutting of the tree into log is known as
(a) Lumber (c) Felling
(b) Limbing (d) Bucking

157. The process of removing branches from the stem of the felled tree is known as
(a) Limbing (c) Both (a) and (b)
(b) Chasing (d) None of these

158. Minor transportation is also known as
(a) Off road transportation
(b) Water transportation
(c) Skyline
(d) None of the above

159. Higher conversion percentage is obtained through
(a) Hand sawing
(b) Machine sawing
(c) Both (a) and (b)
(d) None of the above

160. In North-western Himalayas the most useful method for extracting sleepers and scantlings is
(a) Gravity ropes
(b) Power ropeway
(c) Donald portable gravity ropeway
(d) All of the above

161. Floating carried out in small side nallas is known as
(a) Sweeping floating
(b) Rafting
(c) Wet slides
(d) Telescopic floating

162. Obstructions made to catch freely floating river timber by barricading across the river is known as
(a) Pathru (c) Booms
(b) Depots (d) None of these

163. The V-shaped boom is known as
(a) Two-way boom
(b) Single arm boom
(c) Double arm boom
(d) All of the above

164. In two way boom, the arms make an angle of.................at the vertex.
(a) 45° (b) 40 ° (c) 50° (d) 55°

165. The place where wood and other forest produce is stored is known as
(a) Depots
(b) Channel
(c) Mandi
(d) All of the above

166. Construction over or all around the obstructions in telescopic floating is known as
(a) Booms
(b) Pathru
(c) Both (a) and (b)
(d) None of the above

167. Method of extracting timber from areas where quantity of water is less by means of
(a) Telescopic floating
(b) Sweeping
(c) Wet slides
(d) None of the above

168. A place where the timber and other forest produce stored for sale is called
(a) Transit depot (c) Land depot
(b) Check depot (d) Sale depot

169. If the two pulley are equal in diameter and belt is tight, than the arc of contact is

(a) Equal to 180°
(b) More than 180°
(c) Less than 180°
(d) None of the above

170. If the size of two pulleys are unequal, the arc of contact of smaller pulley will be
(a) More than 180°
(b) Less than 180°
(c) Equal to 180°
(d) None of the above

171. The methods of joining the end of belt is/are
(a) Butt jointing
(b) Spliced jointing
(c) Staggered jointing
(d) All of the above

172. Swage set is also known as
(a) Square dress
(b) Bewelled dress
(c) Briar dress
(d) None of these

173. Spring set is also known as
(a) Briar dress
(b) Bewelled dress
(c) Both (a) and (b)
(d) None of the above

174. A log with a large check should sawn............to grain
(a) Parallel
(b) Perpendicular
(c) Tangential
(d) All of the above

175. The intelligent choice of timber for specific purpose is known as
(a) Choice (c) Both (a) and (b)
(b) Grading (d) None of these

176. First written set of grading rules introduced by
(a) Germany (c) Sweden
(b) France (d) India

177. Based on the physical and mechanical properties of timbers test, the grading rule for structural timber were first reported in 1924 by
(a) SH Howard
(b) LN Seaman
(c) CR Ranganathan
(d) All of the above

178. In India, grading is done on the basis of
(a) Dimension and appearance of logs
(b) Use of log or converted material
(c) Based on defect & estimate of out-turn
(d) All of the above

179. .... is measured by acute angle with the axis.
(a) Twist of the tree
(b) Spiral grain
(c) Both (a) and (b)
(d) None of the above

180. In India, structural timber for building purpose are classified into .... grade
(a) Four (c) Six
(b) Three (d) Ten

181. In selected grade of timber for building purpose, the defect should not exceed
(a) 25% (c) 37.5%
(b) 5% (d) 12.5%

182. Exact quantity of produce is not known at the time of sale
(a) Payment of outturn
(b) Lump-sum sale
(c) Both (a) and (b)
(d) None of the above

183. Most satisfactory system of felling and extraction adopted by
(a) Government agency
(b) Both Government and purchasers combined
(c) Purchasers alone
(d) All of the above

184. Sale of the forest produce based on the individual consultation of
(a) Private negotiation
(b) Highest bidder
(c) Both (a) and (b)
(d) None of the above

185. ....is a price below which no bids are accepted.
(a) Upset price
(b) Upper price
(c) Reserve price
(d) None of the above

186. The total usable rough timber from the average tree is
(a) 20–70%
(b) 39–45%
(c) 37–52%
(d) None of the above

187. The total estimated loss from high stump logging is about
(a) > 50% (c) 10–15%
(b) 2–5% (d) 50–60%

188. ...... are more than 4 m in length and cross-section not less than 15 cm.
(a) Waney baulks (c) Half baulks
(b) Scantlings (d) Beam

Answer Keys

1. (c)	**2.** (a)	**3.** (c)	**4.** (b)	**5.** (a)	**6.** (d)	**7.** (b)	**8.** (a)	**9.** (b)	**10.** (c)
11. (a)	**12.** (d)	**13.** (d)	**14.** (b)	**15.** (b)	**16.** (d)	**17.** (c)	**18.** (a)	**19.** (c)	**20.** (d)
21. (b)	**22.** (a)	**23.** (c)	**24.** (a)	**25.** (d)	**26.** (b)	**27.** (c)	**28.** (d)	**29.** (d)	**30.** (c)
31. (c)	**32.** (a)	**33.** (b)	**34.** (b)	**35.** (d)	**36.** (c)	**37.** (a)	**38.** (b)	**39.** (b)	**40.** (d)
41. (c)	**42.** (d)	**43.** (b)	**44.** (b)	**45.** (a)	**46.** (b)	**47.** (d)	**48.** (d)	**49.** (b)	**50.** (a)
51. (b)	**52.** (c)	**53.** (a)	**54.** (d)	**55.** (c)	**56.** (b)	**57.** (d)	**58.** (c)	**59.** (d)	**60.** (b)
61. (a)	**62.** (c)	**63.** (d)	**64.** (a)	**65.** (c)	**66.** (d)	**67.** (d)	**68.** (a)	**69.** (c)	**70.** (b)
71. (c)	**72.** (a)	**73.** (a)	**74.** (c)	**75.** (d)	**76.** (b)	**77.** (b)	**78.** (a)	**79.** (c)	**80.** (a)
81. (d)	**82.** (d)	**83.** (b)	**84.** (d)	**85.** (a)	**86.** (c)	**87.** (d)	**88.** (b)	**89.** (c)	**90.** (a)
91. (d)	**92.** (b)	**93.** (c)	**94.** (a)	**95.** (b)	**96.** (a)	**97.** (c)	**98.** (c)	**99.** (d)	**100.** (c)
101. (a)	**102.** (b)	**103.** (d)	**104.** (a)	**105.** (c)	**106.** (a)	**107.** (c)	**108.** (c)	**109.** (d)	**110.** (a)
111. (c)	**112.** (b)	**113.** (b)	**114.** (d)	**115.** (b)	**116.** (d)	**117.** (b)	**118.** (c)	**119.** (a)	**120.** (d)
121. (b)	**122.** (c)	**123.** (a)	**124.** (d)	**125.** (c)	**126.** (d)	**127.** (b)	**128.** (a)	**129.** (a)	**130.** (b)
131. (b)	**132.** (a)	**133.** (d)	**134.** (b)	**135.** (c)	**136.** (c)	**137.** (b)	**138.** (d)	**139.** (b)	**140.** (a)
141. (a)	**142.** (b)	**143.** (c)	**144.** (a)	**145.** (b)	**146.** (a)	**147.** (d)	**148.** (a)	**149.** (c)	**150.** (b)
151. (a)	**152.** (c)	**153.** (c)	**154.** (a)	**155.** (b)	**156.** (d)	**157.** (c)	**158.** (a)	**159.** (b)	**160.** (c)
161. (d)	**162.** (c)	**163.** (a)	**164.** (b)	**165.** (a)	**166.** (b)	**167.** (c)	**168.** (d)	**169.** (a)	**170.** (b)
171. (d)	**172.** (a)	**173.** (c)	**174.** (d)	**175.** (b)	**176.** (c)	**177.** (b)	**178.** (d)	**179.** (c)	**180.** (a)
181. (d)	**182.** (b)	**183.** (c)	**184.** (a)	**185.** (a)	**186.** (c)	**187.** (b)	**188.** (d)		

10

Rangeland Management

1. The Dabadghao & Shankar Narayan has classified the Indian grasslands into

(a) 4 types (c) 5 types
(b) 6 types (d) 7 types

2. The most abundant grassland type in India is

(a) *Phragmites-Saccharum-Imperata*
(b) *Schima- Dicanthium*
(c) *Dicanthium- Cenchrus*
(d) *Themeda- Arundinella*

3. The main reason for decline in range condition is/are

(a) Overgrazing
(b) Climatic variation
(c) Invasion of poisonous plants
(d) All of the above

4. The most productive rangeland in the world is

(a) Savana (c) Pastures
(b) Grassland (d) Woodland

5. Indian Grassland and Fodder Research Institute is located at

(a) Bhopal (c) Jhansi
(b) Bangalore (d) Jodhpur

6. The science and art of planning and directing range use so as to obtain maximum livestock production and conservation of range resources is/are

(a) Rangeland management
(b) Forest management
(c) Forest Resource management
(d) All of the above

7. The land primarily composed of management of native grass, forbs, shrubs valuable forage is known as

(a) Grassland (c) Pasture
(b) Rangeland (d) Forest

8. A low level land covered with low vegetation or scattered patches of tree is found is called

(a) Prairie (c) Savanna
(b) Steppe (d) Grassland

9. Rangeland is mainly dominated by

(a) Grasses (c) Trees
(b) Forbs (d) Both (a) and (b)

10. All the vegetation consumed by grazing animals including fruits and twigs of trees and shrubs is termed as

(a) Browsing (c) Grazing
(b) Pasturage (d) Both (a) and (c)

11. The temperate grassland biome in central Asia is known as

(a) Steppe (c) Prairie
(b) Pampas (d) Savanna

12. The temperate grasslands of Argentina is known as

(a) Steppe (c) Prairie
(b) Pampas (d) Savanna

13. Range condition is an indicator of

(a) Forage crop production
(b) Quality of livestock products

(c) Both (a) and (b)
(d) None of the above

14. The maximum number of animals which can graze each year on a given area of range for specific number of days without deteriorating the forage production, quality and soil is
(a) Pasteurizing capacity
(b) Carrying capacity
(c) Stocking capacity
(d) None of the above

15. The number of animals grazing a unit area at a particular time is known as
(a) Grazing rate
(b) Standing rate
(c) Browsing rate
(d) Stocking rate

16. The relationship between the demand for forage by animals and combination of daily herbage increment and standing crop of vegetation is called
(a) Grazing pressure
(b) Carrying capacity
(c) Grazing intensity
(d) None of the above

17. One Goat is equal to how many cow units?
(a) 2 (b) 8 (c) 1/2 (d) 3/4

18. The total utilization of a species after the full use of range has been made is
(a) Forage factor
(b) Forage index
(c) Proper use factor
(d) Forage use factor

19. The proper use factor varies according to
(a) Species
(b) Season
(c) Kind of stock
(d) All of the above

20. ICAR conducted the grassland survey of India in
(a) 1958 (c) 1954
(b) 1976 (d) 1854

21. Dabadghao and Shankar Narayan classified the grasslands of India in
(a) 1973 (c) 1976
(b) 1954 (d) 1970

22. The practice of grazing in which the number of animals allowed to graze per unit area is fixed according to the carrying capacity is
(a) Continuous grazing
(b) Deferred grazing
(c) Controlled grazing
(d) Rotational grazing

23. Seeding rate in rangeland depends on
(a) Seed size
(b) Germination percentage
(c) Purity of seed
(d) All of the above

24. The seed rate/ha for *Cenchrus ciliaris* is
(a) 45 (b) 89 (c) 56 (d) 79

25. The optimum depth for sowing seeds of grasses is
(a) 2-3 cm (c) 0.5-4 cm
(b) 1-2 cm (d) 5-10 cm

26. Conservation of green forage into dry matter without affecting the quality of original matter is
(a) Hay
(b) Silage
(c) Forage
(d) All of the above

27. The species best suited for making hay is/are
(a) *Cynodon dactylon*
(b) *Schima* species
(c) *Cenchrus* species
(d) All of the above

28. The green plant material allowed to ferment under anaerobic conditions is
(a) Hay
(b) Ensilage
(c) Silage
(d) All of the above

29. The process of preserving green fodder under anaerobic conditions
(a) Hay
(b) Ensilage
(c) Silage
(d) None of the above

30. Silage have
(a) High palatability and low nutrition value
(b) Low palatability and high nutrition value
(c) Low palatability and low nutrition value
(d) High palatability and high nutrition value

31. Cattle Trespass Act was enacted in
(a) 1871 (c) 1976
(b) 1978 (d) 1876

32. Which of the following plays major role in grass cover distribution?
(a) Temperature
(b) Relative humidity
(c) Rainfall
(d) Both (a) and (c)

33. The success of plantation program depends upon
(a) Technical factors
(b) Social factors
(c) Economic factors
(d) All of the above

34. Pitting is quite effective mechanical measure to reduce runoff in
(a) Semi-arid areas
(b) Alpine areas
(c) Dry areas
(d) All of the above

35. The most common grazing system practiced in India is
(a) Deferred grazing system
(b) Control grazing system
(c) Open grazing system
(d) Stall feeding

36. Which of the following is the best method for controlling erosion in grasslands?
(a) Mechanical methods
(b) Vegetative methods
(c) Both (a) and (b)
(d) None of the above

37. The range condition is grouped into five different classes by
(a) Dabadghao and Shankar Narayan
(b) Stoddart
(c) Both (a) and (b)
(d) None of the above

38. Controlled burning of grassland leads to
(a) Increase in soil fertility
(b) Decrease in soil fertility
(c) Both (a) and (b)
(d) None of the above

39. The central objective of rangeland management is
(a) Manage land
(b) Produce forage

(c) Both (a) and (b)
(d) None of the above

40. Range system consist of
(a) Interacting environment forces
(b) Local combination of organism
(c) The impacts of management
(d) All of the above

41. The cut and carry system is also known as
(a) Zero Grazing
(b) Over Grazing
(c) Both (a) and (b)
(d) None of the above

42. Range improvement include
(a) Brush control
(b) Terracing
(c) Seeding
(d) All of the above

43. Grazing animal an influence upon the productive rangeland system by their defoliation of plants through
(a) Eating and physical damage
(b) By their digestive process
(c) By their movement
(d) All of the above

44. Grazing does not include defoliation in
(a) Intensity
(b) Frequency
(c) Seasonality and selectivity
(d) None of the above

45. The degree to which herbage has been removed is called
(a) The intensity of defoliation
(b) Frequency of defoliation
(c) Season of defoliation
(d) None of the above

46. The ratio of living leaf area to ground surface is called
(a) Leaf Area Index
(b) Ground Surface Index
(c) Leaf Ground Surface Index
(d) None of the above

47. Grasses have a segmented growth form and each unit is known as
(a) Internode (c) Node
(b) Phytomer (d) Tiller

48. Overharvesting reduces
(a) Non-structural carbohydrates
(b) Perennating stem bases
(c) Both (a) and (b)
(d) None of the above

49. Primary productivity transforms radiant energy from sun into
(a) Physical energy
(b) Chemical energy
(c) Both (a) and (b)
(d) None of the above

50. Large herbivores animal trend to concentrate near
(a) Water
(b) Salt
(c) Feeding areas
(d) All of the above

51. Adaptation of seeds and other plant parts which facilitate their dispersal by animals include
(a) Burrs
(b) Barbs
(c) Mucilaginous covering
(d) All of the above

52. The number of animals over a long time period is called
(a) Grazing capacity
(b) Stocking rate
(c) Animal unit
(d) Carrying capacity

53. The amount of plant material consumed by herbivores, expressed as percentage of the current herbage crop is known as
(a) Range utilization
(b) Degree of use
(c) Percentage use
(d) All of the above

54. The planned distribution of the amount of salt require by livestock for the grazing period is
(a) Salting
(b) Herding
(c) Animal distribution
(d) Grazing plan

55. The unrestricted grazing throughout the whole year of the grazing season is known as
(a) Repeated seasonal grazing
(b) Continuous grazing
(c) Deferred grazing
(d) None of the above

56. Which of the following species on rangeland are difficult to control with chemicals?
(a) *Artemisia* species
(b) *Opuntia* species
(c) *Juniper* species
(d) All of the above

57. The term fire type and fire species refer to vegetation and species that are
(a) Not favored by burning
(b) Favored by burning
(c) Both (a) and (b)
(d) None of the above

58. The grassland types are
(a) *Arundinella* type
(b) *Brothriochloa* type
(c) *Cymbopogon* type
(d) All of the above

59. The intermediate feeders are
(a) Domestic sheep
(b) Burros
(c) Caribou
(d) All of the above

60. Smaller animal typically have
(a) Lower metabolic rate
(b) Medium metabolic rate
(c) High metabolic rate
(d) None of the above

61. Vitamin that is deficient in most of the range livestock diets is
(a) Vitamin A
(b) Vitamin C
(c) Vitamin B
(d) Vitamin D

62 Heavy grazing generally cause problems in
(a) Insects
(b) Jack rabbits
(c) Prairie dogs
(d) All of the above

63. Factors influencing grazing distribution of forest rangeland is
(a) Stock water distribution
(b) Slope steepness
(c) Trails and access routes
(d) All of the above

64. "Cheat grass" is
(a) *Bromus tectorus*
(b) *Juniper* species
(c) *Rhus trilobata*
(d) None of the above

65. Introduced or native plant can be protected from grazing until late in dry season is known as
(a) Stored feed

(b) Forage reserves
(c) Supplemental feed
(d) Complementary forage

66. Perennial grass is
(a) *Cenchrus ciliaris*
(b) *Indigofera cardifolia*
(c) *Justicia simplex*
(d) None of the above

67. Grasslands commonly found in sub-tropical and semi-arid region is
(a) *Cymbopogon*
(b) *Arundinella*
(c) *Sehima-Dicanthium*
(d) *Dicanthium-Cenchrus-Lasiurus*

68. Pasture irrigation depends upon
(a) Time of irrigation
(b) Amount of irrigation
(c) Both (a) and (b)
(d) None of the above

69. "Buffel grass" is
(a) *Cenchrus ciliaris*
(b) *Agropyron cristatum*
(c) *Eragrostis* species
(d) None of the above

70. Rangeland productivity is determined by
(a) Soil
(b) Topography
(c) Climate
(d) All of the above

71. Rangeland are the primary habitat for nearly all the land-dwelling wild animals highly valued for
(a) Meat
(b) Hunting
(c) Aesthetic viewing
(d) All of the above

72. Range management decision making process considers
(a) Social aspects
(b) Economic aspects
(c) Cultural and technological aspect
(d) All of the above

73. Drought is defined as prolonged dry weather, generally when precipitation is less than
(a) 75%
(b) 85%
(c) 95%
(d) None of the above

74. Savanna woodlands are dominated by
(a) Scattered
(b) Short growing trees less than 12 m
(c) Both (a) and (b)
(d) None of the above

75. Tundra refers to
(a) Level region
(b) Tree less plain in Arctic region
(c) High elevation (cold) region
(d) All of the above

76. Alpine tundra occurs at
(a) High elevation (more than 2800 m)
(b) Low elevation
(c) Medium elevation
(d) All of the above

77. Rotational grazing system is/are more effective in
(a) Rugged terrain
(b) Humid areas
(c) Heterogeneity place
(d) All of the above

78. Cattle use is generally diminished for slopes with
(a) More than 10% gradient
(b) More than 20% gradient
(c) More than 30% gradient
(d) More than 50% gradient

79. Forage intake of grazing animal varies with
(a) Body weight
(b) Forage quality
(c) Availability of forage
(d) All of the above

80. Rangeland vegetation includes
(a) Shrublands
(b) Grasslands
(c) Open forests
(d) All of the above

81. An extensive formation dominated by grasses and largely composed of bunch grasses is known as
(a) Prairie
(b) Steppe
(c) Savanna
(d) All of the above

82. A track of levelled land, covered with low vegetation, tree less or dotted with trees or patches of wood is known as
(a) Savanna (c) Steppe
(b) Prairie(d) Pampas

83. Pasturage term is synonym to
(a) Rangeland
(b) Forage
(c) Savanna
(d) All of the above

84. Cymbopogon grassland types grows in
(a)Black soil
(b) High mountains
(c) Sandy loam
(d) Low hills

85. Perennial species is
(a) *Themeda triandra*
(b) *Cynodon dactylon*
(c) *Arstida setacea*
(d) All of the above

86. In popular sense "range condition" refers to a comparison of the
(a) Current year forage production
(b) Previous year forage production
(c) Both (a) and (b)
(d) None of the above

87. Criteria for range classification depends upon
(a) Soil factor
(b) Plant composition
(c) Forage value
(d) All of the above

88 Forage acre is
(a) Forage acre factor × Surface area
(b) Range forage × Surface area
(c) Dry matter demand × dry matter available
(d) None of the above

89. Surface irrigation system includes
(a) Flooding
(b) Border irrigation
(c) Basin and furrow irrigation
(d) All of the above

90. Moist tropical grasses is/are
(a) *Brachiaria mutica*
(b) *Panicum maximum*
(c) *Pennisetum purpureum*
(d) All of the above

91. Ravine grasses is/are
(a) *Dicanthium annulatum*
(b) *Brothrichloa ischaemum*
(c) *Cenchrus ciliaris*
(d) All of the above

92. "Buffalo grass" is also known as
(a) *Brachiaria mutica*
(b) *Cenchrus ciliaris*
(c) *Cynodont dactylon*
(d) None of the above

93. "Sundan grass" is also known as
(a) Chloris
(b) Sorghum
(c) Panicum
(d) None of the above

94. PCB herbicide stand for
(a) Parequat
(c) Picloram
(c) Pentachlorophenol
(d) All of the above

95. Palatability is usually reduced by the presence of
(a) Awns
(b) Spines
(c) Hairs
(d) All of the above

96. Soil compaction
(a) Increases specific gravity
(b) Decreases pore space
(c) Packing soil particles
(d) All of the above

97. Trampling with an artificial hoof increases overland flow of water in the
(a) *Agropyron spicatum*
(b) *Bromus tectorum*
(c) Both (a) and (b)
(d) None of the above

98. AUM stands for
(a) An Animal Unit Month
(b) An Animal Unit Mission
(c) Agriculture Unit Mode
(d) None of the above

99. The factors affecting grazing behavior is/are
(a) Palatability
(b) Topography
(c) Temperature
(d) Both (a) and (b)

100. Ideal crop for silage making is
(a) Rich in soluble carbohydrates
(b) Low to medium in protein content
(c) Both (a) and (b)
(d) None of the above

Answer Keys

1. (c)	**2.** (a)	**3.** (d)	**4.** (b)	**5.** (c)	**6.** (a)	**7.** (b)	**8.** (c)	**9.** (d)	**10.** (b)
11. (a)	**12.** (b)	**13.** (c)	**14.** (b)	**15.** (d)	**16.** (a)	**17.** (c)	**18.** (c)	**19.** (d)	**20.** (c)
21. (a)	**22.** (c)	**23.** (d)	**24.** (a)	**25.** (b)	**26.** (a)	**27.** (d)	**28.** (c)	**29.** (b)	**30.** (d)
31. (a)	**32.** (d)	**33.** (d)	**34.** (a)	**35.** (c)	**36.** (b)	**37.** (b)	**38.** (a)	**39.** (c)	**40.** (d)
41. (a)	**42.** (d)	**43.** (d)	**44.** (d)	**45.** (a)	**46.** (a)	**47.** (b)	**48.** (c)	**49.** (b)	**50.** (d)
51. (d)	**52.** (a)	**53.** (d)	**54.** (a)	**55.** (b)	**56.** (c)	**57.** (b)	**58.** (d)	**59.** (d)	**60.** (c)
61. (a)	**62.** (d)	**63.** (d)	**64.** (a)	**65.** (b)	**66.** (a)	**67.** (d)	**68.** (c)	**69.** (a)	**70.** (d)
71. (d)	**72.** (d)	**73.** (a)	**74.** (c)	**75.** (d)	**76.** (a)	**77.** (d)	**78.** (a)	**79.** (d)	**80.** (d)
81. (a)	**82.** (c)	**83.** (b)	**84.** (d)	**85.** (d)	**86.** (c)	**87.** (d)	**88.** (a)	**89.** (d)	**90.** (d)
91. (d)	**92.** (a)	**93.** (b)	**94.** (d)	**95.** (d)	**96.** (d)	**97.** (c)	**98.** (a)	**99.** (d)	**100.** (c)

11

Wildlife

1. In WPA 1972, "Pitcher Plant" is included in
(a) Schedule IV (c) Schedule V
(b) Schedule VI (d) Schedule VII

2. The community reserves are included in ………..section of WPA, 1972.
(a) 36a (c) 36b
(b) 36c (d) 36d

3. Worlds first halophyte botanical garden is located at
(a) Maharashtra (c) Tamil Nadu
(a) Gujarat (d) West Bengal

4. Vulture breeding centre is located at
(a) Jharkhand (c) Bihar
(b) West Bengal (d) Assam

5. Silent valley National park is situated in the state of
(a) Tamil Nadu
(b) Andhra Pradesh
(c) Kerala
(d) Karnataka

6. National park for conservation of insectivorous plant is
(a) Dachigam, J&K
(b) Keibul Lamjo.
(c) Balpakram, Meghalaya
(d) None of the above

7. ……………. is a marine biosphere.
(a) Gulf of Mannar
(b) Sunderbans
(c) Rann of Kutch
(d) None of these

8. Name the only state where all three species of crocodile are found in wild
(a) UP (c) Bihar
(b) West Bengal (d) Odisha

9. What is the zoological name of African elephant?
(a) *Elephas maximus*
(b) *Loxodanta africana*
(c) *Elephas africana*
(d) None of the above

10. Zoological name of lion tailed macaque
(a) *Macaca silensus.*
(b) *Hominoidea* species
(c) *Macaca fasicularis*
(d) *Hylobatidae* species

11. Wildlife conservation strategy 2002 announced during
(a) 5th meeting of IBWL
(b) 10th meeting of IBWL
(c) 21st meeting of IBWL
(d) None of the above

12. The IBWL set up in
(a) 1950 (c) 1956
(b) 1948 (d) 1952

13. The chairman of IBWL is
(a) Prime minister
(b) Chief wildlife warden

(c) Director of ICFRE
(d) None of the above

14. The Central Zoo Authority (CZA) comes under ministry of
(a) Ministry of Environment and Forest
(b) Ministry of Wildlife
(c) Ministry of Environment, Forest and Climate Change
(d) None of the above

15. Zoological Survey of India is located at
(a) Lucknow (c) Chennai
(b) Kolkata (d) New Delhi

16. The term "MIKE" denotes
(a) Monitoring of Illegal Killing of Endangered and Rare species
(b) Monitoring of Illegal Killing of Eagle
(c) Monitoring of Illegal Killing of Elephants
(d) None of the above

17. The Central Zoo Authority (CZA) established in the year
(a) 1952 (c) 1958
(b) 1880 (d) 1992

18. The BNHS was started in the year
(a) 1992 (c) 1975
(b) 1883 (d) 1969

19. India became the member of CITES in
(a) 1976 (c) 1975
(b) 1958 (d) 1852

20. The hunting of wild animal is totally prohibited in
(a) Section-7 (c) Section-9
(b) Section-16 (d) Section-35

21. The National park declared by state government under
(a) Section-18(1)
(b) Section-38(1)
(c) Section-17(1)
(d) Section-35(1)

22. The Wildlife sanctuary declared by state government under
(a) Section-18(1
(b) Section-35(1)
(c) Section-38(1)
(d) All of the above

23. The National park and Wildlife sanctuary declared by central government under
(a) Section-18(1)
(b) Section-35(1)
(c) Section-38(1)
(d) All of the above

24. The Central Zoo Authority (CZA) comes under section
(a) 4a (b) 4c (c) 4b (d) 4d

25. The Vermins are included in
(a) Schedule-II (c) Schedule-III
(b) Schedule-IV (d) Schedule-V

26. The protection of certain plants in WPA, 1972 comes under
(a) Section-9 (c) Section-3
(b) Section-3a (d) Section- 7

27. The trade in commerce in wild animals, animal articles and trophies in WPA, 1972 comes under
(a) Section-9 to 17
(b) Section-18 to 38
(c) Section-39 to 49
(d) Section-50 to 66

28. The National Tiger Conservation Authority (NTCA) was constituted under WPA-1972, amended in
(a) 2005 (c) 2010
(b) 2000 (d) 2006

29. The Tiger Task Force was constituted on recommendation of NBWL on
(a) 17th March, 2005
(b) 17th May, 2006
(c) 17th March, 2010
(d) None of the above

30. Elephant Task Force constituted under ministry of MOEF&CC on
(a) January, 2005
(b) February, 2010
(c) February, 2012
(d) March, 2009

31. Which of the following is not a viral disease ?
(a) FMD (c) Rinder pest
(b) Rabies (d) Anthrax

32. The scientific name of green turtle is
(a) *Chelonia olivaceae*
(b) *Lepidochelys olivaceae*
(c) *Chelonia mydas*
(d) *Cuon alpinus*

33. Scientific name of dhole is
(a) *Canus lupus* (c) *Vulpus vulpus*
(b) *Axis axis* (d) *Cuon alpinus*

34. Indian python is
(a) *Python morulus*
(b) *Naja naja*
(c) *Ophiophagus hanna*
(d) None of the above

35. Gharial is
(a) *Chrocodylus palustris*
(b) *Chrocodylus porosus*
(c) *Gravialus gangeticus*
(d) None of the above

36. The four horned antelope is
(a) *Gazella benetii*
(b) *Tetracerus quadricornis*
(c) Both (a) and (b)
(d) None of the above

37. Cheetah is found in
(a) Gujarat (c) Karnataka
(a) West Bengal (d) Extinct

38. Scientific name of snow leopard is
(a) *Uncia uncia*
(b) *Panthera pardus*
(c) *Panthera leo*
(d) *Panthera tigris*

39. Indian cobra is
(a) *Ophiophagus hanna*
(b) *Bungaris* species
(c) *Naja naja*
(d) None of the above

40. Territory is ...... than Home range.
(a) Larger.
(b) Smaller
(c) Equal
(d) All of the above

41. Pampas are found in
(a) Europe (c) America
(b) Asia (d) Argentina

42. Which Biome is found in the colder regions of Europe?
(a) Tundra (c) Taiga
(b) Parries (d) Pampas

43. ......is a large terrestrial community characterised by typical flora and fauna.
(a) Ecosystem (c) Biosphere
(b) Biome (d) Niche

44. Bergman's rule is
(a) Colder climate has large sized animals
(b) Due to heat animal changes colour to dark

(c) Animal tail, ear, neck, etc... become short in cold climate.
(d) Due to increase in temperature the number of vertebrate decreases

45 Jordan's rule is
(a) Colder climate has large sized animals
(b) Due to heat animal changes colour to dark
(c) Animal tail, ear, neck, etc... become short in cold climate
(d) Due to increase in temperature the number of vertebrate decrease

46. Which of the following shows poikilothermic behaviour?
(a) Fish (c) Crocodile
(b) Birds (d) Both (a) and (b)

47. Generally insects and lower vertebrates during extreme conditions undergo
(a) Hibernation (c) Both (a) and (b)
(b) Aestivation (d) None of these

48. Monotremes are
(a) Egg laying mammals
(b) Give birth to young ones
(c) Highly developed animals
(d) None of the above

49. Several species in an ecosystem which profoundly influence other species is known as
(a) Indicator species
(b) Umbrella species
(c) Keystone species
(d) None of the above

50. The Wildlife Institute of India (WII) is located at
(a) Delhi (c) Mumbai
(b) Bangalore (d) Dehradun

51. The Ramsar wetland convention came into force in
(a) 1971 (c) 1970
(b) 1975 (d) 1984

52. The total number of Tiger Reserves in India (2019) are
(a) 48 (b) 52 (c) 35 (d) 50

53. The total protected area network constitutes.......% of total geographic area.
(a) 5.02% (c) 4.31%
(b) 6.3% (d) 4%

54. The total number of National Parks in India are
(a) 103 (b) 100 (c) 104 (d) 200

55. The total number of Biosphere Reserves in India are
(a) 20 (b) 4 (c) 15 (d) 18

56. First Marine National Park is situated in
(a) Gujarat (c) West Bengal
(c) Goa (d) Orrisa

57. The total number of Biosphere Reserves in world are
(a) 721 in 120 countries
(b) 686 in 122 countries
(c) 701 in 124 countries
(d) None of the above

58. In which of the following, limited human activities are allowed
(a) National Park
(b) Wildlife Sanctuary
(c) Biosphere Reserve
(d) All of the above

59. Biggest National Park in India
(a) Desert National Park
(b) Jim Corbett National Park
(c) Kanha National Park
(d) Hemis National Park

60. The smallest National Park is
(a) South button Island National Park
(b) Kaziranga National Park.
(c) Mandla plant fossil National Park
(d) Manas National Park

61. The first National Park declared in India is
(a) Rajaji National Park
(b) Keoladeo National Park
(c) Jim Corbett National Park
(d) Manas National Park

62. Wildlife week is celebrated during
(a) 2nd week of July
(b) Last week of October
(c) 3rd Week of December
(d) 1st week of October

63. Mammal which can swim
(a) Whale
(b) Bat
(c) Hen
(d) All of the above

64. Gibbon Wildlife sanctuary is located in
(a) Manipur
(b) Andhra Pradesh
(c) Assam
(d) Kerala

65. Royal Bengal Tiger is found in
(a) Sunderban
(b) Gujarat
(c) Karnataka
(d) Madhya Pradesh

66. Lion tailed macaque is key faunal species of
(a) Nanda Devi Biosphere Reserve
(b) Norek Biosphere Reserve
(c) Simlipal Biosphere Reserve
(d) Nilgiri Biosphere Reserve

67. The emblem of BNHS is
(a) Penguin (c) Felicon
(b) Hornbill (d) Fish

68. The headquarter of IUCN is at
(a) Gland, Switzerland
(b) Nairobi, Kenya
(c) Japan
(d) Rome, Italy

69. A taxon restricted to a particular geographic location is termed as
(a) Rare
(b) Endemic
(c) Vulnerable
(d) All of the above

70. Biosphere Reserve has
(a) Buffer zone
(b) Core zone
(c) Command zone
(d) All of the above

71. The linear strip of habitat that facilitates the movement of organism between larger habitat patches is termed as
(a) Habitat corridor
(b) Habitat fragment
(c) Both (a) and (b)
(d) None of the above

72. The heaviest flying bird in India is
(a) Indian peacock
(b) Siberian crane
(c) The great Indian Bustard
(d) None of the above

73. Which of the following is a Biosphere Reserve?
(a) Nilgiri Biosphere Reserve
(b) Gulf of Mannar
(c) Nanda Devi Biosphere Reserve
(d) All of the above

74. *Ex-situ* conservation is done by
(a) Cryopreservation
(b) Gene Banks
(c) Botanical Gardens
(d) All of the above

75. Which of the following is not in-situ conservation?
(a) Biosphere Reserve
(b) Botanical Gardens
(c) National Parks
(d) Wildlife Sanctuary

76. Silent valley is famous for
(a) Lion tailed macaque
(b) Golden langoor
(c) Hanuman langoor
(d) All of the above

77. Bharatpur is famous for
(a) Lion (c) Tiger
(b) Birds (d) Swamp deer

78. Ranthambore National Park is in
(a) Gujarat (c) Tamil Nadu
(b) Haryana (d) Rajasthan

79. Which of the following is not found in Indian desert?
(a) Wild ass (c) Ibex
(c) Blackbuck (d) Both (a) and (c)

80. Mammal which can fly
(a) Rat (c) Whale
(b) Hen (d) Bat

81. Tiger Census is done by
(a) Pug mark (c) Camera traps
(b) Water hole (d) Total count

82. Project Tiger was started in the year
(a) 1972 (c) 2006
(b) 1973 (d) 1990

83. The Red list of Threatened species is published by
(a) CITES (c) WII
(b) MOEF & CC (d) IUCN

84. The National Aquatic Animal recognized by Government of India is
(a) Crocodile
(b) Seal
(c) Ganges River Dolphin
(d) Turtle

85. The total number of states which have implemented Project Tiger in India are
(a) 18 (b) 20 (c) 16 (d) 30

86. Endemic primate species of Western Ghats is
(a) Nilgiri langur
(b) Lion tailed macaque
(c) Rhesus macaque
(d) None of the above

87. Pigmy hog was rediscovered from
(a) Manas (c) Assam
(b) Kaziranga (d) Both (a) and (b)

88. Which of the following is not an antelope?
(a) Hangul (c) Nilgai
(b) Gazelle (d) Chiru

89. Which among the following is key faunal species conserved and monitored in 'Dachigam National Park' ?
(a) Musk deer
(b) Golden oriole
(c) Yellow-throated Marten
(d) Hangul

90. Western Ghats are found in .... states of India.
(a) 4 (c) 5
(b) 6 (d) 7

91. The project Elephant was launched in

(a) 1994 (c) 1993
(b) 1990 (d) 1992

92. The most widely distributed bear in India is

(a) Black bear (c) Sloth bear
(b) Polar bear (d) Brown bear

93. Which of the following is not a goat antelope?

(a) Takin (c) Serow
(b) Goral (d) Chausingha

94 The total tiger population in India (2019), according to the "All India Tiger Estimation Report" is

(a) 2022 (c) 2145
(b) 2967 (d) 3074

95. International Tiger Day is celebrated on

(a) 29th July (c) 31st December
(b) 20th August (d) 22nd September

96. Indirect method of animal census includes

(a) Counting of burrows
(b) Pugmark census
(c) Track count
(d) All of the above

97. The Elephant catching method adopted in India is/are

(a) Pit method
(b) Water hole method
(c) Kheda system
(d) Both (a) and (b)

98. The space defended by the wild animal is

(a) Home range (c) Kingdom
(b) Territory (d) Both (a) and (b)

99. The Project Snow leopard was launched in

(a) 2000 (c) 2009
(b) 2010 (d) 1970

100. The Vulture conservation and breeding centre is located at

(a) Haryana (c) West Bengal
(b) Punjab (d) Gujarat

101. Dodo is

(a) Critically Endangered
(b) Rare
(c) Endangered
(d) Extinct

102. Salim Ali Centre for Ornithology and Natural History is located at

(a) Bombay (c) Coimbatore
(b) Goa (d) Kerala

Answer Keys

1. (b)	**2.** (a)	**3.** (c)	**4.** (d)	**5.** (c)	**6.** (c)	**7.** (a)	**8.** (d)	**9.** (b)	**10.** (a)
11. (c)	**12.** (d)	**13.** (a)	**14.** (c)	**15.** (b)	**16.** (c)	**17.** (d)	**18.** (b)	**19.** (a)	**20.** (c)
21. (d)	**22.** (a)	**23.** (c)	**24.** (a)	**25.** (d)	**26.** (b)	**27.** (c)	**28.** (d)	**29.** (a)	**30.** (b)
31. (d)	**32.** (c)	**33.** (d)	**34.** (a)	**35.** (c)	**36.** (b)	**37.** (d)	**38.** (a)	**39.** (c)	**40.** (b)
41. (d)	**42.** (a)	**43.** (b)	**44.** (a)	**45.** (d)	**46.** (d)	**47.** (b)	**48.** (a)	**49.** (c)	**50.** (d)
51. (b)	**52.** (d)	**53.** (a)	**54.** (c)	**55.** (d)	**56.** (a)	**57.** (c)	**58.** (b)	**59.** (d)	**60.** (a)
61. (c)	**62.** (d)	**63.** (c)	**64.** (c)	**65.** (a)	**66.** (d)	**67.** (b)	**68.** (a)	**69.** (b)	**70.** (d)
71. (a)	**72.** (c)	**73.** (d)	**74.** (d)	**75.** (b)	**76.** (a)	**77.** (b)	**78.** (d)	**79.** (c)	**80.** (d)
81. (a)	**82.** (b)	**83.** (d)	**84.** (c)	**85.** (a)	**86.** (b)	**87.** (a)	**88.** (c)	**89.** (d)	**90.** (b)
91. (d)	**92.** (c)	**93.** (d)	**94.** (b)	**95.** (a)	**96.** (d)	**97.** (d)	**98.** (b)	**99.** (c)	**100.** (a)
101. (d)	**102.** (c)								

12

Tree Seed Technology

1. ISTA stands for
(a) Indian Seed Testing Association
(b) International Seed Testing Association
(c) International Seed Testing Agency
(d) Indian Seed Testing Association

2. The certificate issued for the genuiness of seed at seed testing station in the country in which the seed is located at the time of sampling and sealing is
(a) Blue (c) Purple
(b) Yellow (d) Orange

3. The certificate issued for the genuiness of seed in the country to the tested sample
(a) Blue (c) Orange
(b) Green (d) Pink

4. Which of the following is classified as vigour of seed germination?
(a) Seed index
(b) Seed quality index
(c) Genetic purity
(d) None of the above

5. The optimum temperature for seed germination is
(a) 30–40% (c) 10–15%
(b) 15–30% (d) 50–60%

6. The ability of seed to germinate is
(a) Germination energy
(b) Seed vigour
(c) Seed viability
(d) Germination capacity

7. Hiltner test is used to test
(a) Seed purity (c) Seed moisture
(b) Seed vigour (d) Seed viability

8. Seed viability is tested by using
(a) Tetrazolium test
(b) Electrical conductivity test
(c) X-ray radiography
(d) All of the above

9. In tetrazolium test, the viable seeds become red due to formation of
(a) Chloride
(b) Forman
(c) Formazan
(d) All of the above

10. The classification of seed germination is proposed by
(a) ISTA
(b) Robert
(c) Haper and Robert
(d) Ewart

11. Germination inhibitors are generally found in
(a) Seed coat
(b) Endosperm
(c) Embryo
(d) Seed

12. Embryo dormancy is generally found in
(a) *Fraxinus excelsior*
(b) *Terminalia* species

(c) Teak
(d) *Ginkgo biloba*

13. Double dormancy is found in
(a) *Fraxinus floribunda*
(b) *Fraxinus excesior*
(c) *Albizzia lebbeck*
(d) *Betula pendula*

14. Secondary dormancy is found in
(a) *Xanthium pennsylvanicum*
(b) *Fraxinus* species
(c) *Pinus* species
(d) Sal

15. Tetrazolium test was given by
(a) Robert (c) Clement
(b) Lakon (d) Mayr

16. Alternative wetting and drying treatment is given to
(a) Sal
(b) *Acacia* species
(c) Teak
(d) *Ecalyptus*

17. Scarification is effective in breaking seed dormancy of
(a) Temperate zone species
(b) Dry zone species
(c) Tropical zone species
(d) Alpine zone species

18. The seeds in which dormancy terminated on exposure to light is
(a) Positive photoblastic
(b) Negative photoblastic
(c) Non-photoblastic
(d) Dormant

19. Which of the following is positively photoblastic?
(a) *Dalbergia sissoo*
(b) Sal
(c) Teak
(d) *Albizzia procera*

20. Which of the following is not used to break seed dormancy
(a) Hydrogen peroxide
(b) Gibberellins
(c) $HgCl_2$
(d) Thiourea

21. The microbiotic, mesobiotic, macrobiotic classification was given by
(a) Stuart (c) Robert
(b) Ewart (d) Henry

22. Mesobiotic seeds having viability of
(a) 3-15 years
(b) less than 3 years
(c) More than 15 years
(d) None of the above

23. Orthodox seeds
(a) 15–20% moisture content
(b) 10–12% moisture content
(c) Thousand years old tree
(d) Old seeds

24. Recalcitrant seeds
(a) Stored for long period
(b) 10-12% moisture content
(c) Stored for short period
(d) None of the above

25. The seeds of Dipterocarpaceae and Lauraceae belongs to
(a) Orthodox seeds
(b) Recalcitrant seeds
(c) Both (a) and (b)
(d) None of the above

26. Grow out test is conducted to determine
(a) Purity percentage
(b) Seed germination energy
(c) Germination capacity
(d) Genetic purity

27. The development of seed without sexual fusion of male and female gamete is known as
(a) Vegetative reproduction
(b) Self incompatibility
(c) Apomixes
(d) Parthenogenesis

28. The development of embryo without fertilisation is termed as
(a) Parthenogenesis
(b) Pseudogamy
(c) Parthenocarpy
(d) Apomixes

29. The most important factor responsible for seed germination
(a) Temperature
(b) Relative humidity
(c) Water
(d) All of the above

30. The deep buried seed fails to germinate due to
(a) Induced dormancy
(b) Secondary dormancy
(c) Innate dormancy
(d) Enforced dormancy

31. Winged seeds are characteristics of which family
(a) Lamiaceae
(b) Dipterocarpaceae
(c) Lauraceae
(d) Malvaceae

32. The exposure of seeds to low temperature for breaking seed dormancy is called
(a) Stratification (c) Vernalisation
(b) Scarification (d) Seed viability

33. The headquarters of International board of plant genetic resources is
(a) Australia (c) Austria
(b) New Delhi (d) USA

34. The seeds from the trees of proven genetic superiority is termed as
(a) Selected seeds
(b) Certified seeds
(c) Source identified seeds
(d) Registered seeds

35. Which label is given to source identified reproductive material?
(a) Green (c) Pink
(b) Blue (d) Yellow

36. Which label is given to genetically tested reproductive material?
(a) Blue (c) Green
(b) Yellow (d) Pink

37. In India for seed certification, ... rules are adopted.
(a) ISTA
(b) AOSA
(c) OECT
(d) All of the above

38. The reproductive material refers to
(a) Cones
(b) Fruits
(c) Seeds
(d) All of the above

39. As the seed matures its dry weight
(a) Decreases (c) Increases
(b) Constant (d) None of these

40. The examination of the development of the embryo and endosperm can be done by using
(a) X- rays (c) IR- rays
(b) UV rays (d) Gamma rays

41. In *Dalbergia sissoo* the brown seeds have.............. germinative capacity than the black seed
(a) Half (c) Triple
(b) Double (d) Equal

42. Chemicals used for treating seed coat
(a) $ZnSO_4$
(b) $MgSO_4$
(c) NaCl
(d) All of the above

43. The germination in which the cotyledons remain in the ground during the germination is
(a) Hypogeal germination
(b) Epigeal germination
(c) Both (a) and (b)
(d) None of the above

44. The germination in which the cotyledons appear above the ground during the germination is
(a) Epigeal germination
(b) Hypogeal germination
(c) Both (a) and (b)
(d) None of the above

45. The Epigeal germination is common in
(a) *Pinus* species
(b) *Terminalia* species
(c) *Tamarindus indica*
(d) All of the above

46. The hypogeal germination is common in
(a) *Tamarindus indica*
(b) *Azadirachta indica*
(c) *Pongamia pinnata*
(d) All of the above

47. The stalk of the ovule is known as
(a) Micropyle
(d) Funicle
(c) Hilum
(d) Tagmen

48. Seed is also known as
(a) Matured ovary
(b) Matured ovule
(c) Matured fruit
(d) All of the above

49. Seed collection in natural forest is difficult due to
(a) Mountainous terrain
(b) Inaccessibility
(c) Climbing on the trees
(d) All of the above

50. Caruncle seed is commonly found in
(a) *Acacia auriculiformis*
(b) *Acacia nilotica*
(c) *Prosopis cineraria*
(d) *Ficus bengalensis*

51. The seed coat is popularly known as
(a) Tegmen (c) Pericarp
(b) Testa (d) Cotyledon

52. Naked seeds are commonly found in
(a) Angiosperms
(b) Gymnosperms
(c) Both (a) and (b)
(d) None of the above

53. Seeds of some plants germinate while attached to mother plant is known as
(a) Ovipary (c) Hypogeal
(b) Epigeal (d) Vivipary

54. Vivipary is commonly found in
(a) Mangroves
(b) Teak
(c) Acacia
(d) Sal

55. Seed fail to germinate when exposed to suitable environment conditions is called as
(a) Apomixes

(b) Dormancy
(c) Abnormality
(d) Hibernation

56. The common inhibitor of seed dormancy is
(a) Ethylene
(b) Gibberellic acid
(c) Abscisic acid
(d) Alcohol

57. Hard seed coat can be softened by
(a) Mechanical aberrations
(a) Alternate freezing
(c) Acid scarification
(d) All of the above

58. Equipment's involved in tree seed collection from standing trees is/are
(a) Extension ladder
(b) Tree cycle
(c) Tarpaulin
(d) All of the above

59. Very minute seeds found in
(a) Casuarina (c) Neem
(b) Teak (d) Sal

60. Large and heavy fruits are col lected from
(a) Standing tree
(b) Felled tree
(c) Ground
(d) All of the above

61. Chemical treatment of concentrated H_2SO_4 is used in
(a) *Pinus* seeds (c) *Acacia* seeds
(b) Teak seeds (d) Bamboo seeds

62. In India first seed testing laboratory for forestry was established in the year
(a) 1960 (c) 1961
(b) 1956 (d) 1962

63. Forest tree testing laboratory was established at
(a) Dehradun (c) New Delhi
(b) Mumbai (d) Coimbatore

64. Tree seeds which give multiple seedlings from single seed is called as
(a) Polyembryony
(b) Polycarpic
(c) Both (a) and (b)
(d) None of the above

65. The breeder seed is the progeny of
(a) Foundation seed
(b) Registered seed
(b) Nucleus seed
(d) Certified seed

66. Which of the following seed has highest genetic purity?
(a) Breeder seed
(b) Foundation seed
(c) Nucleus seed
(d) Certified seed

67. The genetic purity of foundation seed is
(a) 98% (c) 99%
(b) 99.5% (d) 99.9%

68. The qualitative test for finding the seed viability is/are
(a) Potassium permanganate test
(b) Tetrazolium test
(c) Both (a) and (b)
(d) None of the above

69. What do you mean by tumbling?
(a) Extraction of cones
(b) Extraction of fruits
(c) Extraction of seeds from cones
(d) None of the above

70. Collection of seed by shaking is efficient when humidity is

(a) High
(b) Low
(c) Both (a) and (b)
(d) None of the above

71. The seed of Bamboos is
(a) Drupe (c) Pod
(b) Samara (d) Caryopsis

72. In flowering plants, ... is known as second seed coat.
(a) Tegamen
(b) Integument
(c) Testa
(d) Pericarp

73. The rudimentary shoot emerging from the seed is known as
(a) Rachis
(b) Radicle
(c) Plumule
(d) All of the above

74. The rudimentary root of the seed is
(a) Hilum (c) Radicle
(b) Plumule (d) Rachis

75. Which of the following instrument is used for grading seeds based on density or weight?
(a) Mechanical sieve
(b) Specific gravity separator
(c) Both (a) and (b)
(d) None of the above

76. Physiological dormancy is commonly observed in seeds of
(a) Tropical areas
(b) Waterlogged areas
(c) Temperate areas
(d) Desert areas

77. Seed dispersal by gravity is common in
(a) *Ficus* species
(b) *Tectona grandis*
(c) *Bombax ceiba*
(d) *Juglans regia*

78. Maturity of seeds is measured through
(a) 100 gm/seed weight
(b) X- ray radiography
(c) Seed moisture percent
(d) Germination percent

79. X-ray technique is used for testing
(a) Viability of seed
(b) Availability of seed
(c) Dormancy of seed
(d) Germination of seed

80. Epigeal formation is common in
(a) Monocots
(b) Dicots
(c) Both (a) and (b)
(d) None of the above

81. Seed processing helps in improving
(a) Moisture
(b) Genetic purity
(c) Physical purity
(d) All of the above

82. The genetic deterioration in seeds occurs due to
(a) Mutation
(b) Developmental variation
(c) Diseases
(d) All of the above

83. In angiosperms, the endosperm is
(a) Triploid
(b) Diploid
(c) Haploid
(d) None of the above

84. The hardness of seed coat is due to
(a) Parenchyma cells

(b) Collenchyma cells
(c) Sclerenchyma cells
(d) None of the above

85. The media used for seed germination is/are
(a) Sand
(b) Paper
(c) Vermiculite
(d) All of the above

86. The various methods for measuring moisture in seed is/are
(a) Universal moisture meter
(b) Digital moisture meter
(c) Oven method
(d) All of the above

87. Microbiotic seeds have life span of
(a) Above 15 years
(b) Not exceeding 3 years
(c) 3-15 years
(d) None of the above

88. Which of the following tree seeds have lowest viability period?
(a) *Acacia nilotica*
(b) *Casuarina equisetifolia*
(c) *Shorea robusta*
(d) None of the above

89. Seed is
(a) Fertilized mature ovule
(b) Mature ovule
(c) Endosperm
(d) None of the above

90. Seed is also described as
(a) Miniature plant
(b) Dormant plant
(c) Propagative material
(d) All of the above

91. Seed vigour is the sum total of those properties of seed which determine
(a) Level of activity
(b) Performance of the seed
(c) Seed germination and seedling emergence
(d) All of the above

92. The pollen is enclosed in
(a) Microsporophyll
(b) Anther
(c) Gynoecium
(d) None of the above

93. Pollination is marked by arrival of pollen close to
(a) Micropyle of the ovule
(b) Archesporial cells
(c) Microstrobili
(d) All of the above

94. The nutritive tissue surrounding the embryo called as
(a) Perisperm (c) Endosperm
(b) Testa (d) Hilum

95. Forest seeds possess very thick seed coat
(a) *Emblica officinalis*
(b) *Tamarindus indica*
(c) *Tectona grandis*
(d) All of the above

96. Germination of all the seeds occurs over relatively brief period is called as
(a) Quasi-simultaneous
(b) Continious
(c) Intermittent
(d) Periodic

97. Recalcitrant seed is/are
(a) Quercus
(b) Jamun
(c) Chestnut
(d) All of the above

98. Orthodox seed is/are
(a) *Acacia* species

(b) *Cassia* species
(c) *Prosopis* species
(d) All of the above

99. The ovule completely bend along the funicle and get fused with the latter known as
(a) Orthotropus
(b) Amphitropous
(c) Anatropus
(d) Campylotropous

100. Embryo sac is 8 nucleate, formed after a single past-meiotic mitosis and its organisation is similar to the polygonum type found in
(a) Adoxa (c) Drusa
(b) Penaca (d) Plumbago

101. Plane winged seed is found in
(a) *Alnus nepalensis*
(b) *Betula utilis*
(c) *Sequia sempervirens*
(d) All of the above

102. The area around the parent tree in which its progeny seed falls is known as
(a) Seed rain (c) Seed handling
(b) Seed shadow (d) Seroting

103. A set of physical conditions which constitutes a germination establishment site is called
(a) Pre-conditioning
(b) Masting effect
(c) Re-generation niche
(d) Escape time

104. The light sensitive inhibitors are present in the intact seeds of
(a) *Betula pubescens*
(b) *Acacia nilotica*
(c) *Albizzia lebbeck*
(d) All of the above

105. Mechanical scarification is required in
(a) *Fraxinus excelsior*
(b) *Acacia catechu*
(c) *Cassia fistula*
(d) *Ziziphus jujuba*

106. Vertical antenna like branch in conifers outside the normal branch whorl known as
(a) Epicormic branches
(b) Ramicorn
(c) Both (a) and (b)
(d) None of the above

107. The colour of seed after maturity in *Ceiba pentandra* is
(a) Yellowish green
(b) Yellow
(c) Black
(d) Brown

108. Which of the following seed have short viability?
(a) *Syzygium cumini*
(b) *Quercus* species
(c) *Bombax ceiba*
(d) *Ceiba pentandra*

109. Macrobiotic seeds have life span of
(a) Above 15 years
(b) Less than 15 years
(c) 1 year
(d) None of the above

110. The maximum seed germination occurs in
(a) Blue light
(b) Far red light
(c) Red light
(d) Both (a) and (c)

Answer Keys									
1. (b)	**2.** (d)	**3.** (a)	**4.** (b)	**5.** (b)	**6.** (c)	**7.** (d)	**8.** (d)	**9.** (c)	**10.** (c)
11. (a)	**12.** (d)	**13.** (b)	**14.** (a)	**15.** (b)	**16.** (c)	**17.** (b)	**18.** (a)	**19.** (d)	**20.** (c)
21. (b)	**22.** (a)	**23.** (b)	**24.** (c)	**25.** (b)	**26.** (d)	**27.** (c)	**28.** (a)	**29.** (c)	**30.** (d)
31. (b)	**32.** (a)	**33.** (c)	**34.** (b)	**35.** (d)	**36.** (a)	**37.** (c)	**38.** (d)	**39.** (c)	**40.** (a)
41. (b)	**42.** (d)	**43.** (a)	**44.** (a)	**45.** (d)	**46.** (c)	**47.** (b)	**48.** (b)	**49.** (d)	**50.** (a)
51. (b)	**52.** (b)	**53.** (d)	**54.** (a)	**55.** (b)	**56.** (c)	**57.** (d)	**58.** (d)	**59.** (a)	**60.** (c)
61. (c)	**62.** (d)	**63.** (a)	**64.** (a)	**65.** (c)	**66.** (c)	**67.** (d)	**68.** (a)	**69.** (c)	**70.** (b)
71. (d)	**72.** (a)	**73.** (c)	**74.** (c)	**75.** (b)	**76.** (c)	**77.** (d)	**78.** (b)	**79.** (a)	**80.** (b)
81. (c)	**82.** (d)	**83.** (b)	**84.** (c)	**85.** (d)	**86.** (d)	**87.** (b)	**88.** (b)	**89.** (a)	**90.** (d)
91. (d)	**92.** (b)	**93.** (a)	**94.** (c)	**95.** (d)	**96.** (a)	**97.** (d)	**98.** (d)	**99.** (c)	**100.** (a)
101. (d)	**102.** (b)	**103.** (c)	**104.** (a)	**105.** (b)	**106.** (b)	**107.** (c)	**108.** (a)	**109.** (a)	**110.** (c)

13

Tree Physiology

1. C_4 plants normally produce more biological yield than C_3 because of
(a) More photorespiration
(b) Less photorespiration
(c) More photophosphorylation
(d) Less photophosphorylation

2. In plants, enzyme responsible for the synthesis of mallic acid
(a) Rubisco
(b) PEP Carboxylase
(c) Kinase
(d) None of the above

3. Photorespiration rate is highest in
(a) C_4 plants (c) C_3 plants
(b) CAM plants (d) Both (a) and (b)

4. CO_2 accepter in C_4 plants is
(a) PGA
(b) RuBP
(c) OAA
(d) None of the above

5. The net gain of ATP in glycolysis is
(a) 12 ATP's (c) 24 ATP's
(b) 16 ATP's (d) 8 ATP's

6. Sulphur containing amino acid is
(a) Methionine
(b) Lycine
(c) Glutamine
(d) Glycine

7. Photosynthesis is more in
(a) Blue light (c) Black light
(b) White light (d) Red light

8. The splitting of water molecules in plant cells under sunlight is called
(a) Phosphorylation
(b) Photorespiration
(c) Photolysis
(d) Photophosphorylation

9. The following condition is favourable when the ratio between Auxin and Cytokinin is less than one
(a) Root growth
(b) Shoot growth
(c) Callus growth
(d) All of the above

10. Sucrose synthesis takes place in
(a) Cytoplasm
(b) Stroma of chloroplast
(c) Mitochondria
(d) None of the above

11. Conversion of starch to organic acid is essential for stomatal
(a) Closure
(b) Initiation
(c) Opening
(d) None of these

12. Which of the following is a 4C compound?
(a) Phosphoglyceric acid
(b) Phosphoenolpyruvate
(c) Ribulose biphosphate
(d) Oxaloacetic acid

13. Hill reaction occurs in
(a) Absence of water

(b) Total darkness
(c) Presence of ferricyanide
(d) High altitude plants

14. The end product of glycolysis is
(a) Pyruvic acid
(b) Citric acid
(c) Glyceraldehyde
(d) Phospho-glyceraldehyde

15. Water in plants transported by
(a) Cambium (c) Phloem
(b) Xylem (d) Epidermis

16. Root hair occurs in
(a) Meristematic zone
(b) Cell maturation zone
(c) Cell elongation zone
(d) None of the above

17. The loss of water from leaves is known as
(a) Guttation (c) Exudation
(b) Evaporation (d) Transpiration

18. Plasmolysis occurs due to
(a) Absorption (c) Osmosis
(b) Endosmosis (d) Exosmosis

19. Stomata remains closed during day time in
(a) Succulents
(b) Pteridophytes
(c) Xerophytes
(d) All of the above

20. In plants, the phenomena of water absorption takes place through
(a) Tap root
(b) Root hair
(c) Lateral root
(d) All of the above

21. The hormone responsible for closure of stomata is
(a) Auxin (c) Abscisic acid
(b) Gibberellins (d) Cycocel

22. Which of the following helps in ascent of sap?
(a) Root pressure.
(b) Transpiration pull.
(c) Both (a) and (b)
(d) None of these

23. Which of the following is a synthetic growth inhibitor?
(a) Morphactin (c) Cycocel
(b) IBA (d) IAA

24. Who proposed the "Transpiration Pull Theory"?
(a) JC Bose (c) Godlewski
(b) Boehm (d) Dixon & Jolly

25. Photosystem II is absent in
(a) C_3 plants.
(b) C_4 plants.
(c) CAM plants.
(d) None of the above

26. Kranz anatomy is shown by
(a) C_2 plants (c) C_4 plants
(b) C_3 plants (d) CAM plants

27. Light reaction of photosynthesis takes place in
(a) Thylakoid
(b) Grana
(c) Chloroplast
(d) All of the above

28. "Pressure flow" theory for movement of food was given by
(a) Funk
(b) Stephen hales
(c) Kurien
(d) Munch

29. Photosynthesis is a
(a) Oxidation process
(b) Reduction process
(c) Oxidation-Reduction process
(d) None of the above

30. Krebs & ETC takes place in
(a) Mitochondria (c) Grana
(b) Chloroplast (d) Cytoplasm

31. Dark reaction takes place in
(a) Grana of chloroplast
(b) Thylakoid of chloroplast
(c) Stroma of chloroplast
(d) All of the above

32. Water use efficiency is highest in
(a) CAM plants
(b) C_3 plants
(c) C_4 plants
(d) All of these

33. The translocation of sugar from source to sink pass through
(a) Companion cells
(b) Xylem
(c) Sieve tubes
(d) Fibres

34. The term photorespiration was coined by
(a) Garner & Stout
(b) Arnan & Stout
(c) Boehm
(d) Garner & Allard

35. ETS and TCA cycle takes place in
(a) Mitochondria under anaerobic condition
(b) Thylakoid under anaerobic condition
(c) Mitochondria under aerobic condition
(d) Thylakoid under aerobic condition

36. One molecule of glucose give rise to ....molecules of pyruvic acid
(a) One (c) Three
(b) Two (d) Four

37. Respiration occurs in
(a) Cytoplasm (c) Protoplasm
(b) Mitochondria (d) All of these

38. Which of the following is an energy consuming process?
(a) C_2 cycle (c) C_3 cycle
(b) C_4 cycle (d) CAM

39. Where does photorespiration take place?
(a) Chloroplast
(b) Peroxisomes
(c) Mitochondria
(d) All of the above

40. Transpiration rate is highest in
(a) C_3 plants (c) C_4 plants
(b) CAM plants (d) All of these

41. The source of oxygen during the photosynthesis is
(a) Water
(b) Carbon dioxide
(c) Both (a) and (b)
(d) None of the above

42. The centre of porphyrine chlorophyll a and b contains
(a) Phosphorus (c) Potassium
(b) Manganese (d) Magnesium

43. Which of the following chlorophyll pigment occurs universally in all green plants?
(a) Chlorophyll b
(b) Chlorophyll a
(c) Both (a) and (b)
(d) None of the above

44. Beta-carotene is a precursor of
(a) Vitamin D (c) Vitamin C
(b) Vitamin B (d) Vitamin A

45. Carotenoids are found in
(a) Chloroplast
(b) Chromoplast

(c) Both (a) and (b)
(d) None of the above

46. Guttation takes place through
(a) Hydathodes (c) Stomata
(b) Pores (d) Cracks

47. Roots are well developed in
(a) Succulents (c) Mesophytes
(b) Xerophytes (d) Hydrophytes

48. The absorption of water by dry roots is known as
(a) Osmosis (c) Diffusion
(b) Hydration (d) Imbibition

49. Which of the following nutrient is involved in the biosynthesis of IAA?
(a) Boron (c) Manganese
(b) Zinc (d) Phosphorus

50. Photosynthesis in C_3 plants takes place in
(a) Dark
(b) Light
(c) Both (a) and (b)
(d) None of the above

51. Which of the following has the highest water use efficiency?
(a) CAM plants
(b) C_3 plants
(c) C_4 plants
(d) C_2 plants

52. Which of the following is associated with filamentous N-fixing bacteria?
(a) *Acacia catechu*
(b) *Casuarina equisetifolia*
(c) *Alnus nepalensis*
(d) *Pongamia pinnata*

53. The enzymes involved in respiration are activated by
(a) N & K (c) Mg & Mn
(b) Mg & Ca (d) P & Mn

54. The concentration of organic acids in CAM plants
(a) Increases during day time
(b) Decreases during day time
(c) Decreases during night time
(d) Increases during night time

55. Defoliation of a tree
(a) Reduce growth directly
(b) Reduce growth indirectly by reducing the amount of photosynthesis
(c) Both (a) and (b)
(d) None of the above

56. The conversion of food into new protoplasm, cell wall and other substances is called
(a) Assimilation
(b) Photosynthesis
(c) Metabolism
(d) Accumulation

57. A tree consist of six organs under
(a) Vegetative structure
(b) Reproductive structure
(c) Both (a) and (b)
(d) None of the above

58. The pith or central core is made up of parenchymatous tissue that arise from
(a) Cambium cell
(b) Cell wall
(c) Apical meristem
(d) None of the above

59. The part of annual increment produced later in the growing season is called
(a) Late wood
(b) Summer wood
(c) Both (a) and (b)
(d) None of the above

60. Heartwood provide mechanical support but does not participate in
(a) Physiological process
(b) Metabolic process
(c) Cell division
(d) All of the above

61. The term "meristem" was first coined in 1858 by
(a) Jhon Ray
(b) Asa Gray
(c) Conrad
(d) Karl Wilhelm

62. Lateral meristem include
(a) Vascular cambium
(b) Secondary cambium
(c) Cork cambium
(d) All of the above

63. Cork cambium is a
(a) Primary meristem
(b) Secondary meristem
(c) Vacuolated cell
(d) All of the above

64. Cork cambium
(a) Highly suberized cellulose wall
(b) Closely packed with no intercellular spaces
(c) Both (a) and (b)
(d) None of the above

65. During maturity time, vessels, tracheid's and fibres lose their
(a) Protoplast
(b) Vacuole
(c) Gum canals
(d) All of the above

66. Growing root usually consist of
(a) Root cap
(b) Root meristem
(c) Epidermis
(d) All of the above

67. Lateral roots arise in the
(a) Meristematic cell
(b) Pericycle
(c) Hypodermis
(d) None of the above

68. Photosynthesis is affected by
(a) Light intensity
(b) Light quality
(c) Duration of exposure
(d) All of the above

69. The component of chlorophyll molecule is/are
(a) Magnesium
(b) Nitrogen and Magnesium
(c) Calcium
(d) All of the above

70. Carbohydrate include
(a) Sugar
(b) Polysaccharides
(c) Compound carbohydrates
(d) All of the above

71. Disaccharides sucrose is
(a) Very abundant in tree
(b) Less abundant in tree
(c) Both (a) and (b)
(d) None of the above

72. Leaves of trees usually contain
(a) Fairly low concentration of carbohydrates
(b) Fairly high concentration of carbohydrates
(c) Both (a) and (b)
(d) None of the above

73. Proteins are the principle constituents of
(a) Protoplasm
(b) Cell wall
(c) Granules
(d) None of the above

74. Fats are ester of
(a) Fatty acids and glycerol
(b) Trihydric alcohol
(c) Both (a) and (b)
(d) None of the above

75. Species showing much fat throughout the stem tissue
(a) *Quercus* species
(b) *Robinia* species
(c) *Alnus* species
(d) *Tilia* species

76. Fat confined to bark in
(a) *Castanea* (c) *Tilia*
(b) *Alnus* (d) *Pines*

77. Both fat and starch confined in wood of
(a) *Pinus* (c) *Alnus*
(b) *Robinia* (d) *Quercus*

78. "Starch tree" is known as
(a) *Pinus densiflora*
(b) *Castanea pubinervis*
(c) *Populus* species
(d) All of the above

79. The tree containing very low-fat content in winter as well as in summer is
(a) Oak and Maple
(b) Alder
(c) Elm
(a) Ash

80. Which of the following plant is found to have minimum transpiration
(a) Hydrilla (c) Mango
(b) Nerium (d) Guava

81. Waxes mainly occur on
(a) Leaf cuticles
(b) Fruit cuticles
(c) Seed coat covers
(d) All of the above

82. Number of high energy phosphate bonds are formed in the aerobic oxidation of a molecule glucose is/are
(a) 38 (b) 40 (c) 39 (d) 50

83. Respiratory quotient is
(a) Ratio of O_2 produced to CO_2 consumed
(b) Ratio of CO_2 produced to O_2 consumed
(c) Both (a) and (b)
(d) None of the above

84. Generally plant absorbed N_2 in the form of
(a) NO_2 (c) NO_3
(b) HNO_2 (d) N_2O

85. The trees of swampy habitats have specialized root systems are called
(a) Aerated roots
(b) Pneumatophores
(c) Both (a) and (b)
(d) None of the above

86. Respiration inhibitors is/are
(a) Cyanide
(b) Azide
(c) Flouride
(d) All of the above

87. Mass flow hypothesis was proposed by
(a) Munch (1930)
(b) Clement (1940)
(c) Mason (1936)
(d) Clark (1950)

88. Calcium occurs in considerable quantities as
(a) Cystine
(b) Calcium oxalate crystals
(c) Amino acids
(d) All of the above

89. Mineral absorption takes place through
(a) Passive mechanism
(b) Active mechanism
(c) Both (a) and (b)
(d) None of the above

90. Diffusion pressure deficit (DPD)
(a) DPD = Osmotic pressure (OP – Wall Pressure (WP)
(b) DPD = Osmotic pressure (OP + Wall Pressure (WP)
(c) DPD = Wall pressure (WP – Osmotic Pressure (OP)
(d) None of the above

91. Cohesion theory of ascent of sap depend upon
(a) High tensile strength
(b) High adhesive forces of cell wall
(c) Imbibition forces of cell wall
(d) All of the above

92. Botanically seed are
(a) Matured ovules
(b) Matured ovary
(c) Matured endosperm
(d) None of the above

93. Seed susceptible to injury from dehydration is/are
(a) Magnolia
(b) Silver maple
(c) Oak
(d) All of the above

94. Actual wound healing begins with the appearance of soft parenchymatous tissue is known as
(a) Cortex
(b) Phloem cell
(c) Callus
(d) None of the above

95. Lignin is present in
(a) Primary plant cell wall
(b) Secondary plant cell wall
(c) Primary animal cell wall
(d) Secondary animal cell wall

96. Which of the following can be measured using potometer?
(a) Rate of photosynthesis
(b) Rate of respiration
(c) Both (a) and (b)
(d) None of the above

97. The rate of transpiration depends upon
(a) Opening and closing of stomata
(b) Position of stomata
(c) Both (a) and (b)
(d) None of the above

98. Alcohol is produced during
(a) Aerobic respiration
(b) Photosynthesis
(c) Anaerobic respiration
(d) Both (a) and (c)

99. Energy generation currency of cell is?
(a) NADP (c) NAD
(b) ATP (d) ADP

100. In stomata, closed guard cells are
(a) Flaccid
(b) Turgid
(c) Both (a) and (b)
(d) None of the above

Answer Keys									
1. (b)	**2.** (a)	**3.** (c)	**4.** (c)	**5.** (d)	**6.** (a)	**7.** (d)	**8.** (c)	**9.** (b)	**10.** (a)
11. (c)	**12.** (d)	**13.** (c)	**14.** (a)	**15.** (b)	**16.** (b)	**17.** (d)	**18.** (d)	**19.** (a)	**20.** (b)
21. (c)	**22.** (c)	**23.** (a)	**24.** (d)	**25.** (b)	**26.** (c)	**27.** (a)	**28.** (d)	**29.** (c)	**30.** (a)
31. (c)	**32.** (a)	**33.** (c)	**34.** (d)	**35.** (c)	**36.** (b)	**37.** (a)	**38.** (a)	**39.** (d)	**40.** (a)
41. (a)	**42.** (d)	**43.** (b)	**44.** (d)	**45.** (c)	**46.** (a)	**47.** (b)	**48.** (d)	**49.** (b)	**50.** (b)
51. (a)	**52.** (b)	**53.** (c)	**54.** (d)	**55.** (b)	**56.** (a)	**57.** (c)	**58.** (c)	**59.** (c)	**60.** (a)
61. (d)	**62.** (d)	**63.** (b)	**64.** (c)	**65.** (a)	**66.** (d)	**67.** (d)	**68.** (d)	**69.** (b)	**70.** (d)
71. (a)	**72.** (b)	**73.** (a)	**74.** (c)	**75.** (d)	**76.** (a)	**77.** (c)	**78.** (b)	**79.** (a)	**80.** (b)
81. (d)	**82.** (a)	**83.** (b)	**84.** (c)	**85.** (b)	**86.** (d)	**87.** (a)	**88.** (b)	**89.** (c)	**90.** (a)
91. (d)	**92.** (a)	**93.** (d)	**94.** (c)	**95.** (b)	**96.** (d)	**97.** (a)	**98.** (c)	**99.** (b)	**100.** (a)

14

Forest Protection

1. Sal root borer is
(a) *Lophostermis hugelii*
(b) *Hoplocerambyx spinicornis*
(c) *Ganoderma lucidum*
(d) None of the above

2. Which of the following is not a bio-pesticide?
(a) Bioneem (c) Carbaryl
(c) Biolap (d) Dipel

3. What are rodenticides?
(a) that kill fish
(b) that kill insects
(c) that kill rats
(d) All of the above

4. The root and shoot borer of casuarina is
(a) *Zeuzera coffeae*
(b) *Celosterna scabrator*
(c) *Sitophilus rugicollis*
(d) None of the above

5. The larva having no legs is
(a) Polypod (c) Apod
(b) Scarabaeiform (d) Protopod

6. Directorate of Plant Protection Quarantine and Storage is situated at
(a) Faridabad (c) Bangalore
(b) New Delhi (d) Hyderabad

7. Project Directorate of Biological Control is located at
(a) Allahabad (c) Bangalore
(b) Lucknow (d) Kolkata

8. *Hyblea purea* is known as
(a) Sal defoliator
(b) Teak defoliator
(c) Tendu defoliator
(d) Ailanthus defoliator

9. Root rot in Sal in caused by
(a) Ganoderma
(b) Heterobasidium
(c) Polyporus
(d) Fomes

10. Scientific name of Indian Gypsy Moth is
(a) *Lymantria obfuscate*
(b) *Ascotis infixaria*
(c) *Laspeyresia jaculatrix*
(d) None of the above

11. Trichoderma species is a
(a) Fungal antagonistic
(b) Parasitoid
(c) Viral antagonistic
(d) NPV

12. *Ailanthus defoliater is*
(a) *Plectoptera reflexa*
(b) *Atteva fabricella*
(c) *Clostera fulgurita*
(d) *Calopepla leyana*

13. The scientific name of the bark eating caterpillar is
(a) *Indarbela quadrinotata*
(b) *Holotrichia consanguinea*
(c) *Sahaydrassus malabaricus*
(d) None of the above

14. Wings are generally found on
(a) Metathorax (c) Prothorax
(b) Mesothorax (d) Both (a) and (b)

15. Insects comes under phylum
(a) Hemichordata (c) Arthropoda
(b) Vertebrata (d) Mamalia

16. Mealy bugs can be controlled by using
(a) Coccinellids (c) Both (a) and (b)
(b) Sticky bands (d) Pheromones

17. In which order usually larval stage is absent
(a) Hymenoptera (c) Coleoptera
(b) Hemiptera (d) Lepidoptera

18. The scientific name of the Sal heartwood borer is
(a) *Sahyadrassus malabaricus*
(b) *Dioryctria abietella*
(c) *Lophostermis hugelii*
(d) *Hoplocerambyx spinicornis*

19. ………… are popularly known as "White Ants"
(a) Cut worms (c) White grub
(b) Termites (d) Crickets

20. Which of the following is "C-shaped"?
(a) White grub (c) Termites
(b) Cut worms (d) Both (a) and (b)

21. The pest population should be controlled, when it reached to
(a) EIL (c) ETL
(b) ETP (d) PBL

22. Central Insecticide Laboratory (CIL) is situated at
(a) Ranchi (c) Manipur
(b) Ghaziabad (d) Faridabad

23. Pine shoot borer is
(a) *Dinoderus minutum*
(b) *Dioryctria abietella*
(c) *Celosterna scabrator*
(d) None of the above

24. Asymmetric mandibles are generally found in
(a) Butterflies (c) Thrips
(b) Wasps (d) Housefly

25. Frenulum type of wings are found in
(a) Butterflies (c) Bees
(b) Grasshopper (d) Housefly

26. The pest whose General Equilibrium Position (GEP) lies above the Economic Threshold Level (EIL) are
(a) Major pest (c) Regular pest
(b) Sporadic pest (d) Key pest

27. *Eutectona machaeralis* is a serious pest of
(a) Sal (c) Deodar
(b) Teak (d) Shisham

28. The bio-control agents used to control Teak defoliator is/are
(a) *Cedria paradoxa*
(b) *Apanteles* species
(c) *Trichogrammatoidea* species
(d) All of the above

29. The casting of skin in insects during various stages of growth and development is termed as
(a) Ecdysis
(b) Metamorphosis
(c) Exuivae
(d) Both (a) and (b)

30. The larvae having two pair of prolegs is
(a) Caterpillar (c) Semilooper
(b) Looper (d) Nymph

31. The type of mouth part of Grasshopper is
(a) Chewing and lapping type

(b) Piercing and sucking type
(c) Sponging type
(d) Biting and chewing type

32. The father of Nematology is
(a) Stanley (c) Cobb
(b) Robert Hartig (d) Brandis

33. Insects which feed on only one species are termed as
(a) Monophagous (c) Oligophagous
(b) Polyphagous (d) Both (a) and (c)

34. Saltorial type of leg is found in
(a) Praying mantis (c) Cricket
(b) Grasshopper (d) Honey bee

35. The first stage of development in all the insects is
(a) Egg (c) Larvae
(b) Pupa (d) Nymph

36. Which of the following is a cultural method of pest management?
(a) Trapping
(b) Handpicking
(c) Flooding and drainage
(d) Pruning and thinning

37. Anomorphosis is a characteristic feature of ....order
(a) Collembola (c) Diplura
(b) Protura (d) Isoptera

38. Destructive Insect Pest Act was enacted in the year
(a) 1982 (c) 1914
(b) 1814 (d) 1954

39. The presence of propodium is a characteristic feature of
(a) Hymenoptera (c) Hemiptera
(b) Coleoptera (d) Isoptera

40. Which of the following is the largest insect order?
(a) Lepidoptera (c) Coleoptera
(b) Diptera (d) Hemiptera

41. The largest commercially cultivated honey bee is
(a) *Apis florea* (c) *Apis mellifera*
(b) *Apis dorsata* (d) *Apis cerana*

42. Insects which are active in day light are
(a) Diurnal (c) Both (a) and (b)
(b) Nocturnal (d) None of these

43. Pathogens like *Beauveria bassiana* and *Nuclear polyhedrosis virus* (NPV) are used against
(a) *Plectoptera reflexa*
(b) *Eutectona machaeralis*
(c) *Dihammus cervinus*
(d) *Ectropis deodara*

44. Tree banding is useful for controling
(a) Mealy bug (c) Moths
(b) Beetles (d) Both (a) and (c)

45. The bacteria commercially produced to control insect population is
(a) *Trichogramma* species
(b) *Bravaria bassiana*
(c) NPV
(d) *Bacillus thuringien6sis*

46. National Centre for Integrated Pest Management (NCIPM) is located at
(a) Bangalore (c) Faridabad
(b) New Delhi (d) Kolkata

47. Cubicula is found in
(a) Wasps (c) Termites
(b) Beetle (d) Honey bee

48. Aphids are controlled using
(a) Systematic insecticide
(b) Contact insecticide
(c) Fumigant insecticide
(d) All of the above

49. Which of the following is considered as the most destructive group of forest insects?
(a) Seed and cone feeders
(b) Borers
(c) Sap suckers
(d) Gall insects

50. The Black sooty mould or honeydew is excreted by
(a) Aphids (c) Moths
(b) Scales (d) Both (a) and (b)

51. The larvae of some insects resemble the adults such insects are called
(a) Ametabolous insects
(b) Hemimetabolous insects
(c) Homometabolous insects
(d) None of the above

52. Flooding is useful in controlling
(a) Termites
(b) White grubs
(c) Cut worms
(d) All of the above

53. Pupae having non articulated mandibles is
(a) Adecticous pupae
(b) Coarctate pupae
(c) Dectious pupae
(d) Both (a) and (b)

54. The wood boring galleries are often stained by fungus instead of boring dust by
(a) Ambrosia beetles
(b) Carpenter ants
(c) Powder post beetles
(d) None of the above

55. Insects such as aphids and mites feed on growing plant tissues thereby causing abnormal growths is called
(a) Burr formation
(b) Skeletonizing
(c) Gall formation
(d) Honeycombing

56. Oviposition damage is done by
(a) Mites (c) Scales
(b) Tree hoppers (d) Catterpillar

57. Root knot nematode is
(a) Heterodera species
(b) Radopholus species
(c) Meloidogyne species
(d) Xanthomonas

58. The trade name of "Dimethoate" is
(a) Rogor (c) Temik
(b) Furadon (d) Thiodon

59. The insecticide made from the microorganism
(a) Carbaryl (c) Pyrethrum
(b) Endosulfan (d) Vertimac

60. The scientific name of Phassus borer is
(a) *Indarbela quadrinotata*
(b) *Sahyadrassus malabaricus*
(c) *Endoclita undulifera*
(d) *Apriona cinerea*

61. The larva of the houseflies, midges are known as
(a) Caterpillar (c) Maggot
(b) Nymph (d) Grub

62. The base of the insect antennae is
(a) Pedicel (c) Flagellum
(b) Flagella (d) Scape

63. The main structural protein in the insect is
(a) Arthropodin (c) Chitin
(b) Pheromones (d) Glucose

64. Scientific name of large poplar beetle
(a) *Atteva fabricella*

(b) *Apriona cinerea*
(c) *Eutectona mecharalis*
(d) *Chrysomela populi*

65. *Toonica niviferana* is a common insect pest of
(a) Sandalwood (c) Semul
(b) Poplar (d) Tamarind

66. Which of the following insecticide can be applied in the form of granules?
(a) Endosulfan (c) Malathion
(b) Carbaryl (d) Carbofuran

67. Methyl Bromide is a
(a) Fumigant
(b) Contact insecticide
(c) Systematic insecticide
(d) Both (b) and (c)

68. Which of the following helps in grinding of food?
(a) Labrum (c) Mandibles
(b) Maxillae (d) Labium

69. The interspecific chemicals excreted by the insects to attract other insects is/are
(a) Pheromones (c) Allomones
(b) Kairomone (d) Both (b) and (c)

70. The moulting hormone "Ecdysone" is produced by
(a) Prothoracic gland
(b) Copora allata
(c) Neuro secretory cells
(d) Corpora cardiaca

71. The most destructive group of forest insects is
(a) Sucking pests (c) Borers
(b) Defoliators (d) Root feeders

72. Maximum damage is done when the insects are in
(a) Egg stage (c) Larva stage
(b) Pupa stage (d) Adult stage

73. *Odontotermes* species belongs to the order
(a) Orthoptera (c) Hemiptera
(b) Isoptera (d) Thysanoptera

74. Animals lacking body cavity are
(a) Pseudocoelomate
(b) Acoelomate
(c) Both (a) and (b)
(d) None of the above

75. *Trichogramma is a*
(a) Egg parasite (c) Adult parasite
(b) Larval parasite(d) Pupal parasite

76. Biopesticide obtained from *Chrysanthimum* flowers is
(a) Rotenone (c) Carbryl
(b) Dipel (d) Pyrethrum

77. Light traps are used to control
(a) Silverfish
(b) Thrips
(c) Moths and Beetles
(d) Aphids

78. Insect pest control by flooding is
(a) Mechanical control
(b) Physical control
(c) Cultural control
(d) Biological control

79. Malpighian tubule is an
(a) Respiratory organ
(b) Digestive organ
(c) Excretory organ
(d) Locomotory organ

80. Sting in honey bee is a modified part of
(a) Cerci (c) Plumule
(b) Style (d) Ovipositor

81. The largest and the strongest part of insect leg is
(a) Coxa (c) Femur
(b) Tibia (d) Trochanter

82. Which of the following plant protection equipment's work on the principle of air blast?
(a) Sprayers (c) Nozzles
(b) Dusters (d) Both (a) and (b)

83. Pheromones are the chemicals used to attract
(a) Male insect (c) Both (a) and (b)
(b) Female insect (d) Mealy bugs

84. Bark blister disease of *Casuarina equisetifolia* is caused by
(a) *Corticum salmonicolor*
(b) *Trichosporium visiculosam*
(c) *Trichoderma* species
(d) *Cronartium rubicola*

85. Causal agent of Nectria canker is
(a) *Nectria galligena*
(b) *Nectria coccinea*
(c) Both (a) and (b)
(d) None of the above

86. Which disease attacks hardwood and softwood and kills shrub vines and forbs shrub vines?
(a) Chestnut blight
(b) *Annosus* root rot
(c) *Armillaria* root rot
(d) All of the above

87. Chest nut blight is
(a) Fungus (c) Bacteria
(b) Algae (d) Virus

88. Symptoms of anthracnose and leaf spot disease are
(a) Leaf dead area (c) Blotches
(b) Dry leaf (d) Both (a) and (b)

89. Root rot disease of conifers in many temperate parts of the world is called
(a) *Annosus* root rot
(b) *Armillaria* root rot
(c) Canker rot
(d) All of the above

90. Slime flux is a major
(a) Root rot
(b) Leaf rot
(c) Bole or trunk rot
(d) None of the above

91. Tree disease are affected by fungi
(a) Slime flux
(b) Beech bark disease
(c) Aspen canker
(d) Blight

92. Canker rot are most important in
(a) Pine (c) Cedrus
(b) White Oak (d) Red Oak

93. Wilting of sissoo caused by
(a) *Fusarium solani*
(b) *Ganoderma lucidium*
(c) *Pythium* species
(d) *Corticum salmonicolor*

94. Sandal spike is caused by
(a) MLO's (c) RLO's
(b) VLO's (d) Virus

95. The species most resistant to decay is
(a) *Bombax* species
(b) *Acacia nilotica*
(c) *Ailanthus* species
(d) All of the above

96. Spike is a common disease in
(a) Deodar (c) Ailanthus
(b) Sandalwood (d) Eucalyptus

97. The healing over or closing of wounds in plants is
(a) Occlusion (c) Canker
(b) Necrosis (d) Die back

98. Father of Forest Pathology is
(a) Robert Herting (c) Robert Hartig
(b) Coccinelidae (d) Lefroy

99. Vector of spike disease in sandalwood is
(a) Psyllids (c) Jassids
(b) Aphids (d) Hoppers

100. Fire blight blight disease is caused by
(a) Fungus (c) Virus
(b) Bacteria (d) MLO's

101. Powdery mildew can be controlled by
(a) Copper fungicides
(b) Sulphur fungicides
(c) Zinc fungicides
(d) Mercury fungicides

102. Abnormal growth around the feeding sites which forms the galls on leaves and petioles is common in
(a) *Pongamia pinnata*
(b) *Tectona grandis*
(c) *Acacia nilotica*
(d) None of the above

103. Vector of spike disease is
(a) *Cratono thrips*
(b) *Jassus indicus*
(c) *Cacoecia* species
(d) None of these

104. The founder of the concept of "Heart rot" in forest pathology is
(a) Dietrich Brandis
(b) Stebbing
(c) Robert Hartig
(d) Troup

105. Leaf spot in *Ailanthus* species is caused by
(a) *Cercospora ailanthicola*
(b) *Cercospora glandulosa*
(c) *Alternaria* species
(d) All of the above

106. Powdery mildew disease is caused by
(a) Virus (c) Bacteria
(b) Fungus (d) Nematode

107. Causes root rot and collar rot in *Gmelina arborea* is
(a) *Sclerotium rolfsii*
(b) *Fusarium oxysporium*
(c) *Phomes* species
(d) None of the above

108. Bamboo is affected by which disease
(a) Spike disease (c) Pink disease
(b) Broom disease (d) Stem borer

109. Damping off due to *Rizoctonia solani* in nursery mainly spread through
(a) Air borne (c) Water borne
(b) Soil borne (d) Soil borne

110. The killing of young seedlings by certain fungi is called
(a) Culling
(b) Pruning
(c) Damping off
(d) Wilting

111. Most common disease in nursery is
(a) Root rot
(b) Damping off
(c) Wilting
(d) All of the above

112. Pink disease occurs in
(a) *Eucalyptus* species

(b) *Casuarina* species
(c) *Artocarpus* species
(d) *Acacia catechu*

113. Which of the following is obligate parasite?
(a) Rust
(b) Smut
(c) Powdery mildew
(d) All of the above

114. Fungi imperfecti includes in
(a) Deuteromycotina
(b) Basidiomycotina
(c) Ascomycotina
(d) Oomycetes

115. Rust belongs to
(a) Oomycetes
(b) Basidiomycetes
(c) Erysiphales
(d) Ascomycetes

116. Smut belongs to
(a) Ascomycetes
(b) Ustilaginomycetes
(c) Basidiomycetes
(d) Zygomycetes

117. Powdery mildew belongs to
(a) Sachromycetes
(b) Basidiomycetes
(c) Oomycetes
(d) Ascomycetes

118. Downy mildew belongs to
(a) Oomycetes
(b) Basidiomycetes
(c) Zygomycetes
(d) Ascomycetes

119. Mycology is the study of
(a) Bacteria
(b) Virus
(c) Fungi
(d) Algae

120. Father of Indian Mycology is
(a) KC Mehta (c) Hartig
(b) RS Singh (d) EJ Butler

121. Bacterial diseases can be controlled by
(a) Antibiotics
(b) Fungicides
(c) Viricides
(d) All of the above

122. Disease resistance in plants is increased by
(a) Nitrogen (c) Carbon
(b) Potassium (d) Phosphorus

123. Race specific resistance is
(a) Horizontal resistance
(b) Vertical resistance
(c) Qualitative resistance
(d) Both (b) and (c)

124. Sexual spores in downy mildew fungus is
(a) Uredospores (c) Oospores
(b) Zoospores (d) All of these

125. In rust fungi repeating spores are
(a) Uredospores (c) Oospores
(b) Zoospores (d) None of these

126. Which of the following are prokaryotes?
(a) BGA
(b) Bacteria
(c) Mycoplasma
(d) All of the above

127. The relative capability of a pathogen to cause disease is called
(a) Infection
(b) Pathogenicity
(c) Pathogenesis
(d) Invasion

128. Loss of green colour due to prolonged exposure to darkness is
(a) Necrosis
(b) Etiolation
(c) Chlorosis
(d) All of the above

129. Overgrowth due to increase in size of cells is
(a) Hypertrophy (c) Hyperplasia
(b) Canker (d) Sclerotia

130. The special structures produced by fungi during asexual reproduction is/are
(a) Pycnidia (c) Sporodochia
(b) Acervuli (d) All of these

131. VAM is a/an
(a) Insecticide (c) Mycorrhiza
(b) Fungi (d) Virus

132. The study of causal agent of disease and its relation to the susceptible plant is known as
(a) Pathology (c) Etiology
(b) Ecology (d) None of these

133. The Foot rot in *Gmelina arborea* is caused by
(a) *Poria rhizomorpha*
(b) *Sclerotium rolfsii*
(c) *Fusarium oxysporum*
(d) *Phomopsis* species

134. *Trichoderma* is use to control
(a) Foliar diseases
(b) Air-borne diseases
(c) Soil-borne diseases
(d) Water-borne diseases

135. Main vegetative body of fungi is
(a) Spore (c) Hyphae
(c) Mycelium (d) None of these

136. The cell wall of fungi is made up of
(a) Hemi-cellulose
(b) Amino acids
(c) Lipid
(d) Chitin

137. Water blister is a
(a) Physiological disorder
(b) Pathological disorder
(c) Both (a) and (b)
(d) None of the above

138. The cell wall of bacteria is composed of
(a) Lipid (c) Protein
(b) Chitin (d) Lipo-protein

139. Wood decaying fungi which attacks on cellulose only is
(a) White rot
(b) Brown rot
(c) Soft rot
(d) All of the above

140. White rot fungi attacks on
(a) Cellulose
(b) Lignin
(c) Hemi-cellulose
(d) All of the above

141. Ooze test is done to detect
(a) Bacteria (c) Virus
(b) Phytoplasma (d) Fungi

142. Chaubattia paste is a mixture of
(a) Copper carbonate + Sodium carbonate + Water
(b) Copper carbonate + Red lead + Linseed oil
(c) Copper Sulphate + Lime + Water
(d) Copper Carbonate + Sodium Carbonate + Lime

143. Which of the following contains both Mn and Zn?

(a) Thiram (c) Zineb
(b) Mancozeb (d) Vapam

144. The first systematic fungicide is
(a) Captan (c) Carboxin
(b) Zineb (d) Carbaryl

145. In burgundy paste lime is replaced by
(a) Sodium carbamate
(b) Linolin
(c) Read lead
(d) Both (a) and (b)

146. The speed of the spread of fire and its direction depends upon
(a) Wind
(b) Topography
(c) Flammable material
(d) All of the above

147. The distance of the fire trace from the advance fire depends on
(a) Speed of trace fire
(b) Topography
(c) Length of advance fire
(d) Humidity

148. Factors which affect the rate of energy release from forest fuels is
(a) Quantity of moisture in forest fuel
(b) Wind speed and direction
(c) Relative humidity in air
(d) All of the above

149. Which of the following is the most difficult to detect and control?
(a) Underground fire
(b) Ground fire
(c) Surface fire
(d) Crown fire

150. Slash disposal in deodar forest should be done after rains and before snowfall to control
(a) Creeping fire
(b) Crown fire
(c) Ground fire
(d) All of the above

Answer Keys

1. (a)	**2.** (c)	**3.** (c)	**4.** (b)	**5.** (d)	**6.** (a)	**7.** (c)	**8.** (b)	**9.** (c)	**10.** (a)
11. (b)	**12.** (b)	**13.** (a)	**14.** (d)	**15.** (c)	**16.** (c)	**17.** (a)	**18.** (d)	**19.** (b)	**20.** (a)
21. (c)	**22.** (d)	**23.** (b)	**24.** (c)	**25.** (a)	**26.** (d)	**27.** (b)	**28.** (d)	**29.** (a)	**30.** (b)
31. (d)	**32.** (c)	**33.** (a)	**34.** (b)	**35.** (a)	**36.** (d)	**37.** (b)	**38.** (c)	**39.** (a)	**40.** (c)
41. (b)	**42.** (a)	**43.** (d)	**44.** (a)	**45.** (d)	**46.** (b)	**47.** (d)	**48.** (a)	**49.** (b)	**50.** (d)
51. (b)	**52.** (d)	**53.** (d)	**54.** (a)	**55.** (c)	**56.** (b)	**57.** (c)	**58.** (a)	**59.** (d)	**60.** (b)
61. (c)	**62.** (d)	**63.** (a)	**64.** (d)	**65.** (c)	**66.** (d)	**67.** (a)	**68.** (c)	**69.** (d)	**70.** (a)
71. (c)	**72.** (c)	**73.** (b)	**74.** (b)	**75.** (a)	**76.** (d)	**77.** (c)	**78.** (a)	**79.** (c)	**80.** (d)
81. (c)	**82.** (b)	**83.** (a)	**84.** (b)	**85.** (a)	**86.** (c)	**87.** (a)	**88.** (d)	**89.** (a)	**90.** (c)
91. (b)	**92.** (d)	**93.** (a)	**94.** (a)	**95.** (b)	**96.** (b)	**97.** (a)	**98.** (c)	**99.** (d)	**100.** (b)
101. (b)	**102.** (a)	**103.** (b)	**104.** (c)	**105.** (d)	**106.** (b)	**107.** (a)	**108.** (b)	**109.** (b)	**110.** (c)
111. (d)	**112.** (a)	**113.** (d)	**114.** (a)	**115.** (b)	**116.** (b)	**117.** (d)	**118.** (a)	**119.** (c)	**120.** (d)
121. (a)	**122.** (b)	**123.** (d)	**124.** (c)	**125.** (a)	**126.** (d)	**127.** (b)	**128.** (b)	**129.** (a)	**130.** (d)
131. (b)	**132.** (c)	**133.** (c)	**134.** (a)	**135.** (b)	**136.** (d)	**137.** (a)	**138.** (d)	**139.** (c)	**140.** (d)
141. (a)	**142.** (b)	**143.** (b)	**144.** (c)	**145.** (a)	**146.** (d)	**147.** (b)	**148.** (d)	**149.** (a)	**150.** (c)

15

Forest Botany

1. Father of "Botany" is
(a) Aristotle
(b) Theophrastus
(c) Carolus Linnaeus
(d) None of the above

2. Father of "Taxonomy" is
(a) Aristotle
(b) Theophrastus
(c) Carolus Linnaeus
(d) None of the above

3. Plants which produce naked seeds are
(a) Gymnosperms
(b) Angiosperms
(c) Both (a) and (b)
(d) None of the above

4. Which of the following are essential floral organs?
(a) Sepal (c) Petal
(b) Stamen (d) Petiole

5. Stem generally arises from
(a) Radicle (c) Rachis
(b) Hypocotyl (d) Plumule

6. *Pandanus tectorius* has
(a) Prop roots
(b) Haustoria roots
(c) Stilt roots
(d) Clinging roots

7. Lamina of the leaf is attached to stem or branch with
(a) Petiole
(b) Peduncle
(c) Stalk
(d) All of the above

8. The arrangement of veins on the lamina of leaf is called
(a) Phyllotaxy (c) Furcate
(b) Venation (d) Divergent

9. Occurrence of more than one type of leaf on the same plant is called
(a) Phyllotaxy (c) Compound
(b) Heterophylly (d) Phylogeneses

10. Which of the following species is tripinnate?
(a) Acacia (c) Albizzia
(b) Teak (d) Moringa

11. Which of the following has serrate type of leaf margin?
(a) *Melia azedarach*
(b) *Moringa oleifera*
(c) *Shorea robusta*
(d) *Populus ciliata*

12. Fruit of bamboo is
(a) Drupe (c) Samaroid
(b) Caryopsis (d) Lomentum

13. Leaf without petiole is
(a) Petiolate (c) Reticulate
(b) Pinnate (d) Sessile

14. The plants which are well developed and differentiated into stem, root, leaves are
(a) Pteridophyta (c) Bryophyta
(b) Thallophyta (d) Both (a) and (b)

15. A bud present in the apex of stem is known as
(a) Axillary bud (c) Adventitious
(b) Terminal bud (d) Apical bud

16. The plants which produce fruits and seeds only once in a life time are called
(a) Polycarpic (c) Polyembrony
(b) Monocarpic (d) Both (a) and (b)

17. Flower bearing stalk is known as
(a) Peduncle (c) Petiole
(b) Lamina (d) Stamen

18. Stalk of seed is
(a) Hilum (c) Micropyle
(b) Raphe (d) Funicle

19. Triangular scar on the raphe is known as
(a) Funicle (c) Chalaza
(b) Hilum (d) Raphe

20. Outer seed coat is also known as
(a) Testa (c) Pericarp
(b) Tegmen (d) Both (a) and (b)

21. Endosperm is the reserve food stored in the form of
(a) Oils and vitamins.
(b) Oils and proteins.
(c) Proteins and vitamins.
(d) All of the above

22. Root arises from the
(a) Radicle
(b) Stipule
(c) Plumule
(d) None of these

23. *Sonneratia* and *Rhizophora* has
(a) Areal roots
(b) Stilt roots
(c) Respiratory roots
(d) Stilt roots

24. Epigeal germination is common in
(a) Teak
(b) Terminalia
(c) Pines
(d) All of the above

25. Haustorial roots are found in
(a) *Santalum album*
(b) *Acacia catechu*
(c) *Shorea robusta*
(d) *Sonneratia* species

26. Point on the stem where leaf aris-es is
(a) Internode (c) Axil
(b) Node (d) Bud

27. Buds present at the tips of the main stem is
(a) Lateral bud (c) Axillary bud
(b) Apical bud (d) Dormant bud

28. Adventitious buds are found in
(a) *Duranta* (c) *Agave*
(b) *Solanum* (d) *Dalbergia*

29. Which of the following are lianas?
(a) *Tinospora cordifolia*
(b) *Bauhinia vahlii*
(c) *Ipomca*
(d) All of the above

30. Thread like structures which coil in the form of a spiral around the stem are known as
(a) Tendrils (c) Hooks
(b) Thorns (d)Twiners

31. Prostate branching stem running on the ground is known as
(a) Offset (c) Sucker
(b) Runner (d) Climber

32. The upper part of the leaf which is flat and expanded
(a) Petiole (c) Petiole
(b) Leaf blade (d) Lamina

33. Border or edge of lamina is called
(a) Apex (c) Margin
(b) Base (d) Stipule

34. Leaves borne on the branches of stem is common in
(a) *Melia azedarach*
(b) *Ficus religiosa*
(c) *Juglans regia*
(d) Both (a) and (b)

35. Which type of venation is found in grasses?
(a) Reticulate (c) Divergent
(b) Parallel (d) Both (a) and (b)

36. Bipinnate leaves are found in
(a) *Acacia* species
(b) *Moringa oleifera*
(c) *Albizzia procera*
(d) Both (a) and (c)

37. Which of the following is trifoliate?
(a) *Bombax ceiba*
(b) *Acacia catechu*
(c) *Santalum album*
(d) *Aegle marmelos*

38. The leaves of pinus are
(a) Linear (c) Elliptical
(b) Acicular (d) Ovate

39. Cordate type of leaf is found in
(a) *Populus ciliata*
(b) *Ficus bengalensis*
(c) Both (a) and (b)
(d) None of the above

40 "Phyllodes" is a characteristic feature of
(a) *Acacia modesta*
(b) *Acacia catechu*
(c) *Acacia mearnsii*
(d) All of the above

41. A flower having all types of floral leaves is called
(a) Incomplete flower
(b) Complete flower
(c) Perfect flower
(d) Imperfect flower

42. If all the sepals are united than it is called as
(a) Gamopetalous
(b) Polypetalous
(c) Polysepalous
(d) Gamosepalous

43. Fruit type of Oak and Chestnut is
(a) Nut (c) Samara
(b) Caryopsis (d) Achene

44. Lomentum fruit type is found in
(a) Acer (c) Sal
(b) Oak (d) Acacia

45. Coir is obtained from
(a) Endocarp (c) Mesocarp
(b) Endosperm (d) Epicarp

46. The innermost layer of fruit is
(a) Epicarp (c) Endocarp
(b) Mesocarp (d) Testa

47. The winged seeds are common in
(a) Dipterocarpaceae family
(b) Salicaceae family
(c) Meliaceae family
(d) All of the above

48. Jute fibre is composed of........
(a) Parenchyma cells
(b) Sclerenchyma cells
(c) Collenchyma cells
(d) All of the above

49. In flowering plants pollen is released from
(a) Stigma (c) Style
(b) Anther (d) Ovary

50. Epidermis is made up of
(a) Parenchyma cells
(b) Collenchyma cells
(c) Chlorenchyma cells
(d) Sclerenchyma cells

51. The Leaf blade is also known as
(a) Lamina (c) Pulvinus
(b) Stipules (d) Both (a) and (b)

52. Which of the following produces edible seeds?
(a) *Pinus roxburghii*
(b) *Pinus pinastris*
(c) *Pinus gerardiana*
(d) *Pinus wallichiana*

53. *Myristica fragrans* is
(a) Peanut (c) Betelvine
(b) Arecanut (d) Nutmeg

54. Economic part of clove is
(a) Flower (c) Fruit
(b) Aril (d) Flower bud

55. The essential oil of *Cinnamomum zeylanicum* is obtained from
(a) Bark
(b) Stem
(c) Bark and Leaf
(d) Flower

56. Blue gum is obtained from
(a) *Eucalyptus globulus*
(b) *Eucalyptus grandis*
(c) *Eucalyptus tereticornis*
(d) *Eucalyptus hybrid*

57. Floss is obtained from
(a) *Bombax ceiba*
(b) *Ceiba pentandra*
(c) Both (a) and (b)
(d) None of the above

58. Which of the pine have three-needles?
(a) *Pinus wallichiana*
(b) *Pinus roxburghii*
(c) *Pinus radiata*
(d) *Pinus gerardiana*

59. Samaroid is a type of
(a) Fruit
(b) Leaf modification
(c) Inflorescence
(d) Root modification

60. Poplar belongs to the family
(a) Tiliaceae (c) Fabaceae
(b) Salicaceae (d) Meliaceae

61. Bamboo leaves has ....
(a) Parallel venation
(b) Reticulate venation
(c) Divergent venation
(d) All of the above

62. Which of the following has minute seeds?
(a) *Eucalyptus* species
(b) *Anthocephalus cadomba*
(c) *Callistemon lanceolatus*
(d) All of the above

63. Which of the following is known as "Flame of the forest"?
(a) *Erythrina* species
(b) *Butea monosperma*
(c) *Pongamia glabra*
(d) None of the above

64. The nicotine content in tabacco plant is extracted from
(a) Root (c) Leaf
(b) Stem (d) Flower

65. Which of the following oil cake act as nitrogen inhibitor?
(a) Mahua (c) Mustard
(b) Neem (d) Phulwara

66. Vertical growth in tress is termed as
(a) Orthotropic

(b) Plagiotropic
(c) Both (a) and (b)
(d) None of these

67. The chemical used for destroying Fungi in water tank is?
(a) Nitric acid
(b) Zinc sulphate
(c) Copper sulphate
(d) Magnesium sulphate

68. Secondary growth is absent in
(a) Monocot stem (c) Monocot root
(b) Dicot stem (d) Dicot root

69. Epidermis is composed of
(a) Collenchyma cells
(b) Parenchyma cells
(c) Sclerenchyma cells
(d) Chlorenchyma cells

70 The study of algae is known as
(a) Plynology (c) Mycology
(b) Agrostology (d) Phycology

71. The fertilized ovule is
(a) Seed (c) Fruit
(b) Flower (d) Embryo

72. Conjunctive tissue of monocot root is usually composed of
(a) Collenchyma cells
(b) Parenchyma cells
(c) Sclerenchyma cells
(d) None of the above

73. Edible part of *Ziziphus mauritiana* is
(a) Endocarp
(b) Epicarp
(c) Mesocarp
(d) Both (a) and (c)

74. The edible part of the *Aegle marmelos* is
(a) Endocarp (c) Thalamus
(b) Mesocarp (d) Placenta

75. The fruit type of "coffee" is
(a) Nut (c) Drupe
(b) Synconus (d) Pepo

76. Sago is prepared from
(a) Root of Cassava
(b) Stem of Cassava
(c) Leaf of Cassava
(d) Flower of Cassava

77. Germination in which cotyledons appear above the ground is termed as
(a) Hypogeal germination
(b) Epigeal germination
(c) Both (a) and (b)
(d) None of the above

78. An ovary attached to the receptables above attachment of other floral parts is known as
(a) Superior ovary
(b) Inferior ovary
(c) Hypogynous ovary
(d) None of the above

79. The drug "Henbane" is extracted from
(a) *Plantago ovata*
(b) *Cannabis sativa*
(c) *Hyoscyamus niger*
(d) *Abrus precatorius*

80 Scientific name of "Rudraksh" is
(a) *Abrus precatorius*
(b) *Elaeocarpus ganitrus*
(c) *Sapindus mukorossi*
(d) None of the above

81. The scientific name of "Liquorice" is
(a) *Picrorhiza kurroa*
(b) *Dioscorea deltoidea*
(c) *Saussurea lappa*
(d) *Glycyrrhiza glabra*

82. Which are is wild edible plant?
(a) *Bauhinia variegata*
(b) *Pinus gerardiana*
(c) *Madhuca longifolia*
(d) All of the above

83. Which of the following has winged fruit type?
(a) Terminalia (c) Sal
(b) Bamboo (d) Both (a) and (b)

84. The multi seeded fruit which breaks into single dehiscent or indehiscent part is known as
(a) Capsule (c) Schizocarpic
(b) Succulent (d) Achene

85. The group of fruitlets formed from whole inflorescence is known as
(a) Aggregate fruit
(b) Composite fruit
(c) Succulent fruit
(d) Lomentum fruit

86. Which of the following meristems are most abundant in grasses?
(a) Intercalary meristem
(b) Lateral meristem
(c) Apical meristem
(d) All of the above

87. Kinetin is a type of
(a) Auxin (c) Gibberellin
(b) ABA (d) Cytokinin

88. Fruit type of *Ailanthus excelsa* is
(a) Etarrea of achenes
(b) Etarrea of samara
(c) Etarrea of follicles
(d) Etarrea of berries

89. The five-needle pine is
(a) *Pinus wallichiana*
(b) *Pinus radiata*
(c) *Pinus roxburghii*
(d) *Pinus gerardiana*

90. The unbranched stem is known as
(a) Racemose (c) Caudex
(b) Cymose (d) Cull

91. Which of the following is a bio-diesel plant?
(a) *Pongamia pinnata*
(b) *Jatropha curcus*
(c) Both (a) and (b)
(d) None of the above

92. Scientific name of "Sweet flag" is
(a) *Ricinus communis*
(b) *Acorus calamus*
(c) *Artimissia brevifolia*
(d) *Rheum emodi*

93. Which of the following have fissured bark?
(a) Erythrina
(b) Calophyllum
(c) Deodar
(d) All of the above

94. Which of the following tree have papery bark?
(a) Eucalyptus (c) Betula
(b) Sal (d) Neem

95. The scientific name of "Myrobalan" is
(a) *Phyllanthus emblica*
(b) *Terminalia chebula*
(c) *Indigofera tinctoria*
(d) All of the above

96. Scientific name of "Ironwood" tree
(a) *Messua ferrea*
(b) *Casuarina equisetifolia*
(c) *Annogeissus latifolia*
(d) None of the above

97. Keora oil is obtained from
(a) Roots
(b) Stem
(c) Flowers
(d) Heartwood

98. The food in plants is transported in the form of
(a) Glucose (c) Fructose
(b) Sucrose (d) Mamose

99. The pollens are released from
(a) Ovary (b) Filament
(b) Style (d) Anther

100. Which of the following is edible fungi?
(a) Agaricus
(b) *Morchella esculanta*
(c) Both (a) and (b)
(d) None of these

Answer Keys									
1. (b)	**2.** (c)	**3.** (a)	**4.** (b)	**5.** (d)	**6.** (c)	**7.** (a)	**8.** (b)	**9.** (b)	**10.** (d)
11. (a)	**12.** (b)	**13.** (d)	**14.** (a)	**15.** (b)	**16.** (b)	**17.** (a)	**18.** (d)	**19.** (c)	**20.** (a)
21. (b)	**22.** (a)	**23.** (c)	**24.** (d)	**25.** (a)	**26.** (b)	**27.** (b)	**28.** (d)	**29.** (b)	**30.** (a)
31. (b)	**32.** (d)	**33.** (c)	**34.** (d)	**35.** (b)	**36.** (d)	**37.** (d)	**38.** (b)	**39.** (a)	**40.** (c)
41. (a)	**42.** (d)	**43.** (a)	**44.** (d)	**45.** (c)	**46.** (c)	**47.** (a)	**48.** (b)	**49.** (b)	**50.** (a)
51. (a)	**52.** (c)	**53.** (b)	**54.** (d)	**55.** (c)	**56.** (a)	**57.** (c)	**58.** (b)	**59.** (a)	**60.** (b)
61. (a)	**62.** (d)	**63.** (b)	**64.** (a)	**65.** (b)	**66.** (a)	**67.** (c)	**68.** (c)	**69.** (b)	**70.** (d)
71. (a)	**72.** (c)	**73.** (b)	**74.** (d)	**75.** (c)	**76.** (a)	**77.** (b)	**78.** (a)	**79.** (c)	**80.** (b)
81. (d)	**82.** (d)	**83.** (a)	**84.** (c)	**85.** (b)	**86.** (a)	**87.** (d)	**88.** (b)	**89.** (a)	**90.** (c)
91. (c)	**92.** (b)	**93.** (a)	**94.** (c)	**95.** (b)	**96.** (a)	**97.** (c)	**98.** (b)	**99.** (d)	**100.** (a)

16

Forest Ecology and Environment

1. Who coined the term ecosystem in 1935?
(a) Tansley (c) Clement
(b) Troup (d) Shelford

2. An ecotype is
(a) Genetically similar species
(b) Genetically different species
(c) Both (a) and (b)
(d) None of the above

3. The transitional zone representing high diversity is known as
(a) Ecad (c) Ecotone
(b) Ecotype (d) Both (a) and (c)

4. The particular position occupied by an organism in an environment is termed as
(a) Niche (c) Habitat
(b) Ecotype (d) Home

5. The concept of Biome was given by
(a) Clement and shelford
(b) Shelford
(c) Tansley
(d) Ernest Hackel

6. The study of the relationship of individual species with its environment is termed as
(a) Synecology
(b) Habitat ecology
(c) Autecology
(d) Both (a) and (b)

7. Plants that grow well in full sunlight
(a) Facultative Heliophytes
(b) Halophytes
(c) Heliophytes
(d) None of the above

8. The relative length of the day to which plant is exposed is termed as
(a) Photoactive (c) Phototropism
(b) Photoperiod (d) Both (b) and (c)

9. The phenomena of movement of animals in response to light is termed as
(a) Photoaxis (c) Phototropism
(b) Photokinensis (d) Photoperiod

10. The humidity of air is measured using
(a) Hydrometer (c) Hygrometer
(b) Barometer (d) Pycnometer

11. The important events such as cloud formation lightning, thundering takes place in
(a) Stratosphere (c) Mesosphere
(b) Troposphere (d) Inosphere

12. The normal lapse rate is
(a) 6.0°C/Km (c) 6.5°C/Km
(b) 7.0°C/Km (d) 5.5°C/Km

13. Population ecology is called
(a) Autecology (c) None of these
(b) Synecology (d) Both (a) and (b)

14. Plants growing in extremely cold soils are
(a) Psycrophyte
(b) Hydromorphic
(c) Acidophyte
(d) None of the above

15. Phytoremedation is
(a) Use of green plants for *in-situ* reduction for contaminated soil
(b) Use of Microbes for *in-situ* risk reduction for contaminated soil
(c) Use of green plants to remove air pollution
(d) None of the above

16. Major contributor to greenhouse gasses is
(a) Sulphur dioxide
(b) Carbon dioxide
(c) Ozone
(d) Ammonia

17. The ozone layer is found in
(a) Mesosphere
(b) Troposphere
(c) Stratosphere
(d) Inosphere

18. The most abundant gas in the atmosphere is
(a) Oxygen (c) Carbon
(b) Sulphur (d) Nitrogen

19. Abrasion is common in
(a) Saline-alkaline soil
(b) Clayey soil
(c) Waterlogged soils
(d) Sandy soils

20. Forest fires are generally classified as
(a) Ground fire
(b) Surface fire
(c) Crown fire
(d) All of the above

21. Plants which grow while immersed in water are
(a) Hydrophytes (c) Hygrophytes
(b) Mesophytes (d) Xerophytes

22. Plants growing in acidic soils are known as
(a) Psammophytes (c) Oxylophytes
(b) Lithophytes (d) Halophytes

23. Plants growing on wastelands are
(a) Psammophytes (c) Halophytes
(b) Heliophytes (d) Chersophytes

24. The term symbiosis was proposed by
(a) De Vary (c) Clement
(b) De Bary (d) Tansley

25 London smog is also known as
(a) Photochemical smog
(b) Sulphur smog
(c) Both (a) and (b)
(d) None of the above

26. The entry of plant nutrients and fertilizers to fresh water leads to
(a) Eutrophication (c) Nitrification
(b) Denitrification (d) Enrichment

27. Major anthropogenic causes of air pollution is
(a) Volcanoes
(b) Oceanic emission
(c) Fuel burning
(d) Automobiles

28. Kyoto protocol came into force in the year
(a) 1997 (c) 2000
(b) 1998 (d) 2005

29. Zoological Park is example of
(a) *Ex-situ* conservation
(b) *In-situ* conservation
(c) Both (a) and (b)
(d) None of the above

30. Geological long-term storage of carbon is in the form of
(a) Fossil fuel (c) Soil carbon
(b) Deep oceans (d) Trees

31. In agro-ecosystem trophic structure is
(a) Simple (c) Not known
(b) Complex (d) Multi-layered

32. The change in the species composition across the environment gradient is indicated by
(a) Alpha diversity
(b) Gamma diversity
(c) Beta diversity
(d) All of the above

33. Spatial heterogeneity in diversity is measured above
(a) Alpha diversity
(b) Gamma diversity
(c) Beta diversity
(d) All of the above

34. Gamma diversity refers to
(a) Diversity existing in an area
(b) Diversity along time
(c) Diversity along areas
(d) None of the above

35. World environment day is celebrated on
(a) 21st march (c) 5th August
(b) 10th may (d) 5th June

36. World ozone day is observed on
(a) 16th September (c) 21st April
(b) 31st January (d) 5th June

37. WWF stands for
(a) World wide fund
(b) World wide fund for plants
(c) World wide fund for nature
(d) None of the above

38. Alpha diversity refers to
(a) Diversity existing in a given area
(b) Diversity across areas
(c) Diversity along time
(d) None of the above

39. Beta diversity refers to
(a) Diversity existing in a given area
(b) Diversity across areas
(c) Diversity along time
(d) None of the above

40. Blue baby syndrome is caused by
(a) Carbon monoxide
(b) Mercury
(c) Fluoride
(d) Nitrate

41. The rate of energy at consumer level is called as
(a) Primary productivity
(b) Secondary productivity
(c) Net primary productivity
(d) Gross primary productivity

42. The presence of higher number of species in ecosystem is
(a) Dominance effect
(b) Edge effect
(c) Both (a) and (b)
(d) None of the above

43. The soil in an area is wetted and allowed to drain till percolation is stopped, the amount of water thus retained is called
(a) Field capacity
(b) Storage capacity
(c) Capillary capacity
(d) Hygroscopic capacity

44. The pyramid of biomass is inverted in
(a) Forest ecosystem

(b) Green land ecosystem
(c) Aquatic ecosystem
(d) All of the above

45. Which year was declared as international year of biodiversity?
(a) 2002 (c) 2018
(b) 1997 (d) 2010

46. International Biodiversity day is celebrated on
(a) 22nd April (c) 22nd June
(b) 22nd May (d) 23rd December

47. Which of the following is not soluble in both alkali and acid?
(a) Humin
(b) Humic acid
(c) Both (a) and (b)
(d) None of the above

48. The Global Warming Potential is least for which of the following gases
(a) N_2O (c) CFC
(b) CH_4 (d) CO_2

49. Top solar producing country in the world is
(a) India (c) USA
(b) China (d) Japan

50. The transformation of REDD to REDD$^+$ was declared on
(a) COP 3, Kyoto
(b) COP 17, Durban
(c) COP 13, Bali
(d) COP 14, Lima

51. Permissible noise level for residential areas is
(a) 55 dB (c) 85 dB
(b) 80 Db (d) 100 Db

52. Which important greenhouse gas is being provided from the paddy field in agriculture?
(a) Sulphur dioxide
(b) Nitrous oxide
(c) Methane
(d) Ammonia

53. Seismograph is used for the measurement of
(a) Earthquakes (c) Humidity
(b) Light (d) Noise

54. The atmosphere is held to earth by
(a) Clouds (c) Wind
(b) Pressure (d) Gravity

55. Earthquakes in earth crust are result of
(a) Endogenic activity
(b) Exogenic activity
(c) Crust formation
(d) Breaking of rocks

56. The main cause of Eutrophication of lakes is
(a) Organic matter
(b) Water hyacinth
(c) Plants
(d) Chlorides

57. Flow of energy in trophic level is
(a) Multidirectional
(b) Unidirectional
(c) Recycled
(d) Increasing

58. In an ecosystem increase in fauna and decrease in flora would be harmful due to increase in
(a) Carbon dioxide
(b) Environment pollution
(c) Sulphur dioxide
(d) Methane

59. An endemic specie used in measuring and indicating the level of contamination is called

(a) Sentinel species
(b) Keystone species
(c) Indicator species
(d) Invasive species

60. The complete degradation of an organic molecule to inorganic components is called
(a) Decomposition
(b) Mineralisation
(c) Conversion
(d) Immobilisation

61. An International treaty to limit and eventually eliminate the use of CFC's that was signed into law in
(a) Montreal protocol
(b) Kyoto protocol
(c) Delhi declaration
(d) Earth summit

62. Greenhouse gas effect is due to
(a) Net warming of earth
(b) Sunlight infux to earth
(c) Infrared radiation back from earth surface
(d) Infrared eflux from earth surface

63. Bhopal gas tragedy occur in the year
(a) 1983 (c) 1985
(b) 1984 (d) 1986

64. During the year, equinoxis happen
(a) Once (c) Twice
(b) Thrice (d) Four times

65. The soil thermometers are placed in the soil at an angle of
(a) 45° (b) 90° (c) 60° (d) 30°

66. The albedo of wet field (not ploughed) is
(a) 5–24% (c) 35%
(b) 5–14% (d) 25%

67. Entropy of a system signifies
(a) Sensible heat (c) Latent heat
(b) Fission (d) Fusion

68. The ice crystal theory of precipitation was propounded by
(a) Tor Bergeson (c) Bjerkens
(b) Evans (d) Pettersen

69. Thornwaite method of evapo-transpiration estimation considers
(a) Temperature only
(b) Temperature and Rainfall
(c) Rainfall only
(d) Rainfall and Radiation

70. The heat unit concept is the extension of
(a) Stefan's law
(b) Von't Hoff law
(c) Beer's law
(d) Wein's law

71. Species that first invades bare soil is known as
(a) Pioneer (c) Sere
(b) Retrogression (d) Ecesis

72. The culminating stage of plant succession is known as
(a) Preclimax (c) Climax
(b) Proclimax (d) Postclimax

73. The term "Forest succession" was first used by
(a) Clement (c) Dawson
(b) Thoreau (d) Cawles

74. Who developed the idea and elaborated the theory of "Plant succession"
(a) Richard peters
(b) Cawles
(c) Thoreau
(d) Clement

75. Succession initiated in extremely dry conditions is known as
(a) Hydrarch (c) Lithosere
(b) Xerach (d) None of these

76. Non-biodegradable pollutant
(a) DDT (c) Both (a) and (b)
(b) BHC (d) Dipel

77. Which of the following is not correctly matched?
(a) Lithosere-rocky surface
(b) Hydrosere-water
(c) Psammosere-moist places
(d) Xerosere-dry areas

78. Succession that takes place in area having no vegetation previously is known as
(a) Secondary succession
(b) Primary succession
(c) Autogenic succession
(d) Both (b) and (c)

79. The succession that takes place after destruction of whole vegetation is known as
(a) Allogenic succession
(b) Autogenic succession
(c) Both (a) and (b)
(d) None of these

80. The initial causes of Primary succession is/are
(a) Erosion
(b) Physiography
(c) Elevation and subsidence
(d) All of the above

81. The initial cause of Secondary succession is/are
(a) Climate
(b) Physiography
(c) Biotic factors
(d) All of the above

82. The continuing causes of plant succession are
(a) Migration-Grouping-Competition-Reaction
(b) Migration- Ecesis- Grouping-Competition- Reaction
(c) Ecesis-Aggregation-Competition
(d) None of the above

83. Examples of primary succession
(a) Estuarine succession
(b) Riverain succession
(c) Both (a) and (b)
(d) None of the above

84. Which of the following is correctly matched?
(a) Theory of vegetational gradient and climax – Whittaker
(b) Mosiac theory – Aubreville
(c) Monoclimax theory – Clement
(d) All of the above

85 Seeds of *Dalbergia sissoo* is dispersed by
(a) Water (c) Wind
(b) Bird (d) Both (a) and (b)

86. Seeds of *Trewia* species and Mangroove species are dispersed by
(a) Water (c) Birds
(b) Wind (d) Both (a) and (b)

87. The Riverine site in Himalayas are first colonised by
(a) *Alnus* species
(b) *Populus* species
(c) *Pinus wallichiana*
(d) Both (a) and (b)

88. Estuarine deposits are colonized by
(a) *Trewia* species
(b) *Mangrove* species
(c) *Casuarina equisetifolia*
(d) *Holoptelia* species

89. In Tropical evergreen secondary succession starts with the arrival of
(a) *Acrocarpus fraxinifolious*
(b) *Macaranga* species
(c) *Trema* species
(d) All of the above

90. After clear-felling in hills secondary successions start with the arrival of
(a) *Woodfordia fruticosa*
(b) *Indigofera pulchella*
(c) *Berberis lycium*
(d) All of the above

91. Cane breaks in Northern tropical wet evergreen forest is example of
(a) Climatic climax
(b) Edaphic climax
(c) Pre-climax
(d) Biotic climax

92. Tropical wet evergreen forest is example of
(a) Post-climax
(b) Climatic climax
(c) Biotic climax
(d) Edaphic climax

93. Chir pine on the ridges in sub-tropical broadleafed forest is example of
(a) Pre-climax
(b) Edaphic climax
(c) Post-climax
(d) Biotic climax

94. In Sal forest presence of evergreen and semi evergreen species is example of
(a) Biotic climax
(b) Post-climax
(c) Climatic climax
(d) Edaphic climax

95. Example of Sub-climax is
(a) Chir pine on ridges in sub-tropical forest
(b) Chir pine forest in hills
(c) Tropical wet evergreen forest
(d) All of the above

96. The culminating stage in Estuarine succession is
(a) Mangrove trees
(b) Fresh water swamp forest
(c) Evergreen or semi-evergreen forest
(d) None of the above

97. The lichen and moss stages occur in
(a) Lithosere (c) Psammosere
(b) Hydrosere (d) Xerosere

98. In a Sal forest, if patch of *Xylia xylocarpa* is regenerating dominantly, then this is called as
(a) Retro regression
(b) Climatic succession
(c) Climatic climax
(d) Edaphic climax

99. Kyoto protocol does not ban the emission of
(a) CFCs (c) Halon's
(b) VOCs (d) HFCs

100. CFCs are emitted from
(a) Refrigerators
(b) Nuclear reactors
(c) Industries
(d) Vehicle emissions

101. Ozone depletion is mainly due to
(a) Water vapour
(b) Carbon monoxide
(c) CFCs
(d) CO_2

102. Fly ash is mainly emitted from
(a) Vehicle emissions
(b) Coal mining's
(c) Industries
(d) Thermal power plants

103. The excess sulphur dioxide in atmosphere reduces the efficiency of which plant metabolic part
(a) Golgi complex (c) Cell wall
(b) Mitochondria (d) Proteins

104. Tannarie effluents from textile industries causes toxicity to the
(a) Soil (c) Wind
(b) Air (d) Water

105. Scrubbers in smoke stack is used for preventing the vehicle emitted pollutant is
(a) SO_2 (c) NO
(b) CO_2 (d) CO

106. Exotic species introduced in a site which naturally acclimatized for the local ecological and edaphic factors
(a) Edaphic land race
(b) Exotic land race
(c) Geographical land race
(d) Ecological land race

107. Free-Air CO_2 Enrichment (FACE) offers mechanism for
(a) Regulation Carbon emissions under Kyoto protocol
(b) Protecting global biodiversity under CBD
(c) Evaluating the effects of enriched atmospheric CO_2 on ecosystem process
(d) Predicting climate change with coupled atmospheric- ocean general circulation models

108. The United Nation Conference and Environment and Development in June, 1992 held at
(a) Brazil (c) Paris
(b) China (d) India

109. The Kyoto protocol came into force on
(a) 16th February, 2001
(b) 16th January, 2005
(c) 16th June, 2005
(d) 16th February, 2005

110. Montreal protocol is related to
(a) Wetland conservation
(b) Ozone depleting substances
(c) Global warming
(d) None of the above

111. The amount of Solar radiation-reaching the surface of earth is called
(a) Solar energy (c) Solar flux
(b) Radiation (d) Solar light

112. A plant community more exacting than the climatic climax is called
(a) Edaphic climax
(b) Climatic climax
(c) Pre-climax
(d) Post-climax

113. Conference of parties (COP), the supreme body of UNFCCC had its first session at
(a) Berlin (c) Java
(b) Geneva (d) Kyoto

114 There is no absolute climate climax for any area and the climax is a function of the sum of the total of all the factors of mature ecosystem, this theory is given by
(a) Aubreville (c) Whittaker's
(b) Clements (d) Smith

115. In plant root and leaves dangerous metals accumulate by
(a) Acid rain
(b) Disease
(c) Air pollution
(d) Both (a) and (c)

116. Solar constant is equal to (cal/ cm^2/min)
(a) 1.94 (c) 19.4
(b) 190.4 (d) 10.94

117. Which of the following statements is not true?
(a) Photochemical smog always contains ozone
(b) The toxic effect of carbon monoxide is due to its greater affinity for haemoglobin as compared to oxygen
(c) Lead is the most hazardous metal pollutant of automobile exhaust
(d) None of the above

118. Which of the following Environmentalists gave the concept of "Biodiversity hotspot"?
(a) Gaylord Nelson
(b) Norman Myers
(c) John Muir
(d) Tansley

119. Ozone hole was first discovered in
(a) South Asia (c) Australia
(b) South America(d) Antarctica

120. CMD is
(a) Complete Development Mechanism
(b) Carbon dioxide Destructive Mechanism
(c) Clean Development Mechanism
(d) Complete Destructive Mechanism

121. As altitude increases the pressure of the air markedly
(a) Decreases
(b) Increases
(c) Constant
(d) May increase or decrease

122 The injection of air under pressure inside water to increase oxygen concentration and enhance the rate of biological degradation is termed as
(a) Bioventing
(b) Biosparging
(c) Bioaugmentation
(d) None of the above

123. Plants growing in very low temperature throughout the year are termed as
(a) Hekistotherms (c) Macrotherms
(b) Mesotherms (d) Heliophytes

124. Natural selection theory of evolution was put forward by
(a) Wiesmann (c) Robert
(b) Darwin (d) Hugo devrie

125. Transitional zone between the river and marine environment is
(a) Shallow (c) Estuary
(b) Limnitic (d) Ecotone

126. Green India Mission (GIM) comes under
(a) National Action Plan on Climate Change (NAPCC)
(b) Indian Forest Policy
(c) Biodiversity Conservation Act of India
(d) Environment Protection Act

127. World Nature Conservation Day is celebrated on
(a) 5th June (c) 22nd April
(b) 23rd March (d) 28th July

128. Pick the odd one out, Ecological hotspots in India
(a) Eastern Himalayas
(b) Eastern Ghats
(c) Western Ghats
(d) North Eastern regions of India

129. Which of the following type is an obligatory type of relationship?
(a) Parasitism
(b) Commensalism
(c) Predation
(d) Mutualism

130. The Largest Ecosystem on the earth is
(a) Marine ecosystem
(b) Desert Ecosystem
(c) Forest ecosystem
(d) Terrestrial ecosystem

131. Vegetation of tropical deciduous forest falls under which of the following categories?
(a) Microtherm
(b) Mesotherm
(c) Megatherm
(d) All of the above

132. The uppermost part of sea ecosystem contains
(a) Zooplanktons
(b) Benthos
(c) Planktons
(d) All of the above

133. The competition for light, nutrients and space is maximum in
(a) Closely related organism growing in different niches
(b) Closely related organism growing in same niche
(c) Distantly related organism growing in the same habitat
(d) None of the above

134. Greenhouse effect is warming due to
(a) Infra-red rays reaching earth
(b) Moisture layer in atmosphere
(c) Increase in temperature due to high carbon dioxide concentration
(d) Ozone depletion

135. The lichen is an association of Algae and Bacteria
(a) Fungi and Virus
(b) Virus and Algae
(c) Algae and Fungi
(d) None of the above

136. Major aerosol pollutant in jet plane emission
(a) Fluorocarbon
(b) Methane
(c) Carbon monoxide
(d) Sulphur

137. Domestic waste is a
(a) Biodegradable pollution
(b) Effluents
(c) Non-biodegradable pollution
(d) None of the above

138. Homeostasis is
(a) Tendency of biological system to change with change in environment
(b) Tendency of biological systems to resist change
(c) Disturbances of self-regulatory system and natural controls
(d) Biotic materials used in homeopathic medicines

139. Most hazardous metal pollutant of automobile exhaust is
(a) Mercury (c) Cadmium
(b) Copper (d) Lead

140. The UV-Rays coming from sun leads to formation of
(a) Carbon monoxide
(b) Ozone
(c) Nitrous dioxide
(d) Oxygen

141. The population composed of same kind of organisms is a
(a) Species (c) Colony
(b) Community (d) Family

142. In a Lake ecosystem the secondary tropic level is occupied by
(a) Benthos
(b) Zooplanktons
(c) Plankton
(d) Phyto-planktons

143. If the huge amount of sewage waste is dumped into the river, its BOD will
(a) Decrease
(b) Slightly decrease
(c) Increase
(d) Slightly increase

144. If decomposers are completely removed from an ecosystem than the
(a) Minerals will not move
(b) Population of herbivores will decrease
(c) Energy will not transfer
(d) Decomposition rate slows down

145. Therophytes survive extreme conditions as
(a) Corms (c) Seed
(b) Fruit (d) Rhizomes

146. The major causes of Desertification are
(a) Overgrazing
(b) Climate change
(c) Deforestation
(d) All of the above

147. Which of the following ecosystem has highest gross primary productivity?
(a) Tropical rain forest
(b) Grassland
(c) Coral reef
(d) None of the above

148. Which of the following is an indicator of water quality?
(a) *Azospirillum*
(b) *Escherichia crassipes*
(c) *Escherichia coli*
(d) Both (b) and (c)

149. Plant decomposers are
(a) Protista and Virus
(b) Bacillus and Virus
(c) Fungi and Animals
(d) Monera and Fungi

150. Maximum GHG is released by
(a) Russia (c) China
(b) USA (d) Germany

151. *Escherichia coli* is an indicator of water contaminated with
(a) Heavy metals
(b) Litter
(c) Fertilizers
(d) Faecal matter

152. In 1948, Bhopal gas tragedy was due to reaction of methyl isocyanate with
(a) Water
(b) Nitrous oxide
(c) Carbon monoxide
(d) Oxygen

153. Maximum growth rate occurs in
(a) Stationary phase
(b) Senescent phase
(c) Log phase
(d) Lag phase

154. Which of the following is not used for the disinfection of drinking water?
(a) Chlorine (c) Chloramine
(b) Phenol (d) Ozone

155. Biodiversity Act of India was passed in
(a) 2000 (c) 2002
(b) 1894 (d) 2004

156. Niche overlapping leads to
(a) Sharing of resources between two species
(b) Mutual association between the two species
(c) Two organism occupying same place
(d) None of the above

157. Which year is declared as "Inter national year of forests"
(a) 2010 (c) 2008
(b) 2011 (d) 2009

158. Which of the following gas is produced from paddy field?
(a) Methane
(b) Nitrate oxide
(c) Nitrite oxide
(d) Both (a) and (b)

159. The term "Biocoenosis" was proposed by
(a) AG Tansley (c) Karl Mobius
(b) RH Whittaker (d) Robert Hooke

160. The species of plants that help a vital role in controlling the relative abundance of other species in a community are called
(a) Edge species
(b) Link species
(c) Pioneer species
(d) Keystone species

161. Which of the following statement best describes a climax community?
(a) More stable and more diverse
(b) More stable and less diverse
(c) Less stable and more diverse
(d) Less stable less diverse

162. Which of the following is the most abundant molecule in atmosphere?
(a) Carbon dioxide
(b) Water vapour
(c) Methane
(d) Argon

163. The most essential micro nutrient is
(a) Calcium (c) Nickel
(b) Zinc (d) Sodium

164. Which of the following is known as the potential bio-indicator of the air pollution?
(a) Corals (c) Lichens
(b) Mosses (d) Ferns

165. Replacement of existing communities by an external condition is termed as
(a) Primary succession
(b) Secondary succession
(c) Autogenic succession
(d) None of the above

166. Bio-oil can be obtained from lignocellulose through
(a) Pyrolysis
(b) Combustion
(c) Transesterification
(d) Esterification

167. Which of the following explains thermal stratification?
(a) Non viability of seeds at low temperature

(b) Adjust of body temperature according to the environment
(c) Layers of water having different temperature in a pond
(d) All of the above

168. "The extremities of animals are relatively shorter in the cooler parts of a species' range than in warmer part" is known as
(a) Bergmann's rule
(b) Gloger's rule
(c) Rensch's rule
(d) Allen rule

169. In ecology "Biocenosis" specifically means
(a) Ecological study of communities
(b) Ecological study of individual species
(c) A group of many species of plants and animals living together in a natural area
(d) All of the above

170. Maximum biodiversity is seen in
(a) Tropical deciduous forest
(b) Tropical rain forest
(c) Mangrove forest
(d) Alpine forest

171. Expand UNCED?
(a) United Nation Conference on Environment and Development
(b) United Nation Conference on Education Dissemination
(c) United Nation Conference on Environment and Education
(d) None of the above

172. The "Chipko Movement" was started by
(a) Bishnoi community
(b) Amrita Devi
(c) Sunder Lal Bahuguna
(d) Medha Patkar

173. MoEF&CC means
(a) Ministry of Forest, Energy and Climate Conditions
(b) Ministry of Environment, Forest and Climate Change
(c) Management of Environment, Forest and Climate Change
(d) None of the above

174. The biggest award in the field of "Environment" is given in the name of
(a) Indra Gandhi
(b) Mahatma Gandhi
(c) Medha Patkar
(d) Sunder Lal Bahuguna

175. The major components of LPG are
(a) Propane & Methane
(b) Methane & Hexane
(c) Ethane & Methane
(d) Propane & Butane

176. Ozone umbrella is found in
(a) Troposphere (c) Stratosphere
(b) Mesosphere (d) Inosphere

177. The largest brackish water lake is
(a) Wular lake (c) Pulicat lake
(b) Chilika lake (d) Loktak lake

178. Sedimentation is a
(a) Primary water treatment
(b) Secondary water treatment
(c) Tertiary water treatment
(d) None of the above

179. Zooplanktons are
(a) Producers of Seas
(b) Primary consumers of Lakes

(c) Secondary consumers of sea
(d) Primary producers of Lakes

180. The major reservoir of carbon is
(a) Forest (c) Atmosphere
(b) Ocean beds (d) Soil

181. The inherent ability of the organism to reproduce and multiply is known as
(a) Breeding value
(b) Carrying capacity
(c) Biotic potential
(d) None of the above

182. Sulphur bacteria are
(a) Autotrophs (c) Detritivores
(b) Heterotrophs (d) Parasitic

183. The phenomena of accumulation of non-biodegradable pesticides in human body is
(a) Biomagnification
(b) Bioaccumulation
(c) Bioremediation
(d) None of the above

184. Which of the following is a physical pollutant?
(a) Water
(b) Soil
(c) Suspended solids
(d) Air

185. The combustion of waste in the absence of air is called
(a) Combustion (c) Pyrolysis
(b) Hydrolysis (d) Gasification

186. The world wetland day is celebrated on
(a) 2nd February (c) 21st March
(b) 5th September (d) 22nd April

187. Where was Stockholm conference on "Human Environment" was held?
(a) India (c) Sweden
(b) Rio de Janeiro (d) Paris

188. The Environment Protection Act was promulgated in the year
(a) 1986 (c) 1776
(b) 1996 (d) 1896

189. In a food chain of grassland ecosystem the top consumers are
(a) Herbivores (c) Carnivores
(b) Bacteria (d) None of these

190. MAB stands for
(a) Man and Biosphere
(b) Man and Biotic community
(c) Man and Biodiversity
(d) None of the above

191. Species that occur in different geographic areas separated by special barriers are called
(a) Allopatric (c) Both (a) and (b)
(b) Sympatric (d) None of these

192. The pioneers of xerach succession is the
(a) Crustose lichen
(b) Foliose lichen
(c) Mosses
(d) Shrubs

193. The conversion of a pond to a climax forest community is
(a) Xerach succession
(b) Mesarch succession
(c) Hydrarch succession
(d) None of the above

194. The pyramid of number in a single tree is
(a) Upright
(b) Spindle shaped
(c) Inverted
(d) Both (a) and (c)

195. Climate includes
(a) Seasonal variations
(b) Generalized atmospheric conditions
(c) Average weather condition
(d) All of the above

196. Respiratory roots are known as
(a) Pneumatophores
(b) Lenticels
(c) Hydathodes
(d) Stilt root

197. The gradual physiological adjustment to changing environment conditions is termed as
(a) Introduction
(b) Acclimatization
(c) Establishment
(d) None of the above

198. IUCN is also known as
(a) Man and Biosphere programme
(b) World Conservation Union
(c) World Conservation Consortium
(d) World Wide Conservation Union

199. The most important reason for decrease in the Biodiversity is
(a) Habitat population
(b) Introduction of exotic species
(c) Over exploitation
(d) Habitat destruction

200. *In-situ* Conservation is
(a) National parks
(b) Biosphere reserves
(c) Wildlife sanctuaries
(d) All of the above

201. *Ex-situ* conservation is
(a) Pollen banks
(b) Zoo
(c) Botanical gardens
(d) All of the above

202. Hotspots are regions of high
(a) Biodiversity
(b) Endemism
(c) Diversity of exotics
(d) None of the above

203. Endemic species are
(a) Rare species
(b) Localised species of a region
(c) Cosmopolitan in nature
(d) Critically endangered species

204. Viable material of endangered species can be preserved by
(a) Gene bank (c) Tissue culture
(b) Gene pool (d) Herbarium

205. Which group of vertebrates is having highest number of endangered species
(a) Mammals (c) Fishes
(b) Birds (d) Reptiles

206. The unfavourable alteration in environment due to human activities is
(a) Ecological disturbance
(b) Catastrophe
(c) Ecological degradation
(d) Pollution

207. BOD stands for
(a) Biotic Oxygen Demand
(b) Biological Oxidation Demand
(c) Biological Oxygen Demand
(d) Biochemical Oxygen Demand

208. Lead Poisoning
(a) Reduces oxygen carrying capacity of haemoglobin in blood
(b) Increases oxygen carrying capacity of haemoglobin in blood
(c) Reduces oxygen carrying capacity of myoglobin in muscles
(d) Increases oxygen carrying capacity of myoglobin in muscles

209. Earth summit was held in
(a) Stockholm, 1972
(b) Rio de Janeiro, 1992
(c) Montreal, 1992
(d) Tokyo, 1987

210. Which of the following is not a primary pollutant?
(a) Ammonia
(b) Peroxy Acyl Nitrate (PAN)
(c) Sulphur dioxide
(d) Hydrogen sulphide

Answer Keys

1. (a)	**2.** (b)	**3.** (c)	**4.** (a)	**5.** (a)	**6.** (c)	**7.** (c)	**8.** (b)	**9.** (a)	**10.** (c)
11. (b)	**12.** (c)	**13.** (b)	**14.** (a)	**15.** (a)	**16.** (b)	**17.** (c)	**18.** (d)	**19.** (d)	**20.** (d)
21. (a)	**22.** (c)	**23.** (d)	**24.** (b)	**25.** (b)	**26.** (a)	**27.** (c)	**28.** (d)	**29.** (a)	**30.** (b)
31. (a)	**32.** (c)	**33.** (c)	**34.** (b)	**35.** (d)	**36.** (a)	**37.** (c)	**38.** (a)	**39.** (b)	**40.** (d)
41. (c)	**42.** (b)	**43.** (a)	**44.** (c)	**45.** (d)	**46.** (b)	**47.** (a)	**48.** (d)	**49.** (b)	**50.** (c)
51. (a)	**52.** (c)	**53.** (a)	**54.** (d)	**55.** (a)	**56.** (a)	**57.** (b)	**58.** (a)	**59.** (a)	**60.** (a)
61. (a)	**62.** (a)	**63.** (b)	**64.** (c)	**65.** (c)	**66.** (b)	**67.** (a)	**68.** (a)	**69.** (a)	**70.** (b)
71. (a)	**72.** (c)	**73.** (b)	**74.** (d)	**75.** (b)	**76.** (c)	**77.** (c)	**78.** (d)	**79.** (a)	**80.** (d)
81. (d)	**82.** (b)	**83.** (c)	**84.** (d)	**85.** (d)	**86.** (a)	**87.** (d)	**88.** (b)	**89.** (d)	**90.** (d)
91. (b)	**92.** (b)	**93.** (a)	**94.** (b)	**95.** (b)	**96.** (c)	**97.** (a)	**98.** (d)	**99.** (b)	**100.** (a)
101. (c)	**102.** (d)	**103.** (b)	**104.** (d)	**105.** (a)	**106.** (b)	**107.** (c)	**108.** (a)	**109.** (d)	**110.** (b)
111. (c)	**112.** (d)	**113.** (a)	**114.** (c)	**115.** (a)	**116.** (a)	**117.** (d)	**118.** (b)	**119.** (d)	**120.** (c)
121. (a)	**122.** (b)	**123.** (a)	**124.** (b)	**125.** (c)	**126.** (a)	**127.** (d)	**128.** (b)	**129.** (d)	**130.** (a)
131. (b)	**132.** (c)	**133.** (b)	**134.** (c)	**135.** (c)	**136.** (a)	**137.** (c)	**138.** (b)	**139.** (d)	**140.** (b)
141. (a)	**142.** (b)	**143.** (c)	**144.** (a)	**145.** (c)	**146.** (d)	**147.** (a)	**148.** (c)	**149.** (d)	**150.** (b)
151. (d)	**152.** (a)	**153.** (c)	**154.** (b)	**155.** (c)	**156.** (a)	**157.** (b)	**158.** (a)	**159.** (c)	**160.** (d)
161. (a)	**162.** (d)	**163.** (b)	**164.** (c)	**165.** (b)	**166.** (a)	**167.** (c)	**168.** (d)	**169.** (c)	**170.** (b)
171. (a)	**172.** (c)	**173.** (b)	**174.** (a)	**175.** (d)	**176.** (c)	**177.** (b)	**178.** (a)	**179.** (c)	**180.** (b)
181. (c)	**182.** (a)	**183.** (b)	**184.** (c)	**185.** (c)	**186.** (a)	**187.** (c)	**188.** (a)	**189.** (a)	**190.** (a)
191. (a)	**192.** (a)	**193.** (c)	**194.** (b)	**195.** (c)	**196.** (a)	**197.** (b)	**198.** (b)	**199.** (d)	**200.** (d)
201. (d)	**202.** (b)	**203.** (b)	**204.** (a)	**205.** (a)	**206.** (d)	**207.** (c)	**208.** (a)	**209.** (b)	**210.** (b)

17

Forest Survey and Engineering

1. How many chains are in a mile?
(a) 80 (b) 140 (c) 120 (d) 50

2. Planimeter used for measurement of
(a) Angle (c) Diameter
(b) Height (d) Area

3. Contour lines are
(a) Proportional to elevation
(b) Inversely proportional to elevation
(c) Small hill between the contours
(d) None of the above

4. An equipment used in forest chain surveying is
(a) Offset
(b) Plumb bob
(c) Ranging rods
(d) All of the above

5. The diameter of the surveyor compass is
(a) 12 cm (c) 20 cm
(b) 10 cm (d) 30 cm

6. The technique of plotting all the points which are accessible from single set of plane table is
(a) Resection method
(b) Radiation method
(c) Triangulation method
(d) Traversing method

7. Bunds constructed across the slop are
(a) Gully plugging
(b) Terraces
(c) Field bunds
(d) Contour bunds

8. An instrument used for measuring the area from the plan of a plot is
(a) Pedometer
(b) Alidade
(c) Planimeter
(d) Tracer

9. A concrete Structure constructed across the stream bed of a seasonal stream is
(a) Cause way
(b) Suspension bridge
(c) Wire rope bridge
(d) Arch bridge

10. In Brick masonry, if the header and stretcher are laid alternatively in each course, it is called
(a) English bond
(b) Header bond
(c) Flemish bond
(d) None of the above

11. The transverse raising of road in the centre above the edges is called
(a) Crown (c) Camber
(b) Shoulder (d) Foundation

12. Building load is transmitted to the earth through
(a) Roof (c) Floor
(b) Walls (d) Foundation

13. The transverse slope or inward tilt imparted to curves in the road to neutralise the centrifugal force is called
(a) Brest wall
(b) Super elevation
(c) Retaining wall
(d) Gradient

14. Lateral measurement in chain surveying is
(a) Offset
(b) Level
(c) Ranging
(d) None of the above

15. In compass surveying, the fore bearing and back bearing of a line differs by
(a) 90° (c) 180°
(b) 270° (d) 360°

16. Vertical and horizontal angles can be measured using
(a) Clinometer
(b) Cross staff
(c) Prismatic compass
(d) Theodolite

17. The error due to sag in chain surveying is
(a) Positive
(b) Systematic
(c) Negative
(d) Both (a) and (b)

18. The length of the offset depends upon
(a) Method of setting out perpendicular
(b) Nature of ground
(c) Scale of plotting
(d) All of the above

19. Which of the following is/are used for measuring perpendicular offsets?
(a) Cross staff
(b) Tape
(c) Optical square
(d) All of the above

20. The local attraction in compass surveying is due to
(a) Presence of magnetic substances
(b) Gravity
(c) Friction in needle
(d) None of the above

21. The object of forest surveying is to prepare
(a) Drawing (c) Sketch
(b) Cross section (d) Map

22. The size of a plane table is
(a) 600 mm × 750 mm
(b) 500 mm × 800 mm
(c) 650 mm × 750 mm
(d) None of the above

23. Which of the following instrument is used for centring in plane table surveying?
(a) Trough compass
(b) Spirit level
(c) Plumbing fork
(d) All of the above

24. The closely spaced contour lines represent
(a) Plane surface
(b) Gentle slope
(c) Uniform depth
(d) Steep slope

25. Intersection method of surveying is most suitable for
(a) Plains (c) Hills
(b) Forest (d) Urban areas

26. Which of the following plane table surveying method is useful in locating the inaccessible points?
(a) Radiation
(b) Intersection
(c) Traversing
(d) All of the above

27. Contour interval is
(a) Horizontal distance between two consecutive contours
(b) Vertical distance between two consecutive contours
(c) Vertical distance between two points on same contour level
(d) Horizontal distance between two points on same contour level

28. The rise and fall method of levelling involves
(a) Back-sight
(b) Fore-sight
(c) Intermediate-sight
(d) All of the above

29. Which of the following method involves the least measurement on ground?
(a) Perpendicular offset method
(b) Oblique offset method
(c) Both (a) and (b)
(d) None of the above

30. Small gullies in forest areas are controlled by
(a) Brushwood dams
(b) Strip cropping
(c) Contour strip cropping
(d) Mixed cropping

31. The smallest horizontal angle between the true meridian and survey line is known as
(a) Declination
(b) Dip
(c) Azimuth
(d) None of the above

32. The check lines (Proof lines) in chain surveying are used for
(a) Plotting chain lines
(b) Plotting offsets
(c) Both (a) and (b)
(d) None of the above

33. The line normal to the plumb line is known as
(a) Horizontal line
(b) Vertical line
(c) Datum line
(d) Level line

34. In Geodetic surveys high accuracy can be achieved, if
(a) Curvature of the earth surface is ignored
(b) Curvature of earth surface is taken into consideration
(c) Both (a) and (b)
(d) None of the above

35. If the whole bearing of a line is 270° than the reduced bearing of a line is
(a) N 90° W (c) S 90° E
(b) 90° W (d) W 90°

36. Surveys which are carried out to depict mountains, rivers, water bodies, cultural details etc are known as
(a) Topographical surveys
(b) Cadastral surveys
(c) City surveys
(d) None of the above

37. In chain surveying
(a) Angular measurements are taken
(b) Linear measurements are taken
(c) Both (a) and (b)
(d) None of the above

38. Direct method of contouring is a
(a) Quick method
(b) Most accurate method
(c) Both (a) and (b)
(d) None of the above

39. Clinometer is used for
(a) Measuring angle of slope
(b) Setting out angles
(c) Both (a) and (b)
(d) None of the above

40. Short offsets are measured using
(a) Steel tape
(b) Metallic tape
(c) Ordinary chain
(d) None of the above

41. The instrument used for accurate centring in plane table surveying is
(a) Spirit level
(b) Plumbing fork
(c) Alidade
(d) Trough compass

42. Planimeter is used for measuring
(a) Slope
(b) Angle
(c) Area
(d) Volume

43. The first reading taken from the levelled station is known as
(a) Fore-sight
(b) Back-sight
(c) Intermediate sight
(d) None of the above

44. The bearing of a line is also known as
(a) Magnetic bearing
(b) Back bearing
(c) Reduced bearing
(d) True bearing

45. The length of Gunter (Surveyor) chain is
(a) 66 feet (c) 60 feet
(b) 100 feet (d) 80 feet

46. The length of Engineer chain is
(a) 200 feet (c) 66 feet
(b) 100 feet (d) 70 feet

47. Deep arrows are used for measurment
(a) In chain surveys
(b) In contour surveys
(c) Both (a) and (b)
(d) None of the above

48. In surveying the instrument used for counting the number of steps in pacing long lines is
(a) Passometer
(b) Pedometer
(c) Odometer
(d) Nanometer

49. Surveying includes
(a) Field work
(b) Office work
(c) Both (a) and (b)
(d) None of the above

50. Primary classification of survey deals with
(a) Plane surveying
(b) Geodetic surveying
(c) Trignometric surveying
(d) All of the above

51. Classification based on instrument used
(a) Cadastral survey
(b) Secondary survey
(c) Technometric survey
(d) City survey

52. Land survey is classified into
(a) Topographical survey

(b) Cadastral survey
(c) City survey
(d) All of the above

53. Classification based on object of survey is
(a) Geological and soil survey
(b) Mineral survey
(c) Archeological survey and military survey
(d) All of the above

54. Vernier scale includes
(a) Double vernier
(b) Extended vernier
(c) Both (a) and (b)
(d) None of the above

55. Town survey includes
(a) 2 cm = 1 Km or 1/50000
(b) 1 cm = 50 m or 1/5000
(c) Both (a) and (b)
(d) None of the above

56. Forest working plan maps include
(a) 4″ = 1 mile
(b) 1 cm = 150 m
(c) 1/15000
(d) All of the above

57. Forest stock maps include
(a) 8″ = 1 mile
(b) 1 cm = 50 m
(c) 1/5000
(d) All of the above

58. Simple device which is attached to the wheel of any vehicle is called
(a) Wheel pedometer
(b) Odometer
(c) Passometer
(d) Speedometer

59. Cleavage planes occur in
(a) Gypsum
(b) Calcite
(c) Flourite
(d) All of the above

60. Fracturing of mineral includes
(a) Conchoidal
(b) Hackly
(c) Earthy
(d) All of the above

61. Transparency includes
(a) Opaque
(b) Translucent
(c) Transparent
(d) All of the above

62. The rocks which are finely grinded are known as
(a) Argillaceous
(b) Arenaceous
(c) Rudaceous
(d) None of the above

63. Purest argillaceous rock is
(a) Shale
(b) Kaolin
(c) Limestone
(d) Clay

64. Good bricks have
(a) Uniform size and colour
(b) No fissures
(c) Hardness
(d) All of the above

65. Cement produced by heating to a clinkering temperature is
(a) Normal portland cement
(b) High alumina cement
(c) Rapid hardening system
(d) Quick setting cement

66. Structure of colums in main bars and binders covers
(a) 1.5 cm and 0.7 cm
(b) 1.7 cm and 5.1 cm
(c) 3.5 cm and 1.5 cm
(d) 2.5 cm and 1.5 cm

67. The process in which the exposed joints of brickwork or stone work are treated with lime or cement is known as
(a) Leeping
(b) Pointing
(c) Peeble dashing
(d) None of the above

68. Diluting the base to make the point lighter or hevier and to reduce cost includes
(a) Gypsum
(b) China clay
(c) Barytes
(d) All of the above

69. An angle that is not a right angle on a plan is
(a) Squint quoin (c) Bond
(b) Header (d) King closer

70. The process in which joint or gaps in the interior of walls are filled with mortar is called
(a) Toothing (c) Grouting
(b) Quoin (d) Lap

71. The type of bond usually used in forest engineering is
(a) Heading bond
(b) Stretcher bond
(c) English bond
(d) All of the above

72. Load on masonary and foundations is
(a) Dead load
(b) Live load
(c) Wind load
(d) All of the above

73. The walls are usually made of split bamboo which are interlaced in order to form a continuous surface and fastened to the frame work of the houses with the help of strings is
(a) Mat walls
(b) Flush boarding
(c) Sheet walls
(d) Trellis work

74. An arrangement of wedge shaped blocks mutually supporting each other constructed to form a curve and supported at both ends by abutments or pillars is
(a) Centring (c) Arches
(b) Roofing (d) Collar beam

75. Different types of oblique shouldered joints
(a) Birds mouth
(b) Oblique tennon
(c) Bridle joint
(d) All of the above

76. The angle that a cord of 100 feet length substends with the centre of the curve is calculated by formula (where, R-radius, D-Degree of curve)
(a) $D = 100/R$
(b) $D = R/100$
(c) $D = R \times 100$
(d) None of the above

77. The bridge made up of two system of counter poised beams built out from the abutments and supporting the road bearer in the middle is
(a) Suspension bridge
(b) Strutt beam bridge
(c) Cantilever bridge
(d) None of the above

78. To facilitate reading of the chain, each chain is divided into ten equal parts by means of brass tag is called as
(a) Tallie

(b) Teller
(c) Tablet
(d) All of the above

79. Engineers chain is 100 feet long and each one of its 100 link measures
(a) 1 foot (c) 5 feet
(b) 2 feet (d) 4 feet

80. Triangles in which the angles are neither very acute (less than 30°) nor very obtuse (more than 120°) in chain surveying is known as
(a) Well conditioned triangle
(b) Ill conditioned triangle
(c) Both (a) and (b)
(d) None of the above

81. The sizes of the drawing papers commonly used is/are
(a) Demy
(b) Royal
(c) Double elephant
(d) All of the above

82. Instrument capable of measuring upto about 140° is known as
(a) Theodolite
(b) Pocket or Box sextant
(c) Compass
(d) All of the above

83. Plane table include
(a) Alidade
(b) Clinometer
(c) Aneroid
(d) Tracer

84. A small circular level is generally provided for levelling the plane table is known as
(a) Declinator
(b) Mounting
(c) Spirit level
(d) All of the above

85. In plane table, alidade is also called as
(a) Index
(b) Sight rule
(c) Sight vane
(d) All of the above

86. Levelling instrument is
(a) Abney level
(b) Ghat tracer
(c) Tangent clinometer
(d) Barometer

87. The important parts of a telescopic level
(a) The levelling head
(b) The limb
(c) The teleacope and bubble tube
(d) All of the above

88. Contouring over large areas is known as
(a) Topographical survey
(b) Plain table survey
(c) Hachures
(d) None of the above

89. Method of finding North direction from the sun is called
(a) Pole shadow method
(b) Pole star method
(c) Both (a) and (b)
(d) None of the above

90. The instrument invented by professor Wallace to remedy the defect of unsteadiness in the pantograph
(a) Photographic method
(b) Proportional compass
(c) Ediograph
(d) None of the above

91. One degree is equal to
(a) $\frac{p}{180}$ radian
(b) $\frac{p}{10800}$ radian

(c) 1/6,48,00 radian
(d) None of the above

92. Cadestral map scales include
(a) 1:1000
(b) 1:2000
(c) 1:5000
(d) All of the above

93. The preliminary inspection of the area to be surveyed is called
(a) Marking station
(b) Reconnaissance
(c) Reference sketches
(d) None of the above

94. The main principle of the surveying is to work from
(a) Part to whole
(b) Whole to part
(c) Higher to lower level
(d) Lower to higher level

95. A fixed refrence point of known elevation is called as
(a) Line of sight (c) Vertical axis
(b) Bench mark (d) Level line

96. The instrument used for measuring angle of elevation or depression is
(a) Watkin's mirror clinometer
(b) Service pattern clinometer
(c) Both (a) and (b)
(d) None of the above

97. In contour survey, if lines are equally spaced they indicate
(a) Irregular surface
(b) Undulated surface
(c) Uniform slope
(d) None of the above

98. The technique of plotting all the accessible stations with a ...... set of plane table is called radiation
(a) Single (c) Double
(b) Triple (d) Four

99. Resection is a method of
(a) Compass survey
(b) Chain survey
(c) Both (a) and (b)
(d) None of the above

100. The instrument abney level used for levelling purpose is an
(a) Barometer
(b) Clinometer
(c) Hypsometer
(d) Calliper

Answer Keys									
1. (a)	**2.** (d)	**3.** (d)	**4.** (d)	**5.** (a)	**6.** (b)	**7.** (d)	**8.** (c)	**9.** (a)	**10.** (c)
11. (c)	**12.** (d)	**13.** (b)	**14.** (a)	**15.** (c)	**16.** (d)	**17.** (c)	**18.** (d)	**19.** (d)	**20.** (a)
21. (d)	**22.** (a)	**23.** (c)	**24.** (d)	**25.** (c)	**26.** (b)	**27.** (a)	**28.** (d)	**29.** (a)	**30.** (a)
31. (c)	**32.** (d)	**33.** (d)	**34.** (b)	**35.** (d)	**36.** (a)	**37.** (b)	**38.** (b)	**39.** (a)	**40.** (c)
41. (b)	**42.** (c)	**43.** (b)	**44.** (d)	**45.** (a)	**46.** (b)	**47.** (d)	**48.** (a)	**49.** (c)	**50.** (d)
51. (c)	**52.** (d)	**53.** (d)	**54.** (c)	**55.** (c)	**56.** (d)	**57.** (d)	**58.** (b)	**59.** (d)	**60.** (d)
61. (d)	**62.** (a)	**63.** (b)	**64.** (d)	**65.** (a)	**66.** (c)	**67.** (b)	**68.** (d)	**69.** (a)	**70.** (c)
71. (d)	**72.** (d)	**73.** (a)	**74.** (c)	**75.** (d)	**76.** (a)	**77.** (c)	**78.** (d)	**79.** (a)	**80.** (a)
81. (d)	**82.** (b)	**83.** (a)	**84.** (c)	**85.** (d)	**86.** (b)	**87.** (d)	**88.** (a)	**89.** (a)	**90.** (c)
91. (a)	**92.** (d)	**93.** (b)	**94.** (b)	**95.** (b)	**96.** (c)	**97.** (c)	**98.** (a)	**99.** (d)	**100.** (b)

18

Forest Economics

1. The discount rate often used in capital budgeting that makes the net present value of all cash flows from a particular project equal to zero is
(a) Internal rate of return
(b) Gross domestic product
(c) Net present value
(d) Benefit-Cost ratio

2. A farmer performs the following functions by himself
(a) Farm manager
(b) Farm executive
(c) Farm entrepreneur
(d) All of the above

3. The ratio of current assets to current liabilities is
(a) Working ratio
(b) Current ratio
(c) Net capital ratio
(d) None of the above

4. In which place, the AGMARK is situated?
(a) Mumbai
(b) Hyderabad
(c) New Delhi
(d) Nagpur

5. The following markets are the biggest markets from area wise and coverage point of view
(a) Terminal markets
(b) Capital markets
(c) International markets
(d) None of the above

6. The right to keep possession of property belonging to another person until a debt owed by that person is discharged is
(a) Lien
(b) Mortage
(c) Liberties
(d) Trust

7. If the services of some resources such as labour is presently not used cannot be stored are called as
(a) Fixed resources
(b) Variable resources
(c) Flow resources
(d) Stock resources

8. The selection criteria for the mutually exclusive projects is
(a) Internal rate of return
(b) Gross domestic product
(c) Benefit-Cost ratio
(d) Net present value

9. Which of the following statement is true
(a) If total output is maximum, MPP is negative in productive function
(b) The relationship between the input and output can be described in cost function
(c) Both (a) and (b)
(d) None of the above

10. While economically analysing the forest projects we should consider
(a) Market price

(b) Consumer price
(c) Shadow price
(d) Surplus price

11. Inferior goods are known as
(a) Griffin's paradox
(b) Dimond water paradox
(c) Economic goods
(d) Producer goods

12. Agencies set up for the marketing and trade of NWFP's is/are
(a) Tribal co-operatives
(b) MFP Trade and Development Federation
(c) Large Multipurpose societies
(d) All of the above

13. Which of the following is the characteristic of production function?
(a) It tells the relationship between input and output
(b) It is for a given period of time
(c) Describe the ways which are technically possible
(d) All of the above

14. Capital is a
(a) Natural resource
(b) Fixed resource
(c) Cultural resource
(d) Human resource

15. Which of the following is true regarding the Marginal Utility (MU) and Total Utility (TU) curve?
(a) If TU increases at diminishing marginal rate than MU falls
(b) If TU becomes maximum and constant, MU becomes zero
(c) If TU declines, MU becomes negative
(d) All of the above

16. The law of Diminishing Marginal Utility was propounded by
(a) Marshall
(b) Ernest Engel
(c) Ulmer
(d) None of the above

17. The different combinations of the two goods consumers can buy is depicted by
(a) Indifference curve
(b) Law of Demand
(c) Price line
(d) None of the above

18. The demand for the raw material at producer level is
(a) Composite demand
(b) Derived demand
(c) Cross demand
(d) Income demand

19. Luxuries usually have
(a) High income elasticity
(b) Low income elasticity
(c) High demand
(d) Low demand

20. The word "marcatus" meaning market is derived from
(a) Germen word (c) French word
(b) Latin word (d) Arabic word

21. Which of the following has kinky demand curve?
(a) Monopoly market
(b) Duopoly market
(c) Oligopoly market
(d) Monopolistic market

22. Engel's curve is
(a) Demand – Supply curve
(b) Income – Consumption curve
(c) Prize – Demand curve
(d) Indifference curve

23. In case of Inferior goods, the income elasticity of demand is
(a) Positive (c) Negative
(b) Infinity (d) Zero

24. Substitutes generally have
(a) Positive cross elasticity of demand
(b) Negative cross elasticity of demand
(c) Negative price elasticity of demand
(d) Positive price elasticity of demand

25. Movement along the demand curve leads to
(a) Extension & Contraction in demand
(b) Increase & Decrease in demand
(c) Both (a) and (b)
(d) None of the above

26. Which of the following shows different combination of two goods which yield equal level of satisfaction?
(a) Marginal Utility Curve
(b) Price Line
(c) Indifference curve
(d) All of the above

27. Minimum Support Price (MSP) is declared by
(a) CACP
(b) TRIFED
(c) AGMARK
(d) DMI

28. Transportation function of marketing adds
(a) Time utility
(b) Place utility
(c) Possession utility
(d) Form utility

29. Which of the following is/are imperfect market
(a) Monopoly
(b) Oligopoly
(c) Monopolistic
(d) All of the above

30. Minimum Support Price (MSP) is declared
(a) Before harvesting of crop
(b) After harvesting of crop
(c) Before sowing of crop
(d) At any stage of crop life

31. Nature of economics is
(a) Positive (c) Negative
(b) Normative (d) Ecessive

32. AGMARK is involved in
(a) Marketing
(b) Quality control
(c) Packaging
(d) All of the above

33. The conditions for the perfect market are
(a) Homogenous product
(b) Sellers and buyers have full knowledge
(c) Large number of buyers and sellers
(d) All of the above

34. Isoquant is also known as
(a) Product Indifference curve
(b) Iso-resource line
(c) Price line
(d) None of the above

35. The cost that varies with the level of production is
(a) Fixed cost
(b) Variable cost
(c) Relative cost
(d) Marginal cost

36. The characteristic feature of iso-quant curve is
(a) Sloping downward
(b) Sloping upward
(c) Concave to origin
(d) Convex to origin

37. Fixed cost is also known as
(a) Sunk cost
(b) Supplementary cost
(c) Both (a) and (b)
(d) None of the above

38. Pesticides, Fertilizers, Seeds etc. are
(a) Dis-contentious production function
(b) Short run production function
(c) Long run production function
(d) Contentious production function

39. The minimum price buyers are required to pay for a good is called
(a) Price ceiling
(b) Price floor
(c) Price control
(d) Price index

40. Utility is a
(a) Desire
(b) Want
(c) Satisfaction
(d) Demand

41. Rent is a
(a) Return
(b) Investment
(c) Profit
(d) None of the above

42. Economics is
(a) Normative
(b) Positive
(c) Both (a) and (b)
(d) None of the above

43. The type of relationships that are associated with marketable and marketed surplus is
(a) Marketed surplus = Marketable surplus
(b) Marketed surplus $<$ Marketable surplus
(c) Marketed surplus $>$ Marketable surplus
(d) All of the above

44. Procurement Price (PP) is declared
(a) Before harvesting of crop
(b) After harvesting of crop
(c) Before sowing of crop
(d) At any stage of crop life

45. Father of Economics is
(a) Adam Smith
(b) Malthus
(c) A Walker
(d) RA Fisher

46. Farm planning means
(a) Farm budgeting
(b) Cropping pattern
(c) Both (a) and (b)
(d) None of the above

47. If the rate of product substitution is decreasing, production possibility curve is
(a) Concave to origin
(b) Straight
(c) Inverted
(d) Convex to origin

48. Which of the following agroforestry project is considered as worthy?
(a) $BCR = 0$
(b) $BCR < 1$
(c) $BCR > 1$
(d) All of the above

49. A business organisation that is owned and operated by a group of individuals for their mutual benefits is
(a) Cooperative (c) Joint venture
(b) Partnership (d) Both (a) and (c)

50. The value associated with benefits received by retaining the option of using a resource in future is
(a) Bequest value
(b) Existence value
(c) Option value
(d) Total economic value

51. Eco-mark was set up in
(a)1990 (c) 1991
(b) 1982 (d) 1857

52. Farm management is a
(a) Macro approach
(b) Micro approach
(c) Both (a) and (b)
(d) None of the above

53. The new name of ISI (Indian Standards Institution) is
(a) ISO (International Organisation of Standardisation)
(b) ISO (Indian Organisation of Standardisation)
(c) BSI (Bureau of Indian Standards)
(d) BIS (Burcau of Indian Standards)

54. Which of the following is responsible for the marketing of Forest Products in tribal societies?
(a) DMI
(b) TRIFED
(c) CACP
(d) All of these

55. The Theory of inflation was given by
(a) AP Lerner (c) A Walker
(b) Hicks-Hansen (d) Adam Smith

56. VAT is a
(a) Direct tax (c) Interest
(b) Indirect tax (d) Profit

57. The loan offered for the time period of 1 to 5 years is
(a) Long term loan
(b) Medium term loan
(c) Short term loan
(d) Crop loan

58. The factor-product relationship is best explained by using
(a) Law of Equi-marginal returns
(b) Law of Demand
(c) Angles' curve
(d) Law of Diminishing returns

59. The headquarters of WTO is situated at
(a) Geneva, Switzerland
(b) Bogor, Indonesia
(c) Rome, Italy
(d) None of the above

60. If the marginal utility is equal to zero than the total utility will be
(a) Minimum
(b) Zero
(c) Maximum
(d) None of the above

61. Factor-Factor relationship deals with
(a) Resource combination
(b) Input-Output combination
(c) Both (a) and (b)
(d) None of the above

62. Product-Product relationship best explained by
(a) Law of Equi-marginal returns
(b) Law of Demand
(c) Angles' curve
(d) Law of Diminishing returns

63. Factor-Factor relationship depicts
(a) Input-Output relationship
(b) Profit-Loss relationship
(c) Least cost combination
(d) None of the above

64. The market values of all goods and services produced in one year by labour and property supplied by the residents of a country
(a) Net National Product
(b) Gross Domestic Product
(c) Gross National Product
(d) Disposable Income

65. Price spread is
(a) Price paid by the producer-Price received by consumer
(b) Marketing costs
(c) Marketing margins
(d) Marketing cost + Marketing margin

66. Inflation is mostly beneficial to which of the following?
(a) Business class (c) Creditors
(b) Debtors (d) All of these

67. Price rigidity is a feature of
(a) Perfect competition
(b) Monopolistic competition
(c) Oligopoly
(d) Monopoly

68. Line passing through the least cost points in isoquant map is called
(a) Isocline
(b) Ridge line
(c) Expansion path
(d) Isoquant curve

69. A decline in the price of commodity attracts the consumers to purchase more in comparison to other commodities which are relatively high priced
(a) Substitution effect
(b) Cost effect
(c) Income effect
(d) Price effect

70. The line connecting loci of points at which the combination of goods gives the same amount of utility is
(a) Law of Marginal Utility
(b) Isobars
(c) Indifference curve
(d) Iso costs

71. Internal rate of return is a
(a) Capital Budgeting technique
(b) Cash Flow Statement
(c) Balance Sheet
(d) Book-Keeping

72. At a point where a straight line from the origin is tangent to the TC curve, AC is
(a) Minimum
(b) Equal to MC
(c) Both (a) and (b)
(d) None of the above

73. The equilibrium of the monopoly firm is generally determined on that part of demand curve, where price elasticity of demand is
(a) $= 1$ (b) < 1 (c) > 1 (d) $= 0$

74. The cost that has been incurred and cannot be altered or recovered is called
(a) Incremental cost
(b) Historical cost
(c) Social cost
(d) Sunk cost

75. The method of calculating the present value of the future returns is termed as
(a) Compounding (c) Discounting
(b) Budgeting (d) Interpolation

76. Which of the following is not a method to calculate National Income?
(a) Product method
(b) Income method
(c) Expenditure method
(d) Cost method

77. The same demand at higher price is known as
(a) Extension in demand
(b) Contradiction in demand
(c) Increase in demand
(d) Decrease in demand

78. A consumer reaches at maximum satisfaction point for a commodity when marginal utility is
(a) Positive (c) Negative
(b) Maximum (d) Zero

79. Market consisting of single buyer is known as
(a) Monopoly
(b) Duopoly
(c) Oligopoly
(d) All of the above

80. The slope at any point on production possibility curve is known as
(a) MRPS (c) MRTS
(b) MR (d) MC

81. The rate of depreciation remains same in
(a) Sum of years digit method
(b) Diminishing balance method
(c) Straight line method
(d) Reducing fraction method

82. The value of next best alternative use of a resource is known as
(a) Social cost
(b) Opportunity cost
(c) Marginal cost
(d) Resource cost

83. In case of joint products production possibility curve is
(a) Concave to origin
(b) Convex to origin
(c) 'L' shaped
(d) Inverted 'L' shaped

84. Which of the following is not the exception of law of demand
(a) Articles of distinction
(b) Giffen good
(c) Ignorance
(d) Normal goods

85. When the demand curve is elastic, MR is
(a) One (c) Zero
(b) Positive (d) Negative

86. The demand of money for speculative motive depend upon current level of
(a) Income
(b) Rate of interest
(c) Price
(d) Wages

87. The function of management that involves establishing an international structure of roles for people to fulfil in an organisation is called
(a) Planning
(b) Leading
(c) Organising
(d) Controlling

88. A break-even point is the point at which
(a) There is no profit, no loss
(b) Contribution is equal to total fixed cost
(c) Total fixed cost equal to total revenue
(d) All of the above

89. As consumption increases, marginal utility of the consumer will
(a) Remain constant
(b) Decrease
(c) Increase
(d) None of these

90. If the demand is inelastic and price increases
(a) Total revenue will decrease
(b) Total revenue will increase
(c) Total revenue will remain constant
(d) Total revenue will become zero

91. The U-shape of long run average cost curve reflects the
(a) Law of variable proportion
(b) Technological changes
(c) Law of returns to scale
(d) Market changes

92. The demand for the products or factors that are not directly consumed but go into production of final product is
(a) Derived demand
(b) Price demand
(c) Income demand
(d) Outcome demand

93. In short run, marginal cost depends upon value of
(a) Average cost
(b) Total variable cost
(c) Total fixed cost
(d) Average variable cost

94. The concept of elasticity of demand was proposed by
(a) Alferd Marshell
(b) Angel
(c) Mayer
(d) Adam Smith

95. If the demand curve is a horizontal line parallel to x-axis than it is called
(a) Perfectly inelastic
(b) Relative elastic
(c) Perfectly elastic
(d) Unitary elastic

96. RBI was established in
(a) 1945 (c) 1935
(b) 1845 (d) 1970

97. Which of the following is not a farming system?
(a) Co-operative farming
(b) Dry land farming
(c) State farming
(d) None of the above

98. Human wants are
(a) Limited (c) Increasing
(b) Unlimited (d) Decreasing

99. Principle of taxation was propounded by
(a) Hicks
(b) Alferd Marshell
(c) Bohem
(d) Adam Smith

100. WTO replaced the GATT in .... round of multilateral trade organisation
(a) Fifth (c) Eight
(b) Sixth (d) Tenth

Answer Keys

1. (a)	**2.** (d)	**3.** (b)	**4.** (d)	**5.** (c)	**6.** (a)	**7.** (c)	**8.** (d)	**9.** (a)	**10.** (c)
11. (a)	**12.** (d)	**13.** (d)	**14.** (c)	**15.** (d)	**16.** (a)	**17.** (c)	**18.** (b)	**19.** (a)	**20.** (b)
21. (c)	**22.** (b)	**23.** (a)	**24.** (d)	**25.** (a)	**26.** (c)	**27.** (a)	**28.** (b)	**29.** (d)	**30.** (c)
31. (a)	**32.** (b)	**33.** (d)	**34.** (a)	**35.** (b)	**36.** (b)	**37.** (a)	**38.** (d)	**39.** (b)	**40.** (c)
41. (a)	**42.** (c)	**43.** (d)	**44.** (a)	**45.** (a)	**46.** (b)	**47.** (d)	**48.** (c)	**49.** (a)	**50.** (c)
51. (c)	**52.** (b)	**53.** (d)	**54.** (b)	**55.** (a)	**56.** (b)	**57.** (b)	**58.** (d)	**59.** (a)	**60.** (c)
61. (a)	**62.** (a)	**63.** (c)	**64.** (c)	**65.** (d)	**66.** (b)	**67.** (c)	**68.** (c)	**69.** (a)	**70.** (c)
71. (a)	**72.** (c)	**73.** (c)	**74.** (d)	**75.** (c)	**76.** (d)	**77.** (a)	**78.** (d)	**79.** (a)	**80.** (a)
81. (c)	**82.** (b)	**83.** (d)	**84.** (d)	**85.** (b)	**86.** (b)	**87.** (c)	**88.** (d)	**89.** (b)	**90.** (b)
91. (a)	**92.** (d)	**93.** (b)	**94.** (a)	**95.** (c)	**96.** (c)	**97.** (d)	**98.** (b)	**99.** (d)	**100.** (c)

19

Forest Soil

1. Which of the following is completely insoluble in water?
(a) Humic acid
(b) Fulvic acid
(c) Humin
(d) All of the above

2. Which of the following is the most widespread and important root symbiont?
(a) Endomycorrhizae
(b) Ectomycorrhizae
(c) Both (a) and (b)
(d) None of the above

3. Which one of the following clay minerals has the highest CEC?
(a) Vermiculite
(b) Montmorillonite
(c) Kaolinite
(d) Illite

4. Lime is added to reclaim
(a) Alkali soil
(b) Acidic soil
(c) Saline soil
(d) All of the above

5. According to USDA, which of the following is the size of clay
(a) 0.05–0.02 mm
(b) 0.02–0.002 mm
(c) 0.10–0.05 mm
(d) < 0.002 mm

6. The rocks formed by cooling and consolidation of molten magma is called
(a) Igneous (c) Secondary
(b) Sedimentary (d) Metamorphic

7. Lime requirement procedure is given by
(a) Olsen
(b) Walkly and Black
(c) Schoonover
(d) Arnon

8. The acidic igneous rock contains silica
(a) < 65% (c) > 65%
(b) 60–65% (d) < 40%

9. At saturation, soil water potential is
(a) -15 bars (c) -31 bars
(b) -1 bars (d) 0 bars

10. Denitrification occurs only, if
(a) Ammonium is present
(b) Ammonium nitrate is present
(c) Nitrate is present
(d) None of the above

11. Highest porosity is found in
(a) Clay (c) Sand
(b) Gravel (d) Silt

12. Pyrites and Gypsum are the main source of inorganic
(a) Phosphorus (c) Zinc
(b) Sulphur (d) Potash

13. Which of the following is known as solon chalk or white alkali?
(a) Acidic soil
(b) Alkaline soil

(c) Saline soil
(d) Saline- Alkaline soil

14. Which of the following is widely used as bio fertilizer in rice crop?
(a) *Rhizobium*
(b) *Actinomycetes*
(c) *Clostridium*
(d) *Azolla*

15. Which of the following gene is responsible for N- fixation?
(a) Nif gene (c) Sym gene
(b) Nod gene (d) None of these

16. The total pore space is highest in
(a) Sandy soil (c) Loamy soil
(b) Sandy loam soil(d) Clayey soil

17. Soil formed by the deposition of gravity is
(a) Alluvial (c) Colluvial
(b) Aeolian (d) Loess

18. Humus mixed below with the mineral soil is
(a) Mull humus
(b) Sour humus
(c) Mor humus
(d) Raw humus

19. Which of the following is not a sedimentary rock stone?
(a) Sandstone (c) Shale
(b) Dolamite (d) Granite

20. Molasses are applied for the reclamation of
(a) Saline-Alkali soil
(b) Alkali soil
(c) Acidic soil
(d) Waterlogged soil

21. Central Soil Salinity Research Institute is located at
(a) Bangalore (c) West Bengal
(b) Jabalpur (d) Karnal

22. In wind erosion the floating of soil particles of less than 0.1 mm in size is
(a) Suspension
(b) Saltation
(c) Surface creep
(d) All of the above

23. In Sandy loam soil, the percentage of sand is
(a) 15–20%
(b) 80–85%
(c) 50–80%
(d) > 85%

24. Erosion caused by the excessive grazing, deforestation, etc is known as
(a) Wave erosion
(b) Anthropogenic erosion
(c) Saltation
(d) None of the above

25. Wilting coefficient is lowest in
(a) Clayey soil
(b) Loamy soil
(c) Clayey loam soil
(d) Sandy soil

26. Which of the cycle is most important for plant growth?
(a) Sulphur cycle
(b) Carbon cycle
(c) Nitrogen cycle
(d) All of the above

27. Which is the most productive cycle?
(a) Sulphur cycle
(b) Phosphorus cycle
(c) Carbon cycle
(d) Nitrogen cycle

28. In Nitrogen cycle, soil nitrates are transformed into free nitrogen by
(a) Nitrifying bacteria
(b) Ammonifying bacteria

(c) Denitrifying bacteria
(d) Both (a) and (c)

29. Which of the following is 1:1 type of minerals?
(a) Kaolinite
(b) Montmorillonite
(c) Vermiculate
(d) Illite

30. Soil density is measured using
(a) Peizometer
(b) Potometer
(c) Pyranometer
(d) Pycnometer

31. The low mechanical strength soil developed under continuous flooding leads to
(a) Surface crusting
(b) Shallow soils
(c) Fluffy soils
(d) Waterlogged and marshy soils

32. Which of the following are acidic in nature?
(a) Genesis
(b) Granite
(c) Sandstone
(d) All of the above

33. The vulnerability or susceptibility of the soil to get eroded is termed as
(a) Erodibility
(b) Erosivity
(c) Erosion
(d) All of the above

34. The universal soil loss equation is
(a) A = R,K,L,S
(b) A = R,K,L,S,C,P
(c) A = K,S,R,P
(d) All of the above

35. The removal of the very fine particles carried off during suspension is termed as
(a) Efflation
(b) Extrusion
(c) Efflexion
(d) All of the above

36. Which of the following is a sub-surface recharge technique?
(a) Flooding
(b) Stream augmentation
(c) Dug well
(d) Percolation tank

37. The natural geo-hydrological unit having single outlet is
(a) Dam
(b) Catchment area
(c) Watershed
(d) All of the above

38. The drainage area of a sub- catchment is
(a) 500000- 100000
(b) 5000-1000
(c) More than 500000
(d) 100000- 5000

39. The loss of plant nutrients by erosion is
(a) Leaching
(b) Seepage
(c) Infiltration
(d) Drainage

40. The vertical movement of water inside the soil is
(a) Seepage
(b) Percolation
(c) Infiltration
(d) All of the above

41. Gully erosion is the advance stage of
(a) Sheet erosion
(b) Splash erosion
(c) Suspension erosion
(d) Rill erosion

42. In a simplified version of a typical forest soil profile, the D horizon is
(a) Parent material
(b) Eluviation
(c) Leaf litter
(d) Illuviation

43. Under pH 5.5 which are the toxic elements
(a) Copper and Manganese
(b) Zinc and Iron
(c) Lead and Zinc
(d) Aluminium and Iron

44. Which of the mineral soils have a high content of clay particles?
(a) Vertisols (c) Histosols
(b) Oxisols (d) Entisols

45. Which soil suitable for bund construction?
(a) Light soil
(b) Loamy soil
(c) Red soil
(d) All of the above

46. The ideal soil structure helps in
(a) Better movement of air and water
(b) Better growth of soil microorganism
(c) Resist erosion
(d) All of the above

47. The soils rich in organic matter are
(a) Mollisols (c) Alfisols
(b) Histosols (d) Spodosols

48. The organic content of the peat soil is
(a) 10-30% (c) 50-90%
(b) 5-20% (d) 20-40%

49. Suspension is related to
(a) Geological erosion
(b) Water erosion
(c) Splash erosion
(d) Wind erosion

50. Indian Institute of Soil Science is located at
(a) Karnal (c) Bhopal
(b) Nagpur (d) New Delhi

51. The unit of total water content of soil is termed as
(a) Holard
(b) Chraserd
(c) Ecard
(d) None of the above

52. In Munsell colour chart "Hue" denotes
(a) Dominant spectrum
(b) Lightness or Darkness
(c) Intensity
(d) None of the above

53. Available water is held at
(a) Saturation at wilting point
(b) Field capacity to hygroscopic coefficient
(c) Field capacity to Wilting coefficient
(d) Field capacity

54. Mica is a ...... type of clay mineral
(a) 1:1 (c) 1:3
(b) 2:1 (d) 1:2

55. Base saturation point in lateritic soil is
(a) >40% (c) <40%
(b) >30% (d) <30%

56. Ammonia volatilization is purely a
(a) Physical process
(b) Biological process
(c) Chemical process
(d) None of the above

57. Feldspar is a primary mineral that occurs pre-dominantly in
(a) Igneous rock
(b) Sedimentary rock
(c) Metamorphic rock
(d) None of the above

58. Bed rock is commonly not found in
(a) Volcanic soil (c) Forest soil
(b) Alluvial soil (d) Organic soil

59. The honey comb structure is found in
(a) Desert soils
(b) Forest soil
(c) Physiological dry soil
(d) Laterite soil

60. Black colour of the soil is due to
(a) Maghemite
(b) Titaniferous magnetite
(c) Iron and copper
(d) None of the above

61. If Organic matter is added to the soil its bulk density
(a) Increases
(b) Remains same
(c) Decreases
(d) None of these

62. The total number of soil orders found in India are
(a) 4 (b) 10 (c) 6 (d) 12

63. Olsen method is used for phosphorus estimation in
(a) Acidic soil
(b) Saline soil
(c) Neutral to alkali soil
(d) All of the above

64. Which instrument is used for estimation of potash (K)?
(a) Spectrophotometer
(b) Colorimeter
(c) Atomic absorption
(d) Flamephotometer

65. Agar-Agar is obtained from
(a) Fungus
(b) Blue green algae (BGA)
(c) Red sea algae (RSA)
(d) Virus

66. Organic matter contains........% of organic carbon
(a) 32 (b) 68 (c) 18 (d) 58

67. In CAN nitrogen is found in
(a) Nitrate form
(b) Ammonical form
(c) Both (a) and (b)
(d) None of the above

68. Soil is bio-chemically weathered part of
(a) Regolith
(b) Bed rock
(c) Clay minerals
(d) All of the above

69. Ammonia is transferred to nitrate in soil by
(a) Fungi (c) Algae
(b) Bacteria (d) Earthworm

70. The symbiotic association between plant root and fungi is known as
(a) Lichen
(b) Antagonism
(c) Mycorrhiza
(d) Both (b) and (c)

71. The specific surface area of colloidal particle is highest for
(a) Mica
(b) Chlorite
(c) Vermiculite
(d) Kaolinite

72. Humus is composed of
(a) Lignin
(b) Protein
(c) Both (a) and (b)
(d) None of the above

73. The C/N ratio of normal soil is
(a) 12:1 (c) 10:1
(b) 1:10 (d) 20:30

74. The device used for measuring percolation, leaching and evaporation losses from the colum of soil under controlled condition is termed as
(a) Hydrometer (c) Lysimeter
(b) Infiltrometer (d) Thermometer

75. Ideal pH for the growth of fungi in soil is
(a) 6.5 to 8.0 (c) 5.5 to 7.0
(b) 6.0 to 7.0 (d) 4.5 to 6.5

76. The organism which require optimum temperature between 20°C to 40°C for their growth called
(a) Psychrophiles
(b) Thermophiles
(c) Mesophiles
(d) All of the above

77. The term "Soil" is derived from
(a) French word
(b) German word
(c) Latin word
(d) Arabic word

78. The inherent capacity of soil to provide essential nutrients is termed as
(a) Soil fertility
(b) Soil productivity
(c) Soil production
(d) All of the above

79. Which of the following clay mineral is dominant in black soils?
(a) Vermiculite
(b) Montmorillonite
(c) Kaolinite
(d) Illite

80. The nitrogen content of urea is
(a) 33% (c) 46%
(b) 21% (d) 26%

81 Which of the following is a sulphur oxidising bacterium is
(a) *Thiobacillus*
(b) *Bacillus*
(c) *Clostridium*
(d) *Desulfotomaculum*

82. Which of the following mineral is most resistant to weathering?
(a) Feldspar (c) Calcite
(b) Muscovite (d) Quartz

83. The chlorophyll content of leaves can be measured using
(a) SPAD (c) Both (a) and (b)
(b) CNC (d) None of these

84. Which type of soil is most suitable for the cultivation of crops?
(a) Laminar soil (c) Platy soil
(b) Crumb soil (d) Prismatic soil

85. The "B- Horizon" is also known as the zone of
(a) Eluviation
(b) Mineralization
(c) Illuviation
(d) Leaching

86. The "C-Horizon" is known as
(a) Mineral soil (c) Parent rock
(b) Illuvial (d) Solum

87. The book entitled "Nature and Properties of Soil" is written by
(a) NC Brady
(b) Dilip kumar das
(c) Daji AJ
(d) All of the above

88. Bacteria in soil helps in
(a) Nitrogen fixation
(b) Sulphur fixation
(c) Both (a) and (b)
(d) None of the above

89. Which of the following is a fertilizer producing bacteria
(a) *Nostoc*
(b) *Rhizobium*
(c) *Aulosira*
(d) All of the above

90. Syntropism means
(a) Exchange of nutrients among species
(b) Exchange of nutrients between two species
(c) No exchange of nutrients
(d) None of the above

91. Assimilative denitrification is done by
(a) Plants
(b) Eukaryotes
(c) Both (a) and (b)
(d) None of the above

92. The maximum percent of Biuret content in urea should not exceed
(a) 10% (c) 1%
(b) 1.5% (d) 4%

93. Land Capability Classification (LCC) was developed by
(a) China
(b) USA
(c) Japan
(d) Germany

94. The total number of soil orders are
(a) 12 (c) 20
(b) 8 (d) 15

95. Soil is also called as "Early soil"?
(a) Lateritic soil
(b) Red soil
(c) Alluvial soil
(d) All of the above

96. Unit of Electrical conductivity is
(a) dS/m (c) sd/g
(b) Mg/s (d) dS/cm

97. Which of the following soils have highest pore space?
(a) Sandy loam soil
(b) Silt loam soil
(c) Clayey soil
(d) Clayey loam soil

98. The appearance of soil in vertical section with particular reference to the presence of layers that are differentiated is
(a) Soil texture
(b) Soil composition
(c) Soil structure
(d) Soil profile

99. The "O-Horizon" is generally absent in
(a) Virgin soils
(b) Arable soil
(c) Forest soil
(d) Both (a) and (b)

100. Which of the following is the zone of maximum leaching?
(a) A horizon
(b) O horizon
(c) B horizon
(d) C horizon

Answer Keys									
1. (c)	**2.** (a)	**3.** (a)	**4.** (b)	**5.** (d)	**6.** (a)	**7.** (c)	**8.** (b)	**9.** (d)	**10.** (c)
11. (a)	**12.** (b)	**13.** (c)	**14.** (d)	**15.** (a)	**16.** (d)	**17.** (c)	**18.** (a)	**19.** (d)	**20.** (b)
21. (d)	**22.** (a)	**23.** (c)	**24.** (b)	**25.** (d)	**26.** (c)	**27.** (d)	**28.** (c)	**29.** (a)	**30.** (d)
31. (c)	**32.** (d)	**33.** (a)	**34.** (b)	**35.** (a)	**36.** (c)	**37.** (c)	**38.** (d)	**39.** (a)	**40.** (b)
41. (d)	**42.** (a)	**43.** (d)	**44.** (a)	**45.** (d)	**46.** (d)	**47.** (b)	**48.** (c)	**49.** (d)	**50.** (c)
51. (d)	**52.** (a)	**53.** (c)	**54.** (b)	**55.** (a)	**56.** (c)	**57.** (a)	**58.** (b)	**59.** (d)	**60.** (b)
61. (c)	**62.** (d)	**63.** (c)	**64.** (d)	**65.** (c)	**66.** (d)	**67.** (c)	**68.** (a)	**69.** (b)	**70.** (c)
71. (c)	**72.** (c)	**73.** (a)	**74.** (c)	**75.** (a)	**76.** (c)	**77.** (c)	**78.** (a)	**79.** (b)	**80.** (c)
81. (a)	**82.** (d)	**83.** (a)	**84.** (b)	**85.** (c)	**86.** (d)	**87.** (a)	**88.** (c)	**89.** (d)	**90.** (b)
91. (a)	**92.** (b)	**93.** (c)	**94.** (a)	**95.** (b)	**96.** (a)	**97.** (c)	**98.** (d)	**99.** (b)	**100.** (a)

20

Statistics

1. The unit less measure of dispersion is
(a) Co-efficient of variation
(b) Mean
(c) Range
(d) Standard deviation

2. Coefficient of variation is
(a) Mean/SD × 100
(b) SD/Mean × 100
(c) Mean/SD
(d) SD/Mean×100

3. Test between two population variances is done by
(a) F-test (c) Z-test
(b) t-test (d) Chisquare test

4. When the fertility gradient of the field is in two directions, the most appropriate experimental design is
(a) CRD (c) RBD
(b) Split plot (d) LSD

5. The square of standard deviation is
(a) Coefficient of variance
(b) Standard deviation
(c) Variance
(d) Square of variance

6. The principle of use of greater homogeneity in groups of experimental units reduce the experimental error is
(a) Local control
(b) Replication
(c) Randomization
(d) All of the above

7. In local control, blocks are formed .... to the fertility gradient.
(a) Parallel
(b) Perpendicular
(c) Both (a) and (b)
(d) None of the above

8. Missing data can be calculated by using
(a) Field plot technique
(b) Missing plot technique
(c) ANOVA
(d) All of the above

9. Who is the father of field experiment?
(a) JW Leather (c) Bousingault
(b) Jenny (d) Von Liebig

10. Completely randomized designs are mostly used in
(a) Field experiment
(b) Pot experiments
(c) Green houses
(d) Both (b) and (c)

11. A latin square design possess
(a) Two-way classification
(b) One-way classification
(c) Three-way classification
(d) None of the above

12. Who is the father of statistics?
(a) AL Bowley (c) Pearson
(b) RA Fisher (d) None of these

13. Class interval is
(a) Sum of lower and upper limit
(b) Half the sum of upper and lower limit
(c) Half the difference between upper and lower limit
(d) Difference between the upper and lower limit

14. From ogive, we can find
(a) Median (c) Mode
(b) Mean (d) Symmetry

15. Continuous distribution data can be best represented through
(a) Ogive
(b) Frequency polygon
(c) Histogram
(d) Frequency curve

16. Histogram is used when the
(a) Class interval are equal or unequal
(b) Class interval are equal
(c) Cass interval are unequal
(d) None of the above

17. The correct relationship between AM, GM and HM is
(a) AM $\leq$ GM $\leq$ HM
(b) GM $\geq$ AM $\geq$ HM
(c) AM $\neq$ HM $\leq$ GM
(d) HM $\leq$ GM $\leq$ AM

18. Sum of deviations of observation from mean is
(a) Zero (c) Positive
(b) Negative (d) Infinity

19. Correlation of continuity in 2 $\times$ 2 contingency table should be used when expected frequency of any cell is
(a) Less than 10 (c) Less than 2
(b) Less than 5 (d) Less than 8

20. Minimum degree of freedom for error in ANOVA is
(a) 10 (b) 24 (c) 12 (d) 18

21. The purpose of randomisation in field experiment is to
(a) Reduce error
(b) Increase critical difference
(c) Decrease homogeneity
(d) Control variance

22. Which of the following is not a absolute measure of Dispersion?
(a) Coefficient of variation
(b) Range
(c) Quartile deviation
(d) Standard deviation

23. If observed frequency is equal to expected than the value of X^2 static is
(a) More than 1
(b) Less than 1
(c) Equal to 1
(d) None of the above

24. Which of the following test is used for testing the significance of several differences?
(a) T test (c) Z test
(b) F test (d) X^2 test

25. In case of Binominal Distribution
(a) Mean < Variance
(b) Variance < Mean
(c) Variance = Mean
(d) None of the above

26. In Skewed distribution
(a) Mean- Mode = 3 (Mean – Median)
(b) Median – 3 Mean = 2 Mean – Median
(c) 3 Median – 3 Mean = Mode – Mean
(d) Mode – 3 median = 2 Mean – Median

27. The value of first central moment is
(a) Skewness (c) Variance
(b) One (d) Zero

28. The correlation between the price and demand of a commodity is
(a) Positive (c) Both (a) and (b)
(b) Negative (d) None of these

29. The lack of homogeneity in the distribution is called
(a) Skewness (c) Kurtosis
(b) Variance (d) Dispersion

30. The regression coefficient lies between
(a) $-\infty$ to $+\infty$ (c) 0 to $+\infty$
(b) $-\infty$ to 0 (d) 0 to 1

31. The value of X^2 always lies between
(a) $-\infty$ to $+\infty$ (c) 0 to $+\infty$
(b) $-\infty$ to 0 (d) 0 to 1

32. When sample size is small and SD is unknown we use
(a) Z test (c) X^2 test
(b) F test (d) T test

33. The probability of an impossible event is
(a) One
(b) Infinity
(c) Zero
(d) All of the above

34. Indian Agriculture Statistical Institute is located at
(a) Kolkata (c) Bangalore
(b) Guwahati (d) New Delhi

35. The relationship between independent and dependent variables is estimated by
(a) Correlation
(b) Regression
(c) Both (a) and (b)
(d) None of the above

36. The chi square test for goodness of fit and independence of attributes comes under
(a) Non parametric test
(b) Parametric test
(c) Both (a) and (b)
(d) None of the above

37. Normal distribution is attributed to
(a) RA Fisher (c) Spearmen
(b) Demoivre (d) John Grant

38. When both the variables are not normally distributed, in such case we use
(a) Partial correlation
(b) Rank correlation
(c) Regression
(d) Multiple regressions

39. In case of Poisson distribution
(a) Mean < Variance
(b) Variance < Mean
(c) Variance = Mean
(d) None of the above

40. How to estimate the soil fertility of a field before laying out the experiment?
(a) Uniformity trial
(b) Experiment trial
(c) Coordinated trials
(d) None of the above

41. If there are 5 treatment with 4 replications each, the error degree of freedom for CRD will be
(a) 15 (b) 20 (c) 12 (d) 9

42. If there are 8 treatments with 3 replications to each, the error degree of freedom for RBD will be
(a) 16 (b) 10 (c) 26 (d) 14

43. The variation due to uncontrolled factors is referred as
(a) Treatment effects
(b) Standard error
(c) Experimental error
(d) All of the above

44. Concept of ANOVA was given by
(a) WS Gosset (c) RA Fisher
(b) F Galton (d) Spearmen

45. Mean deviation is the least when calculated from
(a) Median (c) Mode
(b) Mean (d) Both (a) and (c)

46. The hypothesis of committing type I error is known as
(a) Test of significance
(b) Level of significance
(c) Both (a) and (b)
(d) None of the above

47. Most appropriate design when all factors are not of equal importance in an experiment is
(a) Field plot design
(b) Strip plot design
(c) Average plot design
(d) Split plot design

48. The critical difference is calculated by
(a) CD = t SE(d) (c) CD= t - SE(d)
(b) CD = t/SE(d) (d) CD= F SE(d)

49. Geometric mean of three numbers 2, 4 and 8 is
(a) 4 (b) 6 (c) 5 (d) 3

50. The regression coefficient is independent of the change of
(a) Scale
(b) Origin
(c) Both (a) and (b)
(d) None of the above

51. Which of the following is not a basic principle of experimental design?
(a) Randomization
(b) Replication
(c) Local control
(d) Cofounding

52. Which of the following sampling technique is preferred when population units are numbered and arranged in order?
(a) Simple Random Sampling
(b) Stratified Sampling
(c) Systematic Sampling
(d) Sequential Sampling

53. When the population consists of heterogeneity, which sampling procedure is preferred?
(a) Stratified Random Sampling
(b) Simple Random Sampling
(c) Systematic Sampling
(d) Double Sampling

54. Sampling error can be reduced by
(a) Non probability sampling
(b) Increasing the population size
(c) Decreasing the population size
(d) Increasing the sample size

55. Mean is a measure of
(a) Dispersion
(b) Skewness
(c) Central tendency
(d) Regression

56. Which of the following represent median?
(a) First Quartile
(b) Second Quartile
(c) Third Quartile
(d) All of the above

57. Geometric mean is used when
(a) Data is positive and negative

(b) Data is in ratio or percentage
(c) Both (a) and (b)
(d) None of the above

58. Which measure of dispersion ensures highest degree of reliability?
(a) Range
(b) Mean deviation
(c) Standard deviation
(d) Quartile deviation

59. If all the values in a given set of data are same, variance is
(a) Zero (c) One
(b) Infinity (d) None of these

60. Local control is a device used to maintain
(a) Homogeneity among blocks
(b) Homogeneity within blocks
(c) Both (a) and (b)
(d) None of the above

61. In sample random sampling with replacement, the sampling unit is included in the sample
(a) Once (c) Twice
(b) Thrice (d) Four time

62. Factor responsible for deciding number of replications in a experiment is
(a) Desire of the experiment
(b) Minimum df for experiment error
(c) Both (a) and (b)
(d) None of the above

63. If Kurtosis of a distribution is less than 3, than it is called
(a) Leptokurtic curve
(b) Platykurtic curve
(c) Mesokurtic curve
(d) None of the above

64. Which of the following design is used in laboratory condition
(a) CRD (c) RBD
(b) LSD (d) Split plot

65. The minimum sample size for using X^2 test is
(a) 20 (b) 30 (c) 50 (d) 100

66. Degree of freedom is related to
(a) Number of observations in a set
(b) Hypothesis under set
(c) Number of independent observations in a set
(d) All of the above

67. The geometric mean of two regression coefficient b_{xy} and b_{yx} is equal to
(a) r (b) 1 (c) r^2 (d) 0

68. The correlation coefficient is independent of
(a) Change of origin
(b) Change of scale
(c) Both (a) and (b)
(d) None of the above

69. If mean is 35 and median is 38, mode will be
(a) 40 (b) 52 (c) 44 (d) 49

70. A statistical measure based on a sample is termed as
(a) Sample parameter
(b) Inference
(c) Population
(d) Statistic

71. Which design is followed in green house experiment?
(a) CRD
(b) Split plot design
(c) LSD
(d) RBD

72. The second central moment is always equal to
(a) Mean deviation
(b) Mean
(c) Zero
(d) Variance

73. Which of the following is the most suitable for calculating the average speed?

(a) AM (c) GM
(b) HM (d) Median

74. How many numbers of treatments is 1 than the two variables are

(a) 3 to 10 (c) 5 to 12
(b) 10 to 12 (d) 1 to 5

75. If the value of correlation coefficient is 1 than the two variables are

(a) Unrelated
(b) Partially related
(c) Highly related
(d) Perfectly related

76. The hypothesis of no difference is

(a) Null hypothesis
(b) Composite hypothesis
(c) Alternate hypothesis
(d) Perfectly related

77. If there are 8 levels of moisture regime and 8 replications each, the error degree of freedom for LSD is

(a) 40 (b) 36 (c) 42 (d) 56

78. Which of the following is an assumption underlying the use of the t – distribution?

(a) The variance of the population is known
(b) The samples are drawn from normally distributed population
(c) Both (a) and (b)
(d) None of the above

79. A coefficient of correlation is computed to be -0.95 means that

(a) The relationship between two variables is weak
(b) The relationship between two variables is strong and positive
(c) The relationship between two variables is strong and negative
(d) Correlation coefficient cannot have this value

80. When bxy is positive, then byx will be

(a) Positive (c) Negative
(b) Zero (d) One

81. The variation due to uncontrolled factors is known as

(a) Standard error
(b) Treatment error
(c) Experimental error
(d) None of the above

82. Two regression lines are parallel to each other if their slope is

(a) Different
(b) Same
(c) Negative
(d) None of the above

83. The mean of a distribution is 23, the median is 24, and the mode is 255 It is most likely that this distribution is

(a) Positively Skewed
(b) Symmetrical
(c) Asymptotic
(d) Negatively Skewed

84. If the standard deviation of a population is nine than the population variance is:

(a) 9 (b) 21 (c) 3 (d) 81

85. In statistics, a sample means

(a) Portion of the sample
(b) A portion of the population
(c) All the items under investigation
(d) None of the above

86. If we reject the null hypothesis, we might be making

(a) Type-I Error

(b) Type-II Error
(c) Unpredictable
(d) Both (a) and (b)

87. Range of multiple correlation co-efficient lies between
(a) -1 to 1
(b) $-\infty$ to $+\infty$
(c) 0 to 1
(d) Both (a) and (c)

88. The repetition of treatment under investigation is known as
(a) Randomisation
(b) Replication
(c) Local control
(d) Both (a) and (c)

89. The accuracy or a measurement signifies the closeness to
(a) True value
(b) Average value
(c) Both (a) and (b)
(d) None of the above

90. The measure of the peakness of a curve is known as
(a) Skewness
(b) Average mean
(c) Kurtosis
(d) Both (a) and (b)

91. When the experimental material is very less and the number of genotypes are very large the most suitable design is
(a) Split plot design
(b) Factorial design
(c) LSD
(d) Augment design

92. The Histogram is a method of
(a) Diagrammatic representation
(b) Graphical representation
(c) Both (a) and (b)
(d) None of the above

93. Indian Agriculture Statistical Research institute is located at
(a) New Delhi (c) Lucknow
(b) Agra (d) Bangalore

94. In chi-square test, theoretically all cell frequency should be at least
(a) 10 (c) 5
(b) 12 (d) 15

95. The statistical procedure for estimating weather the difference under study is significant or non-significant is
(a) Level of significance
(b) Null hypothesis
(c) Standard error
(d) None of the above

96. In chi-square test the goodness of fit is
(a) Non-parametric test
(b) Parametric test
(c) Both (a) and (b)
(d) None of the above

97. The mean of the distribution is 14 and the standard deviation is 5, than the coefficient of variation is
(a) 68%
(b) 35.7%
(c) 40%
(d) 37%

98. A researcher selects a sample out of a population with probability of 50% is
(a) Cluster sample
(b) Stratified sample
(c) Systematic sample
(d) Random sample

99. First quartile is also known as
(a) Lower quartile
(b) Upper quartile
(c) Median
(d) Both (a) and (c)

100. The range of probability lies from
(a) -1 to 1
(b) $-\infty$ to 0
(c) 0 to 1
(d) 0 to $+\infty$

Answer Keys									
1. (a)	**2.** (b)	**3.** (a)	**4.** (d)	**5.** (c)	**6.** (a)	**7.** (b)	**8.** (b)	**9.** (d)	**10.** (d)
11. (a)	**12.** (b)	**13.** (d)	**14.** (a)	**15.** (c)	**16.** (b)	**17.** (d)	**18.** (a)	**19.** (b)	**20.** (c)
21. (d)	**22.** (a)	**23.** (d)	**24.** (b)	**25.** (b)	**26.** (a)	**27.** (d)	**28.** (b)	**29.** (d)	**30.** (a)
31. (c)	**32.** (d)	**33.** (c)	**34.** (d)	**35.** (b)	**36.** (a)	**37.** (b)	**38.** (b)	**39.** (c)	**40.** (a)
41. (a)	**42.** (d)	**43.** (c)	**44.** (c)	**45.** (a)	**46.** (b)	**47.** (d)	**48.** (a)	**49.** (a)	**50.** (b)
51. (d)	**52.** (c)	**53.** (a)	**54.** (d)	**55.** (c)	**56.** (d)	**57.** (b)	**58.** (c)	**59.** (a)	**60.** (b)
61. (a)	**62.** (b)	**63.** (c)	**64.** (a)	**65.** (c)	**66.** (b)	**67.** (a)	**68.** (c)	**69.** (c)	**70.** (d)
71. (a)	**72.** (d)	**73.** (b)	**74.** (c)	**75.** (d)	**76.** (d)	**77.** (c)	**78.** (b)	**79.** (c)	**80.** (a)
81. (c)	**82.** (b)	**83.** (d)	**84.** (d)	**85.** (b)	**86.** (a)	**87.** (c)	**88.** (b)	**89.** (a)	**90.** (c)
91. (d)	**92.** (b)	**93.** (a)	**94.** (c)	**95.** (d)	**96.** (a)	**97.** (b)	**98.** (d)	**99.** (a)	**100.** (c)

21

General Agriculture

1. Khaira disease of rice is caused by .... deficiency.
(a) Mo (c) Fe
(b) Zn (d) K

2. Copper excess indicator plant is
(a) Sugarbeet (c) Barley
(b) Sugarcane (d) Paddy

3. DDT banned for agriculture purposes in India.
(a) 1974 (c) 1962
(b) 1971 (d) 1985

4. Pollen sterility in wheat is due to
(a) Calcium deficiency
(b) Zinc deficiency
(c) Boron deficiency
(d) None of the above

5. The crops that absorb nitrogen directly in the ammonical form is
(a) Paddy (c) Both (a) and (b)
(b) Potato (d) Wheat

6. In rice the ammonical fertilizers are applied in
(a) Reduced zone
(b) Upper zone
(c) Lower zone
(d) All of the above

7. Which of the following crops can tolerate the complete absence of oxygen in the soil?
(a) Wheat (c) Maize
(b) Bajra (d) Rice

8. Which of the following responsible for aerial N- fixation?
(a) *Sesbania aculeata*
(b) *Crotolaria juncea*
(c) *Sesbania rostrata*
(d) All of the above

9. SRI technique to increase the crop production used in
(a) Maize (c) Paddy
(b) Barley (d) Wheat

10. The pollen grains from flower of one plant falls on receptive stigma of another flower on same plant is called
(a) Autogamy (c) Caleistogamy
(b) Getinogamy (d) Chasmogamy

11. Protein revolution is attributed to
(a) MS Swaminathan
(b) Dr Arun Krishnan
(c) V Kurien
(d) None of the above

12. Which of the following is not a cash crop?
(a) Rubber (c) Jute
(b) Oilseeds (d) Sugarcane

13. The National Bamboo Mission was launched in 2006-07 is a
(a) Centrally sponsored scheme
(b) State sponsored scheme
(c) Centre and State sponsored scheme
(d) None of the above

14. Central Rice Research Institute located

(a) Kolkata (c) Chennai
(b) Cuttack (d) Bangalore

15. When more than one gene governs a same character it is termed as

(a) Polygenic inheritance
(b) Polypeptides
(c) Polymorphism
(d) Pleiotropy

16. The first country to use the term extension education

(a) India (c) USA
(b) Africa (d) China

17. Burgundy mixture is a

(a) Insecticide (c) Rodenticide
(b) Fungicide (d) Nematicide

18. The Physiological basis of the life is

(a) Protoplasm (c) Cell
(b) Cytoplasm (d) Both (b) and (c)

19. The Family of Kesar is

(a) Theaceae (c) Compositae
(b) Asteraceae (d) Irridaceae

20. Which country is the second largest producer of Vegetables?

(a) China (c) Brazil
(b) USA (d) India

21. "Rockery" is the main feature of

(a) English garden
(b) Japanese garden
(c) Mughal garden
(d) Persian gardens

22. Which of the following chemicals are used for preservation of food

(a) Salt
(b) KMS
(c) Agar-agar
(d) PMS

23. The word "Agriculture" is derived from

(a) Greek word (c) Latin word
(b) Arabic word (d) French word

24. The oil percentage in the groundnut is

(a) 45% (c) 60%
(b) 20% (d) 35%

25. The "Dapog method" of Rice cultivation was introduced by

(a) Philippines (c) Japanese
(b) Chinese (d) Germany

26 The first manmade cereal "Triticale" is a cross between

(a) Wheat × Bajra
(b) Bajra × Rye
(c) Wheat × Rice
(d) Wheat × Rye

27. The six rowed barley is

(a) *Hordeum distichon*
(b) *Hordeum irregulare*
(c) *Hordeum vulgare*
(d) All of the above

28. Cropping logging is done for

(a) Bajra (c) Wheat
(b) Sugarcane (d) Rubber

29. The origin of Tea is

(a) India (c) China
(b) Indonesia (d) West Asia

30. The edible part of the coconut is

(a) Mesocrap (c) Endocarp
(b) Exocarp (d) Endosperm

31. The development of fruit without fertilization is

(a) Parthenogenesis
(b) Parthenocarpy
(c) Apomixis
(d) Seedless

32. A cross between the inbreed line and an open pollinated variety is
(a) Top cross
(b) Double cross
(c) Back cross
(d) Threeway cross

33. The inflorescence of Rice is
(a) Spike (c) Panicle
(b) Ear (d) Spike

34 .... is popularly known as "Queen of Cereals"
(a) Wheat (c) Rice
(b) Bajra (d) Maize

35. Peg in groundnut is a stalk like structure (originating from the meristematic region at the base of ovary) is known as
(a) Gynophore
(b) Androphore
(c) Perianth
(d) Androgynophore

36. The first State Agriculture University
(a) PAU, Ludhiana
(b) TNAU, Tamil Nadu
(c) IGKV, Raipur
(d) GBPUAT, Pantnagar

37. Moisture content for safe storage of cereals maintained at
(a) 8 – 12% (c) 16 – 20%
(b) 12 – 16% (d) 20 – 25%

38. Diamond black moth is a serious pest of
(a) Rice (c) Wheat
(b) Maize (d) Cabbage

39. ....is the important pest of rice nursery
(a) Hopper (c) Jassid
(b) Thrips (d) Termites

40. The structure of chromosomes can be best observed during
(a) Metaphase (c) Anaphase
(b) Telophase (d) Prophase

41. The minimum thermometer is filled with
(a) Mercury (c) Water
(b) Alcohol (d) None of these

42. Which of the following is a flower and fruit setting hormone?
(a) IBA (c) NAA
(b) IAA (d) CCC

43. Cycocel is a.............. type of anti-transpirant
(a) Stomata closing
(b) Film forming
(c) Reflecting type
(d) Growth retardant

44. Green revolution in India was started by
(a) MS Swaminathan
(b) V Kurien
(c) NE Borlaug
(d) None of the above

45. The intercropping in an area is considered to be beneficial, when
(a) LER = 1 (c) LER < 1
(b) LER > 1 (d) LER > 10

46. According to ICAR, India is divided into......agroecological zones
(a) 10 (c) 8
(b) 15 (d) 21

47. Which of the following is commonly practiced in coffee gardens?
(a) Vegetative barriers
(b) Bench terracing
(c) Vertical mulching
(d) Contour cultivation

48. The line joining points of equal amount of rainfall is
(a) Isohyets (c) Isotherms
(b) Isobar (d) None of these

49. Which of the following is considered useful for choosing cropping pattern?
(a) Short range forecast
(b) Medium range forecast
(c) Long range forecast
(d) None of the above

50. The principal yellow pigment in maize is
(a) Carotene
(b) Zeaxanthin
(c) Anthocyanin
(d) Zeatin

51. Rice is a
(a) Short day plant
(b) Long day plant
(c) Day neutral plant
(d) All of the above

52. Inflorescence of rice is
(a) Aun (c) Spike
(b) Ear (d) Panicle

53. Common bread wheat is
(a) *Triticum durum*
(b) *Triticum aestivum*
(c) *Triticum monococcum*
(d) None of the above

54. Protein found in maize is
(a) Gluten
(b) Oryzein
(c) Zein
(d) All of the above

55. Which of the following has highest protein content?
(a) Sesame (c) Groundnut
(b) Linseed (d) Safflower

56. Ratooning is common in
(a) Sugarcane (c) Wheat
(b) Maize (d) Banana

57. Which of the following is not a part of primary tillage?
(a) Ploughing
(b) Rouging
(c) Harrowing
(d) Weeding

58. Milk is a rich source of
(a) Calcium
(b) Phosphorus
(c) Vitamin A
(d) All of the above

59. The Whiptail of cauliflower is due to the deficiency of
(a) Copper (c) Molybdenum
(b) Zinc (d) Manganese

60. The pungency in pepper is due to
(a) Oleoresin (c) Sinigrin
(b) Oxalic acid (d) Anthocyanin

61. The inflorescence of sugarcane is
(a) Panicle (c) Spike
(b) Ear (d) Arrow

62. Centre of origin of "Tea" is
(a) India (c) Burma
(b) China (d) Brazil

63. The important pest of rice nursery is
(a) Gall midge
(b) Thrips
(c) Spotted ball worm
(d) Ghundi bug

64. Flare square is a characteristic symptom of
(a) Pink ball worm
(b) Whitefly
(c) American boll worm
(d) Spotted ball worm

65 Which of the following is the specific pest of wheat nursery?
(a) Ghujhia weevil
(b) White grub
(c) Termite
(d) Shoot fly

66. Crop logging is common in
(a) Sugarbeet (c) Rice
(b) Maize (d) Sugarcane

67. National Rural Employment Guarantee act (NAREGA) was enacted in the year
(a) 2005 (c) 2006
(b) 2007 (d) 2000

68. Undifferentiated mass of cells is also known as
(a) Explant (c) Burr
(b) Callus (d) Cull

69. .... is known as suicidal bag
(a) Mitochondria
(b) Per-oximes
(c) Lysosomes
(d) Golgi apparatus

70. Glycolysis pathway is the first step of
(a) Photorespiration
(b) Transpiration
(c) Evaporation
(d) Respiration

71. Haploids can be produced through
(a) Anther culture
(b) Meristem culture
(c) Pollen culture
(d) Callus culture

72. Which of the following is a stress hormone
(a) Ethylene (c) Cycocel
(b) ABA (d) IAA

73. Certified seed is produced from
(a) Nucleus seed
(b) Foundation seed
(c) Breeder seed
(d) None of the above

74. At saturation, soil water potential is
(a) -15 bars (c) -31 bars
(b) -1 bars (d) 0 bars

75 The square of standard deviation is
(a) Coefficient of variance
(b) Standard deviation
(c) Variance
(d) Square of variance

76. Powdery mildew can be controlled by
(a) Copper fungicides
(b) Sulphur fungicides
(c) Zinc fungicides
(d) Mercury fungicides

77. Mica is a ...... type of clay mineral
(a) 2:1 (c) 1:1
(b) 1:2:1 (d) 1:2

78. Secondary growth is absent in
(a) Monocot stem
(b) Dicot root
(c) Dicot stem
(d) Monocot root

79. Which of the following is absent in plant cell?
(a) Starch
(b) Fructose
(c) Glucose
(d) None of the above

80. Commonly used surface sterilant in tissue culture is
(a) $KMnO_4$ (c) $CaCl_2$
(b) $HgCl_2$ (d) NaCl

81. Individual propagules originating from ortet are
(a) Ramet
(b) Clone
(c) Explant
(d) Both (a) and (b)

82. Operation flood is related to
(a) Green revolution
(b) Golden revolution
(c) Blue revolution
(d) White revolution

83. In case of Poisson distribution
(a) Mean < Variance
(b) Variance < Mean
(c) Variance = Mean
(d) None of the above

84. The edible part of apple is
(a) Thalamus (c) Endocarp
(b) Epicarp (d) Endosperm

85. The most common disease in nursery is
(a) Root rot
(b) Damping off
(c) Wilting
(d) All of the above

86. The humidity of air is measured using
(a) Hydrometer (c) Barometer
(b) Hygrometer (d) Lycimeter

87. Normal lapse rate is
(a) 6.0°C per kilometre
(b) 6.5°C per kilometre
(c) 7.0°C per kilometre
(d) 5.5°C per kilometre

88. Recently banned pesticide is
(a) Endosulfan
(b) DDT
(c) CAN
(d) All of the above

89. Vitamin B_2 is also known as
(a) Thiamine (c) Retinol
(b) Niacin (d) Riboflavin

90. The concept of zero tillage was first introduced in
(a) USA (c) China
(b) Japan (d) India

91. Which of the following is the most abundant greenhouse gas?
(a) N_2O (c) CFC
(b) CO_2 (d) H_2O

92. Which of the following is the most abundant soil type of India?
(a) Black cotton soil
(b) Alluvial soil
(c) Sandy loam soil
(d) Red soil

93. Family of sesame is
(a) Fabaceae (c) Graminae
(b) Tiliaceae (d) Pedaliaceae

94. "Indian Journal of Agricultural Sciences" is published by
(a) ICAR (c) IARI
(b) MOEF&CC (d) None of these

95. Which of the following is sold under the name of "Sproutstop"?
(a) Cloemequat
(b) BAP
(c) Mallic hydrazide
(d) ABA

96. India is placed at .... position for the production of wheat in world agriculture
(a) First (c) Second
(b) Third (d) Fourth

97. Number of agroclimatic zones in India are
(a) 8 (c) 10
(b) 15 (d) 20

98. Amarpali is a variety of
(a) Mango (c) Guava
(b) Pomegranate (d) Grapes

99. Pusa jwala is a variety of
(a) Capsicum (c) Brinjal
(b) Okra (d) Chilli

100. Triple gene dwarf varieties of wheat is/are
(a) Heera
(b) Moti
(c) Arjun
(d) All of the above

Answer Keys

1. (b)	**2.** (a)	**3.** (d)	**4.** (c)	**5.** (c)	**6.** (a)	**7.** (d)	**8.** (c)	**9.** (c)	**10.** (b)
11. (d)	**12.** (a)	**13.** (a)	**14.** (b)	**15.** (d)	**16.** (c)	**17.** (b)	**18.** (a)	**19.** (d)	**20.** (d)
21. (a)	**22.** (b)	**23.** (c)	**24.** (a)	**25.** (a)	**26.** (d)	**27.** (c)	**28.** (b)	**29.** (c)	**30.** (d)
31. (b)	**32.** (a)	**33.** (c)	**34.** (d)	**35.** (a)	**36.** (d)	**37.** (b)	**38.** (d)	**39.** (b)	**40.** (a)
41. (b)	**42.** (c)	**43.** (d)	**44.** (a)	**45.** (b)	**46.** (c)	**47.** (c)	**48.** (a)	**49.** (c)	**50.** (b)
51. (b)	**52.** (d)	**53.** (b)	**54.** (c)	**55.** (d)	**56.** (a)	**57.** (b)	**58.** (d)	**59.** (c)	**60.** (a)
61. (d)	**62.** (b)	**63.** (b)	**64.** (d)	**65.** (a)	**66.** (d)	**67.** (a)	**68.** (b)	**69.** (c)	**70.** (d)
71. (a)	**72.** (b)	**73.** (b)	**74.** (d)	**75.** (c)	**76.** (b)	**77.** (a)	**78.** (d)	**79.** (d)	**80.** (b)
81. (a)	**82.** (d)	**83.** (c)	**84.** (a)	**85.** (d)	**86.** (b)	**87.** (b)	**88.** (a)	**89.** (d)	**90.** (a)
91. (b)	**92.** (b)	**93.** (d)	**94.** (a)	**95.** (c)	**96.** (c)	**97.** (b)	**98.** (a)	**99.** (d)	**100.** (d)

Previous Year Original Solved JRF Question Paper (2019)

1. The forest conservation act was promulgated in the year
(a) 1982 (c) 1972
(b) 1988 (d) 1980

2. Wood is biodegradable because it can be reduced to its component such as
(a) Fibres of cells
(b) Cellulose and carbohydrates
(c) Simple sugars and lignin elements
(d) Cellulose and Hemi-cellulose

3. A mineral deficiency is likely to affect older leaves more than younger leaves if
(a) The mineral is a macronutrient
(b) The mineral is a micronutrient
(c) The mineral is mobile
(d) The mineral is required for chlorophyll synthesis

4. The relation that yields highest net return in invested capital is called
(a) Financial rotation
(b) Rotation of highest income
(c) Technical rotation
(d) Rotation of maximum volume production

5. Mechanical properties of wood can be known by testing for
(a) Chemical strength & Physical strength
(b) Tension and Compression
(c) Bending and Tatoing
(d) Boring and Splitting

6. Which among the following mentioned bamboo flowers annually?
(a) *Dendrocalamus strictus*
(b) *Arundinaria wightiana*
(c) *Bamboosa vulgaris*
(d) *Dendrocalamus stocksii*

7. Identify the correct statement from the ones mentioned below-
(a) CAI curve is always higher than MAI curve
(b) MAI curve is always higher than CAI curve
(c) CAI curve and MAI curve meets at two points in life time of a crop
(d) The rotation of maximum volume production is when CAI curve is at maximum

8. Tree having a height of $3/4^{th}$ of the tallest tree in forest are referred to as
(a) Pre-dominants
(b) Co-dominants
(c) Dominated trees
(d) Sub-dominated trees

9. The timber which can withstand rapid seasoning are termed as
(a) Highly refractory wood
(b) Non-refractory wood
(c) Durable wood
(d) Non-durable wood

10. Annual yield according to Von mantle formula is
(a) GS/2r
(b) 2 GS/r
(c) GS/3r
(d) GS/r

11. When veneer sides are joined in sequence without flipping the pattern it is called
(a) Book matching
(b) Slip matching
(c) Pleasing matching
(d) Random matching

12. Seasoning defect in wood are
(a) Cross grain, stains and splits
(b) Warping, cracking and shrinking
(c) Loosing lustre, loosing colour and loosing shape
(d) Holes, decay, pitch pockets

13. Quarter girth formula is used to calculate volume of logs. The formula is
(a) $(g/4) \times l$
(b) $(g/4)^2 \times l$
(c) $(dbh/4) \times l$
(d) $(dbh/4) \times l$

14. The theory of plant succession leading to climax was proposed by
(a) Dauson
(b) Liocourt
(c) Odum
(d) Clement

15. The classification of seed into micro, meso and macro biotic was put forth by
(a) Robert
(b) Ewart
(c) Stuart
(d) Pammenter

16. Which among the following is the most natural system in tropical forestry?
(a) Shelterwood system
(b) Clear felling system
(c) Selection system
(d) Uniform system

17. For the manufacture of paper, the pulp should contain more
(a) Lignin
(b) Hemi-cellulose
(c) Pectin
(d) Cellulose

18. Which among the following tree exhibits whorl type of branching?
(a) *Populus nigra*
(b) *Terminalia arjuna*
(c) *Mangifera indica*
(d) *Bombax ceiba*

19. Who among the following can be described as the first entomologist of India?
(a) EP Stebbing
(b) AD Immse
(c) EC Beeson
(d) HM Lefroy

20. While grading timber which of the following defects are taken into consideration?
(a) Log length, Log taper, Log diameter
(b) Check, Decay, Flutes, Knots
(c) Grain, Quality of sapwood, Taper
(d) Straightness, Surface cracks, curvature

21. "Rasaunt" is extracted from
(a) *Picrorhiza kurroa*
(b) *Datura stramonium*
(c) *Cannabis sativa*
(d) *Berberis aristata*

22. Who among the following can be classified as frost hardy species
(a) *Dalbergia sissoo*
(b) *Adina cordifolia*
(c) *Tectona grandis*
(d) *Boswellia serrata*

23. Removal of certain species of high value trees above certain size and certain species without regard to silvicultural treatment is called
(a) Selection felling
(b) Regeneration felling
(c) Improvement felling
(d) Selective felling

24. Quantitative traits exhibit continuous variation as they are controlled by
(a) Single gene
(b) Oligo genes
(c) Poly genes
(d) Special genes

25. Potential capacity of seed to germinate is known as
(a) Vigour of seed
(b) Viability of seed
(c) Dormancy of seed
(d) Germination energy of seed

26. The practise of cutting a stem of a tree in order to obtain flush of shoots, usually above the height to which the browsing animals can reach is
(a) Lopping (c) Pollarding
(b) Pruning (d) Hacking

27. Heating biomass under pressure anaerobically upto 500°C to obtain gaseous and liquid hydrocarbon is known as
(a) Wood distillation
(b) Wood gasification
(c) Pyrolysis
(d) Destructive distillation

28. The first policy of Independent India was initiated in
(a) 1952 (c) 1988
(b) 1980 (d) 2012

29. The rotation which coincides with the natural lease of life of a species at a given site is called
(a) Silvicultural rotation
(b) Technical rotation
(c) Physical rotation
(d) Biological rotation

30. Tree which is called 'Naked lady of the forest' is
(a) *Lagerstromia lanceolata*
(b) *Anogessus latifolia*
(c) *Terminalia arjuna*
(d) *Tectona grandis*

31. Glory lily (*Gloriosa superba*) is a medicinal plant and its chief medicinal properties are that it
(a) Yields aromatic oil which is used for skin care
(b) Yields colchicine, used for treatment of gout and polyploidy breeding
(c) Yields drug used for cardiovascular diseases
(d) Is the primary source of menthol

32. Genetic engineering is a technique in which
(a) DNA of two organisms is joined with each other
(b) DNA is analysed
(c) DNA and RNA is processed
(d) RNA is analysed

33. According to Champion and Seth classification of forests in India, which of the following is not a major group of forest?
(a) Alpine scrub
(b) Mangrove forest
(c) Tropical forest
(d) Montane sub tropical forest

34. Most accurate instrument for measuring atmospheric pressure is
(a) Aneroid barometer
(b) Barograph
(c) Kew pattern barometer
(d) Mercurial barometer

35. If 80 chromatids are present in beginning of meiosis, than chromosomes present in cell after completion of cell division will be
(a) 40 (b) 80 (c) 20 (d) 10

36. What is a particle board?
(a) A board or sheet constituted from fragments of wood or other lignocellulosic materials bounded with organic bounders with the help of heat and pressure
(b) A board having a core of light material faced on both sides, with relatively thin layer having high strength
(c) A board made up of strips of wood of various dimensions
(d) It is a chemically modified wood to make it less hygroscopic

37. "*Ginkgo biloba*" is known as "living fossil" of the world because it
(a) Is extinct
(b) Has fossilized
(c) Has survived the changes on earth for millions of years
(d) Can be used as a fuel

38. A plantation of genetically superior trees isolated to reduce cross pollination by other trees are managed intensively for seed production is called
(a) Seed stand (c) Plus stand
(b) Seed orchard (d) Elite stand

39. Scientific name of "Lion tailed macaque" is
(a) *Macca silensus*
(b) *Macca aratoides*
(c) *Macca assamensis*
(d) *Macca namatoetrina*

40. The vegetation of Southern Tropical Wet Evergreen Forest comprises –
(a) Dipterocarpus, Hopea, Mesua, Artocarpus, etc...
(b) Sal, Teak, Dendrocalamus, Syzygium, etc......
(c) Terminalia,Lagerstromia, Dalbergia etc.
(d) Diospyros, Bamboos, Acacia's, etc.

41. Tussar silk is obtained from
(a) *Antheria milletia*
(b) *Bombyx morii*
(c) *Lacifer lacca*
(d) *Vetiver zizanioides*

42. Red Sanders tree is
(a) *Pinus gerardiana*
(b) *Pterocarpus santalinus*
(c) *Santalum album*
(d) *Aquilaria agallocha*

43. Isabgol (*Plantago ovata*) is important for its
(a) Root leaves
(b) Bark and Roots
(c) Seed and Husk
(d) Leaves and twigs

44. The state in India with maximum mangrove cover (ISFR-2015) is
(a) Orissa
(b) Andhra Pradesh
(c) Gujarat
(d) West Bengal

45. About 50 percent of the dry weight of a plant is made up of
(a) Nitrogen
(b) Hydrogen
(c) Carbon
(d) Oxygen

46. The most important human activity leading to the extinction of wildlife is
(a) Diseases of human spread to wildlife
(b) Hunting of wildlife for valuable products
(c) Alteration and destruction of natural habitats
(d) Introduction of alien species in our ecosystem

47. Chapter IV B of Wildlife Protection Act, 1972 deals with
(a) Central Zoo Authority
(b) National Parks
(c) National Tiger Conservation Authority
(d) Community reserves

48. Selection purely based on phenotypic characters from a population is called as
(a) Pure line selection
(b) Tandem selection
(c) Mass selection
(d) Recurrent selection

49. Specific gravity of wood can be calculated by formula
(a) Weight of wood/Weight of water
(b) Weight of pieces of wood/Weight of volume of water displaced by same piece of wood
(c) Weight of wood in Kg/Volume of water in cubic inches
(d) Weight of water/Weight of wood

50. Onge-tribe group is mainly found in
(a) Orissa
(b) Tripura
(c) Meghalaya
(d) Andaman and Nicobar islands

51. The calorific value of wood is quantitatively expressed as the number of heat units obtained
(a) By complete combustion of a unit mass of wood
(b) To warm water by 10°C
(c) By burning of wood
(d) By combustion with hydrogen

52. A product released in the course of trans-esterification of crude biodiesel is
(a) Methyl amine
(b) Ethylene
(c) Glycerin
(d) Glyceraldehyde

53. The volume of a forest at 70 years was 100 m^3 and the volume 10 years ago was 70 m^3. The increment percent percent of the forest would be
(a) 4.2 m^3
(b) 4.8 m^3
(c) 3.5 m^3
(d) 4.0 m^3

54. Which among the following is not a type of particle board
(a) Chip board
(b) Flake board
(c) Shaving board
(d) Lamin board

55. A forest that is regenerated from seed is
(a) Normal forest
(b) Natural forest
(c) High forest
(d) Regular forest

56. A hygrophyte is a plant that grows in
(a) Moist soils
(b) Waterlogged soils
(c) Aerated soils with sufficient moisture
(d) Aerated soils with deficient moisture

57. "Ecotype" is described as
(a) Temporary
(b) Are genetically fixed
(c) Are genetically not related
(d) Can't be measured

58. In *Atropa belladona,* the commercial drug is obtained from
(a) Roots only
(b) Leaves, flowering tops and roots
(c) Flowering tops only
(d) Bark

59. National Agroforestry Policy was implemented in the year
(a) 2014 (c) 2000
(b) 2012 (d) 2010

60. Which of the following plants among the following can be classified as a total parasite?
(a) Cuscuta
(b) Loranthus
(c) Viscum
(d) Arceuthobium

61. The process of felling all poles of non-commercial species, under growth upto 10 m and girdling trees is not required as seed trees (10-20 m) is called
(a) Improvement felling
(b) Regeneration felling
(c) Canopy lifting
(d) Selective felling

62. In a sample plot of 0.25 hectares, while viewing from a wedge prism with BAF 2, the tree with overlapping images is 8 and trees touching margins is 11, the basal area of the sample plot is
(a) 27×0.25 m^3 (c) 8×0.25 m^3
(b) 19×0.25 m^3 (d) 13.5×0.25 m^3

63. Conservation Reserves is declared using the provisions under which section of Wildlife Protection Act, 1972?
(a) 36a (c) 36c
(b) 36b (d) 36d

64. Poplars and Willows belong to which of the following family?
(a) Meliaceae
(b) Dipterocarpaceae
(c) Salicaceae
(d) Verbenaceae

65. What is a stump planting?
(a) Planting of stump and root
(b) Root cutting
(c) Shoot cutting
(d) Planting of grafts

66. Choro Flouro Carbons (CFC's) are responsible for
(a) Ozone formation
(b) Ozone depletion
(c) Ozone formation
(d) Methane formation

67. Which among the following forests can be classified as a 'Shade bearer' of tropical wet evergreen forest?
(a) *Callophyllum inophyllum*
(b) *Messua ferrea*
(c) *Xylia xylocarpa*
(d) *Artocarpus hetrophyllus*

68. Crop diameter corresponds to
(a) Mean diameter of 250 dominant trees per hectare
(b) Diameter corresponds to mean basal area of the crop
(c) Average diameter of all trees in a stand
(d) Diameter corresponding to average crop height

69. What is the minimum number of codons required for amino acids containing five amino acids (ignore start and stop codons)?
(a) 10 (c) 20
(b) 5 (d) 15

70. Indian Institute of Natural Resins and Gums (IINRG) is located at
(a) Guwahati
(b) Ranchi
(c) Bhubaneshwar
(d) Jabalpur

71. 6-Benzyl Amino Purine (BAP) belongs to category of which plant growth regulator?
(a) Auxin
(b) ABA
(c) Cytokinin
(d) Ethylene

72. Which of the following is not a vegetative propagation?
(a) Coppice
(b) Container raised plants
(c) Grafting
(d) Root sucker

73. If the normal growing stock of a forest is 350 m^3/ha as per yield table method and 800 m^3/ha as per MAI method. The flurry's constant is
(a) 2.2 (c) 47
(b) 22.6 (d) 0.47

74. Commonly used surface sterilant in tissue culture is
(a) $KMnO_4$
(b) $CaCl_2$
(c) $HgCl_2$
(d) NaCl

75. In which method of thinning is the attention concentrated an evenly spaced selected stems so that they are retained until maturity?
(a) Crown thinning
(b) Free thinning
(c) Mechanical thinning
(d) High thinning

76. The hardness is seed coat is due to which type of cells?
(a) Parenchyma
(b) Sclerenchyma
(c) Collenchyma
(d) Xylem

77. The bacteria used in genetic engineering is
(a) *Agrobacterium* species
(b) *Rhizobium* species
(c) *Frankia* species
(d) *Escherichia coli*

78. Free – Air CO_2 Enrichment (FACE) offers mechanism for
(a) Regulation Carbon emissions under Kyoto protocol
(b) Protecting global biodiversity under CBD
(c) Evaluating the effects of enriched atmospheric CO_2 on ecosystem process
(d) Predicting climate change with coupled atmospheric- ocean general circulation models

79. A sampling method in which sampling units composing a sample

selected in such a manner that all possible units of the same size have chance of being chosen is
(a) Multiphase sampling
(b) List sampling
(c) Sequential sampling
(d) Random sampling

80. Pitch of a saw is a
(a) Length of teeth
(b) Thickness of teeth
(c) Number of teeth
(d) Angle between teeth

81. The increment percent according to the Schneider's formula (n- number of annual rings in last inch of tree, D- is D.B.H.)
(a) 200/nD (c) 100/nD
(b) (n/D)*100 (d) 400/nD

82. If a soil test data indicates 12% exchangeable sodium, 0.15% soluble salts and a pH of 8.0, it would be classified as
(a) Saline soil
(b) Alkaline soil
(c) Sodic soil
(d) Saline-Alkaline soil

83. Calculate spacing between rows in which plants are 2.5 m apart if the plantation has 1000 plants per hectare
(a) 5 m (c) 4 m
(b) 3.5 m (d) 6 m

84. A classic example of ring porus wood among Indian timbers is
(a) *Salix alba & Popular ciliata*
(b) *Tectona grandis & Toona ciliata*
(c) *Shorea robusta & Lagerstromia parviflora*
(d) *Cedrus deodara & Pinus wallichiana*

85. Length of the lowest section in stem analysis is
(a) 3.0 m (c) 5.0 m
(b) 13.7 m (d) 2.74 m

86. Which of the following is known as Smalian's formula for calculating the volume of logs ?
(a) $Sm \times l$
(b) $\left(\frac{S_1 + 4S_m + S_2}{6}\right) \times l$
(c) $\left(\frac{S_1 + S_2}{2}\right) \times l$
(d) $\left(\frac{l}{S_1 + S_2}\right) \times g$

87. The edible part of nutmeg is
(a) Leaves (c) Flower
(b) Fruit (d) Aril

88. Major nutrients or macro-nutrients ie. mineral required in large quantities by plants are
(a) Iron, Manganese, Copper, Boron
(b) Nitrogen, Phosphorus, Potassium, Calcium, Magnesium, Sulphur
(c) Zinc, Copper, Molybdenum, Iron
(d) Oxygen, Hydrogen, Nitrogen, Carbon

89. Pneumatophores are
(a) Tree of arid regions
(b) Tree of shola forest
(c) Tree of cold forest
(d) Tree of mangrove forest

90. Moisture content of seed can be determined in a high constant temperature oven at
(a) 130°C ± 2°C
(b) 100°C ± 2°C
(c) 50°C ± 2°C
(d) 200°C ± 2°C

91. Number of seeds that germinate in a sample upto a time when germination reaches its peak is
(a) Germinative capacity
(b) Germinative potential
(c) Germinative capability
(d) Germinative energy

92. The appearance of soil in vertical section with particular reference to the presence of layers that are differentiated is
(a) Soil texture
(b) Soil profile
(c) Soil composition
(d) Soil structure

93. Which two seed physiological qualities are maintained at higher levels from harvesting to actual use in nursery?
(a) Respiration and Transpiration
(b) Heat and Photosynthesis
(c) Viability and Vigour
(d) Moisture content and Embryo quality

94. The Land use factor (L) in shifting cultivation where permanent cultivation is reached (Zero fallow phase) through improved fallow management will be
(a) > 10
(b) = 1
(c) < 1
(d) = 0

95. Radar Remote sensing can be categorized as
(a) Normal Remote sensing
(b) High Remote sensing
(c) Active Remote sensing
(d) Passive Remote sensing

96. African padauk is
(a) *Tectona grandis*
(b) *Pterocarpus soyauxii*
(c) *Pterocarpus marsupium*
(d) *Artocarpus hirsutus*

97. The biochemical test employed to access seed viability is
(a) Formaldehyde test
(b) Tetrazolium test
(c) Ferric chloride test
(d) Mercuric chloride test

98. What is podzolization?
(a) Soil forming process by which igneous rocks, usually of basic composition
(b) A soil forming process of cool, humid climate in which leaching by organic acids remove bases and translocates sesquioxides from A horizon
(c) Disintegration of rocks chemically under the influence of water containing dissolved carbon dioxide
(d) Physical disintegration of rocks resulting in exposure of internal surfaces

99. Seed stand has two areas
(a) Breeding and production area
(b) Buffer and production area
(c) Good and bad area
(d) Same species and different species areas

100. Full Cell Bethel process is a technique used for
(a) Wood preservation technique
(b) Wood seasoning technique
(c) Test for mechanical properties
(d) Test to find out luminescence of wood

101. The amount of visible radiation received in surface of earth as a proportion of total solar radiation is about
(a) 40% (c) 60%
(b) 20% (d) 80%

102. The average height of dominant trees in a stand is called
(a) Crop height (c) Top height
(b) Mean height (d) Canopy height

103. The Institute of Forest Genetics and Tree breeding is located at
(a) Bengaluru (c) Jabalpur
(b) Coimbatore (d) Jhansi

104. Which one of the following is used as fillers during plywood manufacturing?
(a) Clay
(b) Copper sulphate
(c) Cotton
(d) Plastic

105. Root suckers are a reliable method for regeneration of
(a) *Diospyros tomentosa*
(b) *Pinus roxburghii*
(c) *Tectona grandis*
(d) *Acacia auriculiformis*

106. Which among the following is a criteria for selecting a region as a biodiversity hotspot?
(a) Biodiversity and extent of area
(b) Specie endemism and degree of threat
(c) Biodiversity and species richness
(d) Biodiversity and degree of threat

107. A normal forest is described as
(a) Forest having all the species of trees and shrubs
(b) Forest which give annual or periodic yield equal to the increment without endangering future yield.
(c) Forest which grows in the normal forest land
(d) Forest which has tall trees with straight stem

108. Height of trees can be determined without horizontal distance measurement using
(a) Hega altimeter
(b) Blume leiss hypsometer
(c) Laser hypsometer
(d) Ravi multimeter

109. On the basis of position of optical axis of camera, photograph has been taken with optical axis of the camera approximately perpendicular to the horizontal plane, such photography is known as
(a) Oblique photography
(b) Vertical photography
(c) Optical photography
(d) Axis photography

110. Shikakai pods are obtained from
(a) *Acacia concinna*
(b) *Acacia arabica*
(c) *Butea monosperma*
(d) *Caesalpinia sappan*

111. The process of DNA synthesis could be best described as
(a) Winding, Pairing, Joining
(b) Pairing, Joining, Winding
(c) Unwinding, Pairing, Joining
(d) Joining, Pairing, Unwinding

112. When *Eucalyptus* is planted at spacing of 1m × 2m the number of plants will be
(a) 5,000 (c) 4,000
(b) 10,000 (d) 2,500

113. If a forest has crop density 1.1, it indicates that forest
(a) Is overstocked compared to yield table
(b) Is overstocked compared to regular crop
(c) Is incomplete as compared to yield table
(d) Is incomplete as compared to regular crop

114. Hydroxyl citric acid, a natural apetite suppressant is found in
(a) *Garcinia gummigutta*
(b) *Strychnous nuxvomica*
(c) *Saraca osoca*
(d) *Aegle marmeloes*

115. The approximately cubic content of a standing tree with gbh 2m and commercial bole height 20 m is
(a) $4m^3$
(b) $1.5\ m^3$
(c) $250\ m^3$
(d) $2.5\ m^3$

116. Natural variation exist in hierarchial arrangements from species variation followed by genus variation followed by site variation followed by
(a) Country variation
(b) Provinance variation
(c) Tree to tree variation
(d) Micro-level variation

117. Essential oil are obtained from
(a) *Cymbopogon martinii* & *Aquilaria agallocha*
(b) *Zizyphus mauritiana* & *Cajanus cajan*
(c) *Pinus roxburghii* & *Cedrus deodara*
(d) *Alstonoia scholaris* & *Sapium sebiferum*

118. A yield table is described as
(a) A table which gives life data of tree
(b) A tabular statement which minimises on unit area basis all the essential data relating to development of a fully stocked even aged crop
(c) A tabular statement which gives on unit the yield of trees in a particular area
(d) A tabular statement which gives the life data of trees growing in a particular forest

119. The tusk of elephant is
(a) Upper incisors
(b) Pre molars
(c) Canines
(d) Molars

120. In eukaryotes, glycolytic reaction take place in
(a) Mitochondria
(b) Cytoplasm
(c) Ribosomes
(d) Endoplasmic Reticulum

121. LULUCF is strategy under
(a) Bhopal India process
(b) Kyoto treaty
(c) Paris agreement
(d) Montreal process

122. Gasification of wood is a process in which
(a) Wood is burnt down
(b) Wood is broken into compounds
(c) Wood is broken down by use of heat to produce combustible gas
(d) Wood is broken down into useful gases

123. Component of variation are

$$\delta^2P = \delta^2NA + \delta^2A + \delta^2 + \cdots$$

(a) Broad sense heritability (H^2)

(b) Narrow sense heritability (h^2)
(c) Environment (E)
(d) Error (r)

124. Veneer matching done by colour and not by grain is called
(a) Slip Matching
(b) Random matching
(c) Book matching
(d) Pleasing matching

125. Normal series of age gradation refers to presence of all
(a) Ages from pole stage to rotation age
(b) Ages from one year to rotation age
(c) Ages from half rotation age to rotation age
(d) Age classes from 1-5 years class to rotation age

126. Cutch and katha are obtained by boiling in water the heart wood of
(a) *Acacia nilotica*
(b) *Acacia catechu*
(c) *Prosopis cineraria*
(d) *Azadirachta indica*

127. More widely used method of veneering is
(a) Slicing (c) Peeling
(b) Sawing (d) Cutting

128. Aswagandha is
(a) *Withania somnifera*
(b) *Commiphora mukul*
(c) *Cymbopogon flexuosus*
(d) *Jasminum grandiflorum*

129. The nursery operation to condition the nursery stalk by loosening the contact between soil and root of the seedlings in nursery bed is called
(a) Box pruning
(b) Lateral pruning
(c) Under cutting
(d) Wrenching

130. One horn rhinoceros lives in swampy and marshy lands of
(a) Tamil Nadu and Kerela
(b) Assam and WB
(c) UP and Bihar
(d) Maharashtra and Telengana

131. Wetting coefficient of soil is lowest in
(a) Sandy soil
(b) Sandy loam
(c) Clay soil
(d) Clay loam

132. The number of coupes in clear felling is a function of
(a) Area of plantation alone
(b) Rotation of the species alone
(c) Area of the plantation and rotation of the species
(d) Site quality of the plantation and rotation of the species

133. 100 Cubic meter of wood is equivalent to
(a) 3531 cubic feet
(b) 7062 cubic feet
(c) 353.1 cubic feet
(d) 706.2 cubic feet

134. Identify the tpe of sampling in which each unit has the same probability of being sampled
(a) Sequential sampling
(b) Random sampling
(c) Systematic sampling
(d) Stratified sampling

135. Mode in statistics is defined as
(a) Middle observation
(b) Result of adding together a set of values and dividing by number of values

(c) Observation which occurs most frequently
(d) Each value is multiplied by its frequency and divided by total frequency

136. Timber obtained from coniferous tree is
(a) Soft wood (c) Porus wood
(b) Hard wood (d) Sap wood

137. Identify the principle on which all instruments used to measure height of trees is based on
(a) Simailar triangle or Trignometric principles
(b) Parallel lines cant meet
(c) Rectangular principles
(d) Triangle having similar angles are equal

138. which is the family of old-world monkey
(a) Cercopithecidae
(b) Lorisidae
(c) Hylobatidae
(d) Pithecidae

139. Xylem of coniferous timber is made up of
(a) Trachieds and vessels
(b) Trachieds only
(c) Vessels only
(d) Vessels and phloem

140. Universal Testing Machine (UMT) is used to test the
(a) Function of machine
(b) Strength of wood
(c) Strength of teeth of saw
(d) Strength of log

141. A measure of the relative productive capacity of a site for a particular specie is
(a) Site factor
(b) Site quality
(c) Vegetative characters
(d) Tree characters

142. What is the percentage of nitrogen in plants on dry weight basis
(a) 50% (c) 25%
(b) 75% (d) 35%

143. Sawn timber can be obtained in form of
(a) Squares, Poles, Sleepers, Rounds
(b) Logs, Hakries, Trimmed wood
(c) Beams, Scantlings, Planks
(d) Baulks, Stacks, Rafts

144. Flame of forest is
(a) *Lannea coromandalica*
(b) *Spathodiea companulata*
(c) *Butea monosperma*
(d) *Delonix regia*

145. Primary classification of silvicultural system is based on
(a) Mode of regeneration
(b) Pattern of felling
(c) Locality factors
(d) Species composition

146. The tree that occupies more than its fair share of growing space in a woodland is
(a) Buttressed tree
(b) Fluted tree
(c) Wolf tree
(d) Clear bole tree

147. The ideal storage conditions for seed are those in which the following are reduced to minimum, without damaging the inherent vitality of seed
(a) Respiration and Energy
(b) Water loss and Food loss
(c) Colour and Injury
(d) Photosynthesis and Light

148. In an irregular forest, the ideal Silvicultural system that ensures irregular nature of crop
(a) Shelterwood system
(b) Selective system
(c) Uniform system
(d) Selection system

149. Sal, most Dipterocarpus, many Mystaceae and Lauraceae seeds can be stored for
(a) Ten years
(b) Less than 1 year
(c) Decades
(d) Centuries

150. Satpura hypothesis was proposed by
(a) Wallace
(b) Sunder Lal Hora
(c) Robert Haeckel
(d) Wittaker

151. Number of seedlings required to plant an area of one hectare, at a spacing of 2 m × 2 m would be
(a) 25000
(b) 10000
(c) 20000
(d) 15000

152. Process by which carbohydrates in wood is converted to simple sugars through chemical treatment is called
(a) Seasoning
(b) Pyrolisation
(c) Saccharification
(d) Carmalisation

153. Which one of the lipid group is most abundant in chloroplast membranes?
(a) Sphigno lipids
(b) Galacto lipids
(c) Phospho lipids
(d) Sulpho lipids

154. In a regeneration survey, the symbol 'w+' would indicate
(a) Woody shoot, not yet established, but expected to establish soon considering its vigour and size
(b) Whippy unestablished seedling with height < 50 cm
(c) Woody shoot, which has been browsed
(d) Established regeneration whose height is > 25 m and diameter > 10 cm

155. A silvicultural system in which regeneration fellings are carried out in narrow strips extending in east – west direction and advancing from North to South is
(a) Bavarian Femelschlag
(b) Two storied high forest system
(c) Wagner's Blender saumschlag
(d) Dauerwald

156. Insoluble fraction in the pulp related to viscose – rayon process is
(a) Beta – cellulose
(b) Cellulose
(c) Alpha – cellulose
(d) Hemi – cellulose

157. A viable seed which falls to germinate even in suitable conditions is
(a) Vigrous seed
(b) Viable seed
(c) Dormant seed
(d) Live seed

158. The inner seed coat is called
(a) Testa
(b) Tegmen
(c) Raphae
(d) Integument

159. Longest living tree of the world is
(a) *Tectona grandis*
(b) *Platenous orientalis*
(c) *Sequoia sempervirens*
(d) *Pseudosuga taxifolia*

160. Haldridge's climate models are not based on
(a) Bio temperature
(b) Average annual precipitation
(c) Potential evapotranspiration
(d) Solar radiation

Answer Keys

1. (d)	**2.** (c)	**3.** (c)	**4.** (a)	**5.** (b)	**6.** (b)	**7.** (c)	**8.** (c)	**9.** (b)	**10.** (b)
11. (b)	**12.** (b)	**13.** (b)	**14.** (d)	**15.** (b)	**16.** (c)	**17.** (d)	**18.** (d)	**19.** (d)	**20.** (b)
21. (b)	**22.** (a)	**23.** (d)	**24.** (c)	**25.** (b)	**26.** (c)	**27.** (c)	**28.** (a)	**29.** (c)	**30.** (a)
31. (b)	**32.** (a)	**33.** (b)	**34.** (d)	**35.** (c)	**36.** (a)	**37.** (c)	**38.** (b)	**39.** (a)	**40.** (a)
41. (a)	**42.** (b)	**43.** (c)	**44.** (d)	**45.** (c)	**46.** (c)	**47.** (c)	**48.** (c)	**49.** (b)	**50.** (d)
51. (a)	**52.** (c)	**53.** (c)	**54.** (d)	**55.** (c)	**56.** (a)	**57.** (b)	**58.** (b)	**59.** (a)	**60.** (a)
61. (c)	**62.** (a)	**63.** (a)	**64.** (c)	**65.** (a)	**66.** (b)	**67.** (d)	**68.** (b)	**69.** (d)	**70.** (b)
71. (c)	**72.** (b)	**73.** (d)	**74.** (c)	**75.** (b)	**76.** (b)	**77.** (a)	**78.** (c)	**79.** (d)	**80.** (d)
81. (d)	**82.** (a)	**83.** (c)	**84.** (b)	**85.** (d)	**86.** (c)	**87.** (d)	**88.** (b)	**89.** (d)	**90.** (b)
91. (d)	**92.** (b)	**93.** (c)	**94.** (b)	**95.** (c)	**96.** (b)	**97.** (b)	**98.** (b)	**99.** (b)	**100.** (a)
101. (a)	**102.** (c)	**103.** (b)	**104.** (a)	**105.** (a)	**106.** (b)	**107.** (b)	**108.** (c)	**109.** (b)	**110.** (a)
111. (c)	**112.** (a)	**113.** (a)	**114.** (a)	**115.** (a)	**116.** (c)	**117.** (a)	**118.** (b)	**119.** (a)	**120.** (b)
121. (b)	**122.** (c)	**123.** (c)	**124.** (d)	**125.** (b)	**126.** (b)	**127.** (c)	**128.** (a)	**129.** (d)	**130.** (b)
131. (a)	**132.** (c)	**133.** (a)	**134.** (b)	**135.** (c)	**136.** (a)	**137.** (a)	**138.** (a)	**139.** (b)	**140.** (b)
141. (b)	**142.** (c)	**143.** (c)	**144.** (c)	**145.** (a)	**146.** (c)	**147.** (a)	**148.** (d)	**149.** (b)	**150.** (b)
151. (a)	**152.** (c)	**153.** (b)	**154.** (c)	**155.** (c)	**156.** (c)	**157.** (c)	**158.** (b)	**159.** (c)	**160.** (d)

Appendix

ICAR Institutions, Deemed Universities, National Research Centres, National Bureaux and Directorate/Project Directorates:

Deemed Universities – 4	
1	ICAR-Indian Agricultural Research Institute, New Delhi
2	ICAR-National Dairy Research Institute, Karnal
3	ICAR-Indian Veterinary Research Institute, Izatnagar
4	ICAR-Central Institute on Fisheries Education, Mumbai
Institutions – 65	
1	ICAR-Central Island Agricultural Research Institute , Port Blair
2	ICAR-Central Arid Zone Research Institute, Jodhpur
3	ICAR-Central Avian Research Institute, Izatnagar
4	ICAR-Central Inland Fisheries Research Institute, Barrackpore
5	ICAR-Central Institute Brackishwater Aquaculture, Chennai
6	ICAR-Central Institute for Research on Buffaloes, Hissar
7	ICAR-Central Institute for Research on Goats, Makhdoom
8	ICAR-Central Institute of Agricultural Engineering, Bhopal
9	ICAR-Central Institute for Arid Horticulture, Bikaner
10	ICAR-Central Institute of Cotton Research, Nagpur
11	ICAR-Central Institute of Fisheries Technology, Cochin
12	ICAR-Central Institute of Freshwater Aquaculture, Bhubneshwar
13	ICAR-Central Institute of Research on Cotton Technology, Mumbai
14	ICAR-Central Institute of Sub Tropical Horticulture, Lucknow
15	ICAR-Central Institute of Temperate Horticulture, Srinagar
16	ICAR-Central Institute on Post Harvest Engineering and Technology, Ludhiana
17	ICAR-Central Marine Fisheries Research Institute, Kochi
18	ICAR-Central Plantation Crops Research Institute, Kasargod
19	ICAR-Central Potato Research Institute, Shimla
20	ICAR-Central Research Institute for Jute and Allied Fibres, Barrackpore
21	ICAR-Central Research Institute of Dryland Agriculture, Hyderabad
22	ICAR-National Rice Research Institute, Cuttack
23	ICAR-Central Sheep and Wool Research Institute, Avikanagar, Rajasthan
24	ICAR- Indian Institute of Soil and Water Conservation, Dehradun
25	ICAR-Central Soil Salinity Research Institute, Karnal
26	ICAR-Central Tobacco Research Institute, Rajahmundry

27	ICAR-Central Tuber Crops Research Institute, Trivandrum
28	ICAR-ICAR Research Complex for Eastern Region, Patna
29	ICAR-ICAR Research Complex for NEH Region, Barapani
30	ICAR-Central Coastal Agricultural Research Institute, Ela, Old Goa, Goa
31	ICAR-Indian Agricultural Statistics Research Institute, New Delhi
32	ICAR-Indian Grassland and Fodder Research Institute, Jhansi
33	ICAR-Indian Institute of Agricultural Biotechnology, Ranchi
34	ICAR-Indian Institute of Horticultural Research, Bengaluru
35	ICAR-Indian Institute of Natural Resins and Gums, Ranchi
36	ICAR-Indian Institute of Pulses Research, Kanpur
37	ICAR-Indian Institute of Soil Sciences, Bhopal
38	ICAR-Indian Institute of Spices Research, Calicut
39	ICAR-Indian Institute of Sugarcane Research, Lucknow
40	ICAR-Indian Institute of Vegetable Research, Varanasi
41	ICAR-National Academy of Agricultural Research & Management, Hyderabad
42	ICAR-National Institute of Biotic Stresses Management, Raipur
43	ICAR-National Institue of Abiotic Stress Management, Malegaon, Maharashtra
44	ICAR-National Institute of Animal Nutrition and Physiology, Bengaluru
45	ICAR-National Institute of Natural Fibre Engineering and Technology, Kolkata, Kolkata
46	ICAR-National Institute of Veterinary Epidemiology and Disease Informatics, Hebbal, Bengaluru
47	ICAR-Sugarcane Breeding Institute, Coimbatore
48	ICAR-Vivekananda Parvatiya Krishi Anusandhan Sansthan, Almora
49	ICAR-Central Institute for Research on Cattle, Meerut, Uttar Pradesh
50	ICAR-National Institute of High Security Animal Diseases, Bhopal
51	ICAR-Indian Institute of Maize Research, New Delhi
52	ICAR- Central Agroforestry Research Institute , Jhansi
53	ICAR-National Institute of Agricultural Economics and Policy Research, New Delhi
54	ICAR- Indian Institute of Wheat and Barley Research, Karnal
55	ICAR- Indian Institute of Farming Systems Research, Modipuram
56	ICAR- Indian Institute of Millets Research, Hyderabad
57	ICAR- Indian Institute of Oilseeds Research, Hyderabad
58	ICAR- Indian Institute of Oil Palm Research, Pedavegi, West Godawari
59	ICAR- Indian Institute of Water Management, Bhubaneshwar
60	ICAR-Indian Institute of Rice Research, Hyderabad
61	ICAR- Central Institute for Women in Agriculture, Bhubaneshwar

62	ICAR-Central Citrus Research Institute, Nagpur
63	ICAR-Indian Institute of Seed Research, Mau
64	ICAR-Indian Agriculture Institute, Post Box No. 48, Hazaribag, Jharkhand
65	ICAR-National Institute for Plant Biotechnology, New Delhi
	National Research Centres – 14
1	ICAR-National Research Centre for Banana, Trichi
2	ICAR-National Research Centre for Grapes, Pune
3	ICAR-National Research Centre for Litchi, Muzaffarpur
4	ICAR-National Research Centre for Pomegranate, Solapur
5	ICAR-National Research Centre on Camel, Bikaner
6	ICAR-National Research Centre on Equines, Hisar
7	ICAR-National Research Centre on Meat, Hyderabad
8	ICAR-National Research Centre on Mithun, Medziphema, Nagaland
9	ICAR-National Research Centre on Orchids, Pakyong, Sikkim
10	ICAR-National Research Centre on Pig, Guwahati
11	ICAR-National Research Centre on Seed Spices, Ajmer
12	ICAR-National Research Centre on Yak, West Kemang
13	ICAR-National Centre for Integrated Pest Management, New Delhi
14	Mahatma Gandhi Integrated Farming Research Institute, Motihari
	National Bureaux – 6
1	ICAR-National Bureau of Plant Genetics Resources, New Delhi
2	ICAR-National Bureau of Agriculturally Important Micro-organisms, Mau, Uttar Pradesh
3	ICAR-National Bureau of Agricultural Insect Resources, Bengaluru
4	ICAR-National Bureau of Soil Survey and Land Use Planning, Nagpur
5	ICAR-National Bureau of Animal Genetic Resources, Karnal
6	ICAR-National Bureau of Fish Genetic Resources, Lucknow
	Directorates/Project Directorates – 13
1	ICAR-Directorate of Groundnut Research, Junagarh
2	ICAR-Directorate of Soybean Research, Indore
3	ICAR-Directorate of Rapeseed & Mustard Research, Bharatpur
4	ICAR-Directorate of Mushroom Research, Solan
5	ICAR-Directorate on Onion and Garlic Research, Pune
6	ICAR-Directorate of Cashew Research, Puttur
7	ICAR-Directorate of Medicinal and Aromatic Plants Research, Anand
8	ICAR-Directorate of Floricultural Research, Pune, Maharashtra
9	ICAR-Directorate of Weed Research, Jabalpur

10	ICAR-Project Directorate on Foot & Mouth Disease, Mukteshwar
11	ICAR-Directorate of Poultry Research, Hyderabad
12	ICAR-Directorate of Knowledge Management in Agriculture (DKMA), New Delhi
13	ICAR-Directorate of Cold Water Fisheries Research, Bhimtal, Nainital

CGIAR Institutes

Active CGIAR Centres	Headquarters
International Plant Genetics Resources Institute, (IPGRI)	Rome, Italy
Centre for International Forestry Research (CIFOR)	Bogor, Indonesia
International Centre for Tropical Agriculture (CIAT)	Cali, Colombia
International Centre for Agricultural Research in the Dry Areas (ICARDA)	Beirut, Lebanon
International Crops Research Institute for the Semi-Arid Tropics (ICRISAT)	Hyderabad, India.
International Food Policy Research Institute (IFPRI)	Washington, D.C., United States
International Institute of Tropical Agriculture (IITA)	Ibadan, Nigeria
International Maize and Wheat Improvement Centre (CIMMYT)	Texcoco, Mexico State, Mexico
International Rice Research Institute (IRRI)	Los Banos, Laguna, Philippines
World Agroforestry Centre (International Centre for Research in Agroforestry, ICRAF)	Nairobi, Kenya

International Decades

S.No.	Decades	United Nations International Decades
1	2019-2028	United Nations Decade of Family Farming.
2	2018-2028	United Nations Decade for Action on Water for Sustainable Development
3	2016-2025	United Nations Decade of Action on Nutrition
4	2015-2024	International Decade for People of African Descent
5	2014-2024	United Nations Decade of Sustainable Energy for All
6	2011-2020	Decade of Action for Road Safety
7	2010-2020	United Nations Decade for Deserts and the Fight against Desertification
8	2008-2017	Second United Nations Decade for the Eradication of Poverty
9	2006-2016	Decade of Recovery and Sustainable Development of the Affected Regions (third decade after the Chernobyl disaster)
10	2005-2015	International Decade for Action, "Water for Life"
11	1995-2004	Decade for Human Rights Education

S.No.	Decades	United Nations International Decades
12	1994-2004	Decade of the World's Indigenous People
13	1993-2003	Third Decade to Combat Racism and Racial Discrimination
14	1991-2000	United Nations Decade Against Drug Abuse Fourth United Nations Development Decade
15	1990-2000	International Decade for the Eradification of Colonialism
16	1990-1999	United Nations Decade of International Law International Decade for Natural Disaster Reduction
17	1990's	Third Disarmament Decade
18	1988-1997	World Decade for Cultural Development
19	1983-1993	Second Decade to Combat Racism and Racial Discrimination
20	1983-1992	United Nations Decade for Disabled Persons
21	1981-1990	International Drinking Water Supply and Sanitation Decade Third United Nations Development Decade
22	1980-1990	Second Disarmament Decade
23	1980's	Industrial Development Decade for Africa
24	1978-1988	Transport and Communications Decade for Africa
25	1976-1985	United Nations Decade for Women: Equality, Development and Peace
26	1973-1983	Decade to Combat Racism and Racial Discrimination
27	1971-1980	Second United Nations Development Decade
28	1970's	Disarmament Decade
29	1960-1970	United Nations Development Decade

International Years

S.No.	Year	United Nations International Years
1	2019	International Year of Indigenous Languages
2	2017	International Year of Sustainable Tourism for Development
3	2016	International Year of Pulses
4	2015	International Year of Soils
5	2014	International Year of Family Farming
6	2013	International Year of Water Cooperation
7	2012	International Year of Cooperatives
8	2011	International Year of Forests
9	2010	International Year of Biodiversity
10	2009	International Year of Planet Earth, International Year of the Potato
11	2007-2008	International Polar Year
12	2006	International Year of Deserts and Desertification

S.No.	Year	United Nations International Years
13	2005	International Year of Microcredit
14	2004	International Year of Rice
15	2003	International Year of Freshwater
16	2002	International Year of Mountains, International Year of Ecotourism
17	2001	International Year of Volunteers
18	2000	International Year for the Culture of Peace
19	1999	International Year of Older Persons
20	1998	International Year of the Ocean

Revolutions

Revolution	Product Related	Father/Person Associated
Protein Revolution	Higher Production (Technology driven 2nd Green revolution)	Coined by PM Narendra Modi and FM Arun Jaitely.
Yellow Revolution	Oil seed Production (Especially Mustard and Sunflower)	Sam Pitroda
Black Revolution	Petroleum Products	-
Blue Revolution	Fish Production	Dr. Arun Krishnan
Brown Revolution	Leather/Cocoa/Non-Conventional Products	-
Golden Fiber Revolution	Jute Production.	-
Golden Revolution	Fruits/Honey Production/ Horticulture Development	Nirpakh Tutej.
Grey Revolution	Fertilizers	-
Pink Revolution	Onion Production/ Pharmaceuticals/Prawn Production	Durgesh Patel
Evergreen Revolution	Overall Production of Agriculture	Started in 11th 5 year Plan
Silver Revolution	Egg Production/Poultry Production	Indira Gandhi
Silver Fiber Revolution	Cotton	-
Red Revolution	Meat Production/Tomato Production	Vishal Tewari
Round Revolution	Potato	-

Revolution	Product Related	Father/Person Associated
Green Revolution	Food Grains	Norman Borlong M.S. Swaminathan
White Revolution (or, Operation Flood)	Milk Production	Verghese Kurien

Important Acts and Organisations

S.No.	Acts/Organisations	Year
1	Botanical Survey of India	1890
2	Fisheries Act	1897
3	Zoological Survey of India	1916
4	Indian Forest Act	1927
5	Indian Forests Act	1927
6	Mining and Mineral Development Regulation Act	1957
7	The Prevention of Cruelty to Animal Act	1960
8	National Remote Sensing Agency	1973
9	The Water Prevention and Control Pollution Act	1974
10	International Seed Testing Association	1976
11	The Forest (Conservation) Act	1980
12	The Air Prevention and Control of Pollution Act	1981
13	Forest Survey of India	1981
14	National Wasteland Development Board	1985
15	The Environment Protection Act	1986
16	IPCC	1988
17	Hazardous Waste Handling and Management Act	1989
18	The National Environment Tribunal Act	1995
19	Biological Diversity Act	2002
20	Wildlife (Protection) Amendment Act	2002
21	National Biodiversity Authority	2003
22	The ST and Other Traditional Forest Dwellers (Forest Rights) Act	2006
23	The National Green Tribunal Act	2010

Environmentally Important Days

February	
February 2	World Wetlands Day
February 27	International Polar Bear Day
February 28	National Science Day

March	
March 3	World Wildlife Day
March 14	International Day of Action for Rivers
March 20	World Sparrow Day
March 21	World Forestry Day, World Planting Day, World Wood Day
March 22	World Water & Sanitation Day
March 23	World Meteorological Day, World Resources Day
April	
April 7	World Health Day
April 10	World Atmosphere Day
April 18	World Heritage Day
April 22	World Earth Day
May	
May 3	International Energy Day
May 8	World Migratory Bird Day
May 11	National Technology Day
May 14	Endemic Bird Day
May 22	World Biodiversity Day
May 23	World Turtle Day
June	
June 5	World Environment Day
June 8	World Ocean Day
June 9	Coral Triangle Day
June 15	Global Wind Day
June 17	World Day to Combat Desertification and Drought
July	
July 1-July 7	Van Mahotsav Saptah
July 3	World Seabird Day
July 11	World Population Day
July 26	International Mangrove Day
July 29	International Tiger Day
August	
August 10	World Lion Day
August 12	World Elephant Day
August 22	Honey Bee Day
September	
September 8	World Cleanup Day

September 16	World Ozone Day
September 18	World Water Monitoring Day
September 21	Zero Emissions Day
September 26	World Environmental Health Day
October	
October 1 – 7	Wildlife Week
October 3	World Nature Day, World Habitat Day
October 4	World Animal Day
October 6	World Wildlife Day
October 24	International Day of Climate Action
November	
November 6	International Day for Preventing the Exploitation of the Environment in War and Armed Conflict
November 12	World Birds Day
November 14	World Energy Conservation Day
December	
December 5	World Soil Day
December 11	International Mountain Day
December 14	National Energy Conservation Day

(*Source*: wiienvis)

Important Environment Conferences and Conventions

S.No.	Conferences/Conventions	Year
1	RAMSAR Convention on Wetlands	1971
2	Stockholm Conference	1972
3	Convention on Protection of World Cultural and Natural Heritage	1972
4	CITES	1976
5	Convention on Migratory Species of Wild Animals	1979
6	World Conservation Strategy	1980
7	Vienna Convention (Ozone layer)	1985
8	Montreal Protocol (ODS-Ozone Depleting Substances)	1987
9	Earth summit	1992
10	UNFCCC	1992
11	Convention on Biological Diversity (CBD)	1992
12	UN Convention on Desertification	1994
13	Kyoto Protocol	1997

S.No.	Conferences/Conventions	Year
14	Stockholm Convention (POPs-Persistent Organic Pollutants)	2000
15	UN World Summit	2005
16	Bali Summit on Climate Change	2007
17	Paris Agreement	2015

Important Conservation Sites in India (As on February, 2019)

Reserves/Sites	Numbers	Total area (in Sq.Kms.)
Tiger Reserves	50	71027.10
Elephant Reserves	32	69,582.80
Biosphere Reserves	18	87491.6
RAMSAR Wetland Sites	27	11121.31
Natural World Heritage Sites	07	11755.84
Cultural World Heritage Sites	28	--
Mixed World Heritage Sites	01	1784.00
Important Coastal and Marine Biodiversity Areas	107	10773.07
Marine Protected Areas	131	9801.13
Important Bird Areas	467	--
Potential Important Bird Areas	96	--
Key Biodiversity Areas	531	--
Biodiversity Heritage Sites	9	--

Tiger Reserves:

Project Tiger was launched by the Government of India in the year 1973 to save the endangered species of tiger in the country. Starting from nine (9) reserves in 1973-2016 the number is grown up to fifty (50). A total area of 71,027.10 km^2 is covered by these project tiger areas.

S.No.	Name of Tiger Reserve	State	Area of the core/critical tiger habitat (in Sq. Kms.)	Area of the buffer/ peripheral (in Sq. Kms.)	Total area (in Sq. Kms.)
1	Nagarjunsagar Srisailam (part)	Andhra Pradesh	2595.72	700.59	3296.31
2	Namdapha	Arunachal Pradesh	1807.82	245	2052.82
3	Kamlang Tiger Reserve	Arunachal Pradesh	671.00	112.00	783.00

S.No.	Name of Tiger Reserve	State	Area of the core/critical tiger habitat (in Sq. Kms.)	Area of the buffer/ peripheral (in Sq. Kms.)	Total area (in Sq. Kms.)
4	Pakke	Arunachal Pradesh	683.45	515	1198.45
5	Manas	Assam	840.04	2310.88	3150.92
6	Nameri	Assam	200	144	344
7	Orang Tiger Reserve	Assam	79.28	413.18	492.46
8	Kaziranga	Assam	625.58	548	1173.58
9	Valmiki	Bihar	598.45	300.93	899.38
10	Udanti-Sitanadi	Chattisgarh	851.09	991.45	1842.54
11	Achanakmar	Chattisgarh	626.195	287.822	914.017
12	Indravati	Chhattisgarh	1258.37	1540.7	2799.07
13	Palamau	Jharkhand	414.08	715.85	1129.93
14	Bandipur	Karnataka	872.24	584.06	1456.3
15	Bhadra	Karnataka	492.46	571.83	1064.29
16	Dandeli-Anshi	Karnataka	814.884	282.63	1097.514
17	Nagarahole	Karnataka	643.35	562.41	1205.76
18	Biligiri Ranganatha Temple	Karnataka	359.1	215.72	574.82
19	Periyar	Kerala	881	44	925
20	Parambikulam	Kerala	390.89	252.772	643.662
21	Kanha	Madhya Pradesh	917.43	1134.361	2051.791
22	Pench	Madhya Pradesh	411.33	768.30225	1179.63225
23	Bandhavgarh	Madhya Pradesh	716.903	820.03509	1598.1
24	Panna	Madhya Pradesh	576.13	1021.97	1578.55
25	Satpura	Madhya Pradesh	1339.264	794.04397	2133.30797
26	Sanjay-Dubri	Madhya Pradesh	812.571	861.931	1674.502
27	Melghat	Maharashtra	1500.49	1268.03	2768.52
28	Tadoba-Andhari	Maharashtra	625.82	1101.7711	1727.5911
29	Pench	Maharashtra	257.26	483.96	741.22

S.No.	Name of Tiger Reserve	State	Area of the core/critical tiger habitat (in Sq. Kms.)	Area of the buffer/ peripheral (in Sq. Kms.)	Total area (in Sq. Kms.)
30	Sahyadri	Maharashtra	600.12	565.45	1165.57
31	Nawegaon-Nagzira	Maharashtra	653.674	-	653.674
32	Bor	Maharashtra	138.12	-	138.12
33	Dampa	Mizoram	500	488	988
34	Similipal	Odisha	1194.75	1555.25	2750
35	Satkosia	Odisha	523.61	440.26	963.87
36	Ranthambore	Rajasthan	1113.364	297.9265	1411.291
37	Sariska	Rajasthan	881.1124	332.23	1213.342
38	Mukandra Hills	Rajasthan	417.17	342.82	759.99
39	Kalakad-Mundanthurai	Tamil Nadu	895	706.542	1601.542
40	Anamalai	Tamil Nadu	958.59	521.28	1479.87
41	Mudumalai	Tamil Nadu	321	367.59	688.59
42	Sathyamangalam	Tamil Nadu	793.49	614.91	1408.4
43	Kawal	Telangana	893.23	1125.89	2019.12
44	Amrabad	Telangana	2166.37	445.02	2611.39
45	Dudhwa	Uttar Pradesh	1093.79	1107.9848	2201.7748
46	Pilibhit	Uttar Pradesh	602.798	127.4518	730.2498
47	Amangarh (buffer of Corbett TR)	Uttar Pradesh	-	80.6	80.6
	Corbett	Uttarakhand	821.99	466.32	1288.31
48	Rajaji TR	Uttarakhand	255.63	819.54	1075.17
49	Sunderbans	West Bengal	1699.62	885.27	2584.89
50	Buxa	West Bengal	390.5813	367.3225	757.9038
	Total		**40340.12**	**30686.98**	**71027.10**

Elephant Reserves

The Indian elephant *Elephas maximus* occurs in the central and southern Western Ghats, North-east India, eastern India and northern India and in some parts of southern peninsular India. It is included in Schedule I of the Indian Wildlife (Protection) Act, 1972 and in Appendix I of the Convention on International Trade in Endangered Species of Flora and Fauna (CITES). It occurs in 16 of the 28 states in the country and is showing an increasing trend across its distributional range. Its population in 2007 was estimated to be in the range from 27,657 to 27,682, whereas in 2012 the population was estimated to be between 27,785 and 31,368.

Project Elephant

Project Elephant was launched by the Government of India in the year 1992 as a Centrally Sponsored Scheme with following objectives:

1. To protect elephants, their habitat & corridors
2. To address issues of man-animal conflict
3. Welfare of captive elephants

The Project is being mainly implemented in 16 States/UTs , viz Andhra Pradesh, Arunachal Pradesh, Assam, Chhattisgarh, Jharkhand, Karnataka, Kerala, Maharashtra, Meghalaya, Nagaland, Orissa, Tamil Nadu, Tripura, Uttarakhand, Uttar Pradesh, West Bengal.

The Ministry of Environment, Forest and Climate Change provides the financial and technical support to major elephant range states in the country through Project Elephant.

Biosphere Reserve

Biosphere reserves are sites established by countries and recognized under UNESCO's Man and the Biosphere (MAB) Programme to promote sustainable development based on local community efforts and sound science. The programme of Biosphere Reserve was initiated by UNESCO in 1971. The purpose of the formation of the biosphere reserve is to conserve ***in-situ*** all forms of life, along with its support system, in its totality, so that it could serve as a referral system for monitoring and evaluating changes in natural ecosystems. The first biosphere reserve of the world was established in 1979, since then the network of biosphere reserves has increased to 631 in 119 countries across the world.

Presently, there are 18 notified biosphere reserves in India

S.No.	Name	Date of Notification	Area (in Km^2)	Location (State)
1	Nilgiri	01.09.1986	5520 (Core 1240 & Buffer 4280)	Part of Wayanad, Nagarhole, Bandipur and Madumalai, Nilambur, Silent Valley and Siruvani hills (Tamil Nadu, Kerala and Karnataka)
2	Nanda Devi	18.01.1988	5860.69 (Core 712.12, Buffer 5,148.570) & T. 546.34)	Part of Chamoli, Pithoragarh, and Bageshwar districts (Uttarakhand)

S.No.	Name	Date of Notification	Area (in Km^2)	Location (State)
3	Nokrek	01.09.1988	820 (Core 47.48 & Buffer 227.92, Transition Zone 544.60)	Part of Garo hills (Meghalaya)
4	Great Nicobar	06.01.1989	885 (Core 705 & Buffer 180)	Southern most islands of Andaman And Nicobar (A&N Islands)
5	Gulf of Mannar	18.02.1989	10,500 Total Gulf area (area of Islands 5.55)	Indian part of Gulf of Mannar between India and Sri Lanka (Tamil Nadu)
6	Manas	14.03.1989	2837 (Core 391 & Buffer 2,446)	Part of Kokrajhar, Bongaigaon, Barpeta, Nalbari, Kamprup and Darang districts (Assam)
7	Sunderbans	29.03.1989	9630 (Core 1700 & Buffer 7900)	Part of delta of Ganges and Brahamaputra river system (West Bengal)
8	Simlipal	21.06.1994	4374 (Core 845, Buffer 2129 & Transition 1400)	Part of Mayurbhanj district (Orissa)
9	Dibru-Saikhowa	28.07.1997	765 (Core 340 & Buffer 425)	Part of Dibrugarh and Tinsukia Districts (Assam)
10	Dehang-Dibang	02.09.1998	5111.50 (Core 4094.80 & Buffer 1016.70)	Part of Siang and Dibang Valley in Arunachal Pradesh
11	Pachmarhi	03.03.1999	4926	Parts of Betul, Hoshangabad and Chindwara districts of Madhya Pradesh
12	Khangchendzonga	07.02.2000	2619.92 (Core 1819.34 & Buffer 835.92)	Parts of Khangchendzonga hills and Sikkim
13	Agasthyamalai	12.11.2001	1828	Neyyar, Peppara and Shendurney Wildlife Sanctuaries and their adjoining areas in Kerala
14	Achanakamar- - Amarkantak	30.3.2005	3835.51 (Core 551.55 & Buffer 3283.86)	Covers parts of Anupur and Dindori districts of M.P. and parts of Bilaspur districts of Chhattishgarh State

S.No.	Name	Date of Notification	Area (in Km^2)	Location (State)
15	Kachchh	29.01.2008	12,454	Part of Kachchh, Rajkot, Surendra Nagar and Patan Civil Districts of Gujarat State
16	Cold Desert	28.08.2009	7770	Pin Valley National Park and surroundings; Chandratal and Sarchu & Kibber Wildlife Sancturary in Himachal Pradesh
17	Seshachalam Hills	20.09.2010	4755.997	Seshachalam Hill Ranges covering parts of Chittoor and Kadapa districts of Andhra Pradesh
18	Panna	25.08.2011	2998.98	Part of Panna and Chhattarpur districts in Madhya Pradesh

RAMSAR Wetland Sites

The Ramsar Convention is an international treaty for the conservation and sustainable utilization of wetlands, recognizing the fundamental ecological functions of wetlands and their economic, cultural, scientific, and recreational value.

S.No.	Name of Site	State Location	Date of Declaration	Area (in Sq. km.)
1	Asthamudi Wetland	Kerala	19.8.2002	614
2	Bhitarkanika Mangroves	Orissa	19.8.2002	650
3	Bhoj Wetlands	Madhya Pradesh	19.8.2002	32.01
4	Chandertal Wetland	Himachal Pradesh	8.11.2005	0.49
5	Chilka Lake	Orissa	1.10.1981	1165
6	Deepor Beel	Assam	19.8.2002	40
7	East Calcutta Wetlands	West Bengal	19.8.2002	125
8	Harike Lake	Punjab	23.3.1990	41
9	Hokera Wetland	Jammu and Kashmir	8.11.2005	13.75
10	Kanjli Lake	Punjab	22.1.2002	1.83
11	Keoladeo Ghana NP	Rajasthan	1.10.1981	28.73
12	Kolleru Lake	Andhra Pradesh	19.8.2002	901
13	Loktak Lake	Manipur	23.3.1990	266

S.No.	Name of Site	State Location	Date of Declaration	Area (in Sq. km.)
14	Nalsarovar Bird Sanctuary	Gujarat	24.09.2012	120
15	Point Calimere	Tamil Nadu	19.8.2002	385
16	Pong Dam Lake	Himachal Pradesh	19.8.2002	156.62
17	Renuka Wetland	Himachal Pradesh	8.11.2005	0.2
18	Ropar Lake	Punjab	22.1.2002	13.65
19	Rudrasagar Lake	Tripura	8.11.2005	2.4
20	Sambhar Lake	Rajasthan	23.3.1990	240
21	Sasthamkotta Lake	Kerala	19.8.2002	3.73
22	Sunderbans Wetland	West Bengal	30.1.2019	4230
23	Surinsar-Mansar Lakes	Jammu and Kashmir	8.11.2005	3.5
24	Tsomoriri Lake	Jammu and Kashmir	19.8.2002	120
25	Upper Ganga River (Brijghat to Narora Stretch)	Uttar Pradesh	8.11.2005	265.9
26	Vembanad Kol Wetland	Kerala	19.8.2002	1512.5
27	Wular Lake	Jammu & Kashmir	23.3.1990	189
		Total Area (in Sq. Km.)		**11121.31**

(*Source*: Ministry of Environment & Forests, Government of India)

World Heritage Sites

A UNESCO World Heritage Site is a place that is listed by the United Nations Educational, Scientific and Cultural Organization as of special cultural or physical significance.

Natural World Heritage Sites in India (As on July, 2016)

S. No.	Name of WH Site	State Location	Year of Notification	Area (sq.km)
1	Great Himalayan National Park Conservation Area	Himachal Pradesh	2014	905.4
2	Western Ghats	Maharashtra, Goa, Karnataka, Tamil Nadu and Kerala	2012	7,953.15
3	Nanda Devi and Valley of Flowers National Parks	Uttarakhand	1988	630.00 87.50
4	Sundarbans National Park	West Bengal	1987	1,330.10
5	Kaziranga National Park	Assam	1985	429.96

S. No.	Name of WH Site	State Location	Year of Notification	Area (sq.km)
6	Keoladeo National Park	Rajasthan	1985	28.73
7	Manas Wildlife Sanctuary	Assam	1985	391.00

Cultural World Heritage Sites in India (As on January, 2018)

S.No.	Name of WH Site	State Location	Year of Notification
1	Historic City of Ahmadabad	Gujarat	2017
2	Archaeological Site of Nalanda Mahavihara (Nalanda University) at Nalanda	Bihar	2016
3	The Architectural Work of Le Corbusier, an Outstanding Contribution to the Modern Movement	Chandigarh	2016
4	Rani-ki-Vav (the Queen's Stepwell) at Patan	Gujarat	2014
5	Hill Forts of Rajasthan	Rajasthan	2013
6	The Jantar Mantar, Jaipur	Rajasthan	2010
7	Red Fort Complex	Delhi	2007
8	Champaner-Pavagadh Archaeological Park	Gujarat	2004
9	Chhatrapati Shivaji Terminus (formerly Victoria Terminus)	Maharashtra	2004
10	Rock Shelters of Bhimbetka	Madhya Pradesh	2003
11	Mahabodhi Temple Complex at Bodh Gaya	Bihar	2002
12	Mountain Railways of India	West Bengal	1999
13	Humayun's Tomb	Delhi	1993
14	Qutb Minar and its Monuments	Delhi	1993
15	Buddhist Monuments at Sanchi	Madhya Pradesh	1989
16	Elephanta Caves	Maharashtra	1987
17	Great Living Chola Temples	Tamil Nadu	1987
18	Group of Monuments at Pattadakal	Karnataka	1987
19	Churches and Convents of Goa	Goa	1986
20	Fatehpur Sikri	Uttar Pradesh	1986
21	Group of Monuments at Hampi	Karnataka	1986
22	Khajuraho Group of Monuments	Madhya Pradesh	1986
23	Group of Monuments at Mahabalipuram	Tamil Nadu	1984
24	Sun Temple, Konârak	Odisha	1984
25	Agra Fort	Uttar Pradesh	1983
26	Ajanta Caves	Maharashtra	1983
27	Ellora Caves	Maharashtra	1983
28	Taj Mahal	Uttar Pradesh	1983

Mixed World Heritage Sites in India (As on July, 2016)

S. No.	Name of WH Site	State Location	Year of Notification
1	Khangchendzonga National Park	Sikkim	2016

(*Source*: UNESCO World Heritage Convention: http://whc.unesco.org/en/statesparties/in-title=World)

National Park

An area, whether within a sanctuary or not, can be notified by the state government to be constituted as a National Park, by reason of its ecological, faunal, floral, geomorphological, or zoological association or importance, needed to for the purpose of protecting & propagating or developing wildlife therein or its environment. No human activity is permitted inside the national park except for the ones permitted by the Chief Wildlife Warden of the state under the conditions given in CHAPTER IV, WPA 1972. There are 104 existing national parks in India covering an area of 40,501.13 Km^2, which is 1.23% of the geographical area of the country (National Wildlife Database, May, 2019).

State	S. No.	Name of State/ Protected Area	Year of Establishment	Area (Km^2)
Andaman & Nicobar Islands	1	Campbell Bay NP	1992	426.23
Andaman & Nicobar Islands	2	Galathea Bay NP	1992	110
Andaman & Nicobar Islands	3	Mahatama Gandhi Marine (Wandoor) NP	1983	281.5
Andaman & Nicobar Islands	4	Middle Button Island NP	1987	0.44
Andaman & Nicobar Islands	5	Mount Harriett NP	1987	46.62
Andaman & Nicobar Islands	6	North Button Island NP	1987	0.44
Andaman & Nicobar Islands	7	Rani Jhansi Marine NP	1996	256.14
Andaman & Nicobar Islands	8	Saddle Peak NP	1987	32.54
Andaman & Nicobar Islands	9	South Button Island NP	1987	0.03
Andhra Pradesh	1	Papikonda NP	2008	1012.86
Andhra Pradesh	2	Rajiv Gandhi (Rameswaram) NP	2005	2.4

State	S. No.	Name of State/ Protected Area	Year of Establishment	Area (Km²)
Andhra Pradesh	3	Sri Venkateswara NP	1989	353.62
Arunachal Pradesh	1	Mouling NP	1986	483
Arunachal Pradesh	2	Namdapha NP	1983	1807.82
Assam	1	Dibru-Saikhowa NP	1999	340
Assam	2	Kaziranga NP	1974	858.98
Assam	3	Manas NP	1990	500
Assam	4	Nameri NP	1998	200
Assam	5	Rajiv Gandhi Orang NP	1999	78.81
Bihar	1	Valmiki NP	1989	335.65
Chhattisgarh	1	Guru Ghasidas (Sanjay) NP	1981	1440.705
Chhattisgarh	2	Indravati (Kutru) NP	1982	1258.37
Chhattisgarh	3	Kanger Valley NP	1982	200
Goa	1	Mollem NP	1992	107
Gujarat	1	Vansda NP	1979	23.99
Gujarat	2	Blackbuck (Velavadar) NP	1976	34.53
Gujarat	3	Gir NP	1975	258.71
Gujarat	4	Marine (Gulf of Kachchh) NP	1982	162.89
Haryana	1	Kalesar NP	2003	46.82
Haryana	2	Sultanpur NP	1989	1.43
Himachal Pradesh	1	Great Himalayan NP	1984	754.4
Himachal Pradesh	2	Inderkilla NP	2010	104
Himachal Pradesh	3	Khirganga NP	2010	710
Himachal Pradesh	4	Pin Valley NP	1987	675
Himachal Pradesh	5	Simbalbara NP	2010	27.88
Jammu & Kashmir	1	City Forest (Salim Ali) NP	1992	9
Jammu & Kashmir	2	Dachigam NP	1981	141
Jammu & Kashmir	3	Hemis NP	1981	3350
Jammu & Kashmir	4	Kishtwar NP	1981	425
Jharkhand	1	Betla NP	1986	226.33
Karnataka	1	Anshi NP	1987	417.34
Karnataka	2	Bandipur NP	1974	874.2
Karnataka	3	Bannerghatta NP	1974	260.51
Karnataka	4	Kudremukh NP	1987	600.32

State	S. No.	Name of State/ Protected Area	Year of Establishment	Area (Km^2)
Karnataka	5	Nagarahole (Rajiv Gandhi) NP	1988	643.39
Kerala	1	Anamudi Shola NP	2003	7.5
Kerala	2	Eravikulam NP	1978	97
Kerala	3	Mathikettan Shola NP	2003	12.82
Kerala	4	Pambadum Shola NP	2003	1.318
Kerala	5	Periyar NP	1982	350
Kerala	6	Silent Valley NP	1984	89.52
Madhya Pradesh	1	Bandhavgarh NP	1968	448.85
Madhya Pradesh	2	Dinosaur Fossils NP	2011	0.8974
Madhya Pradesh	3	Fossil NP	1983	0.27
Madhya Pradesh	4	Indira Priyadarshini Pench NP	1975	292.85
Madhya Pradesh	5	Kanha NP	1955	940
Madhya Pradesh	6	Madhav NP	1959	375.22
Madhya Pradesh	7	Panna NP	1981	542.67
Madhya Pradesh	8	Sanjay NP	1981	466.88
Madhya Pradesh	9	Satpura NP	1981	585.17
Madhya Pradesh	10	Van Vihar NP	1979	4.45
Maharashtra	1	Chandoli NP	2004	317.67
Maharashtra	2	Gugamal NP	1975	361.28
Maharashtra	3	Nawegaon NP	1975	133.88
Maharashtra	4	Pench (Jawaharlal Nehru) NP	1975	257.26
Maharashtra	5	Sanjay Gandhi (Borivilli) NP	1983	86.96
Maharashtra	6	Tadoba NP	1955	116.55
Manipur	1	Keibul-Lamjao NP	1977	40
Meghalaya	1	Balphakram NP	1985	220
Meghalaya	2	Nokrek Ridge NP	1986	47.48
Mizoram	1	Murlen NP	1991	100
Mizoram	2	Phawngpui Blue Mountain NP	1992	50
Nagaland	1	Intanki NP	1993	202.02
Odisha	1	Bhitarkanika NP	1988	145
Odisha	2	Simlipal NP	1980	845.7

State	S. No.	Name of State/ Protected Area	Year of Establishment	Area (Km^2)
Rajasthan	1	Desert NP	1992	3162
Rajasthan	2	Keoladeo Ghana NP	1981	28.73
Rajasthan	3	Mukundra Hills NP	2006	200.54
Rajasthan	4	Ranthambhore NP	1980	282
Rajasthan	5	Sariska NP	1992	273.8
Sikkim	1	Khangchendzonga NP	1977	1784
Tamil Nadu	1	Guindy NP	1976	2.82
Tamil Nadu	2	Gulf of Mannar Marine NP	1980	6.23
Tamil Nadu	3	Indira Gandhi (Annamalai) NP	1989	117.1
Tamil Nadu	4	Mudumalai NP	1990	103.23
Tamil Nadu	5	Mukurthi NP	1990	78.46
Telangana	1	Kasu Brahmananda Reddy NP	1994	1.43
Telangana	2	Mahaveer Harina Vanasthali NP	1994	14.59
Telangana	3	Mrugavani NP	1994	3.6
Tripura	1	Clouded Leopard NP	2007	5.08
Tripura	2	Bison (Rajbari) NP	2007	31.63
Uttar Pradesh	1	Dudhwa NP	1977	490
Uttarakhand	1	Corbett NP	1936	520.82
Uttarakhand	2	Gangotri NP	1989	2390.02
Uttarakhand	3	Govind NP	1990	472.08
Uttarakhand	4	Nanda Devi NP	1982	624.6
Uttarakhand	5	Rajaji NP	1983	820
Uttarakhand	6	Valley of Flowers NP	1982	87.5
West Bengal	1	Buxa NP	1992	117.1
West Bengal	2	Gorumara NP	1992	79.45
West Bengal	3	Jaldapara NP	2014	216.51
West Bengal	4	Neora Valley NP	1986	159.89
West Bengal	5	Singalila NP	1986	78.6
West Bengal	6	Sunderban NP	1984	1330.1

(*Source*: National Wildlife Database, Wildlife Institute of India)

Wildlife Sanctuary

Any area other than area comprised with any reserve forest or the territorial waters can be notified by the State Government to constitute as a sanctuary if such area is of adequate ecological, faunal, floral, geomorphological, natural or zoological significance, for the purpose of protecting, propagating or developing wildlife or its environment. Some restricted human activities are allowed inside the Sanctuary area details of which are given in CHAPTER IV, WPA 1972. There are 551 existing wildlife sanctuaries in India covering an area of 1,19,775.80 km^2, which is 3.64% of the geographical area of the country (National Wildlife Database, May, 2019).

List of Wildlife Sanctuaries (May, 2019)

State	S. No.	Name of State/ Protected Area	Year of Establishment	Area (Km^2)
Andaman & Nicobar Islands	1	Arial Island WLS	1987	0.05
Andaman & Nicobar Islands	2	Bamboo Island WLS	1987	0.05
Andaman & Nicobar Islands	3	Barren Island WLS	1987	8.1
Andaman & Nicobar Islands	4	Battimalv Island WLS	1987	2.23
Andaman & Nicobar Islands	5	Belle Island WLS	1987	0.08
Andaman & Nicobar Islands	6	Benett Island WLS	1987	3.46
Andaman & Nicobar Islands	7	Bingham Island WLS	1987	0.08
Andaman & Nicobar Islands	8	Blister Island WLS	1987	0.26
Andaman & Nicobar Islands	9	Bluff Island WLS	1987	1.14
Andaman & Nicobar Islands	10	Bondoville Island WLS	1987	2.55
Andaman & Nicobar Islands	11	Brush Island WLS	1987	0.23
Andaman & Nicobar Islands	12	Buchanan Island WLS	1987	9.33
Andaman & Nicobar Islands	13	Chanel Island WLS	1987	0.13

State	S. No.	Name of State/ Protected Area	Year of Establishment	Area (Km²)
Andaman & Nicobar Islands	14	Cinque Islands WLS	1987	9.51
Andaman & Nicobar Islands	15	Clyde Island WLS	1987	0.54
Andaman & Nicobar Islands	16	Cone Island WLS	1987	0.65
Andaman & Nicobar Islands	17	Curlew (B.P.) Island WLS	1987	0.16
Andaman & Nicobar Islands	18	Curlew Island WLS	1987	0.03
Andaman & Nicobar Islands	19	Cuthbert Bay WLS	1997	5.82
Andaman & Nicobar Islands	20	Defence Island WLS	1987	10.49
Andaman & Nicobar Islands	21	Dot Island WLS	1987	0.13
Andaman & Nicobar Islands	22	Dottrell Island WLS	1987	0.13
Andaman & Nicobar Islands	23	Duncan Island WLS	1987	0.73
Andaman & Nicobar Islands	24	East Island WLS	1987	6.11
Andaman & Nicobar Islands	25	East of Inglis Island WLS	1987	3.55
Andaman & Nicobar Islands	26	Egg Island WLS	1987	0.05
Andaman & Nicobar Islands	27	Elat Island WLS	1987	9.36
Andaman & Nicobar Islands	28	Entrance Island WLS	1987	0.96
Andaman & Nicobar Islands	29	Gander Island WLS	1987	0.05
Andaman & Nicobar Islands	30	Galathea Bay WLS	1997	11.44
Andaman & Nicobar Islands	31	Girjan Island WLS	1987	0.16
Andaman & Nicobar Islands	32	Goose Island WLS	1987	0.01

State	S. No.	Name of State/ Protected Area	Year of Establishment	Area (Km²)
Andaman & Nicobar Islands	33	Hump Island WLS	1987	0.47
Andaman & Nicobar Islands	34	Interview Island WLS	1987	133.87
Andaman & Nicobar Islands	35	James Island WLS	1987	2.1
Andaman & Nicobar Islands	36	Jungle Island WLS	1987	0.52
Andaman & Nicobar Islands	37	Kwangtung Island WLS	1987	0.57
Andaman & Nicobar Islands	38	Kyd Island WLS	1987	8
Andaman & Nicobar Islands	39	Landfall Island WLS	1987	29.48
Andaman & Nicobar Islands	40	Latouche Island WLS	1987	0.96
Andaman & Nicobar Islands	41	Lohabarrack (Saltwater Crocodile) WLS	1987	22.21
Andaman & Nicobar Islands	42	Mangrove Island WLS	1987	0.39
Andaman & Nicobar Islands	43	Mask Island WLS	1987	0.78
Andaman & Nicobar Islands	44	Mayo Island WLS	1987	0.1
Andaman & Nicobar Islands	45	Megapode Island WLS	1987	0.12
Andaman & Nicobar Islands	46	Montogemery Island WLS	1987	0.21
Andaman & Nicobar Islands	47	Narcondam Island WLS	1987	6.81
Andaman & Nicobar Islands	48	North Brother Island WLS	1987	0.75
Andaman & Nicobar Islands	49	North Island WLS	1987	0.49
Andaman & Nicobar Islands	50	North Reef Island WLS	1987	3.48
Andaman & Nicobar Islands	51	Oliver Island WLS	1987	0.16

State	S. No.	Name of State/ Protected Area	Year of Establishment	Area (Km^2)
Andaman & Nicobar Islands	52	Orchid Island WLS	1987	0.1
Andaman & Nicobar Islands	53	Ox Island WLS	1987	0.13
Andaman & Nicobar Islands	54	Oyster Island-I WLS	1987	0.08
Andaman & Nicobar Islands	55	Oyster Island-II WLS	1987	0.21
Andaman & Nicobar Islands	56	Paget Island WLS	1987	7.36
Andaman & Nicobar Islands	57	Parkinson Island WLS	1987	0.34
Andaman & Nicobar Islands	58	Passage Island WLS	1987	0.62
Andaman & Nicobar Islands	59	Patric Island WLS	1987	0.13
Andaman & Nicobar Islands	60	Peacock Island WLS	1987	0.62
Andaman & Nicobar Islands	61	Pitman Island WLS	1987	1.37
Andaman & Nicobar Islands	62	Point Island WLS	1987	3.07
Andaman & Nicobar Islands	63	Potanma Islands WLS	1987	0.16
Andaman & Nicobar Islands	64	Ranger Island WLS	1987	4.26
Andaman & Nicobar Islands	65	Reef Island WLS	1987	1.74
Andaman & Nicobar Islands	66	Roper Island WLS	1987	1.46
Andaman & Nicobar Islands	67	Ross Island WLS	1987	1.01
Andaman & Nicobar Islands	68	Rowe Island WLS	1987	0.01
Andaman & Nicobar Islands	69	Sandy Island WLS	1987	1.58
Andaman & Nicobar Islands	70	Sea Serpent Island WLS	1987	0.78

State	S. No.	Name of State/ Protected Area	Year of Establishment	Area (Km^2)
Andaman & Nicobar Islands	71	Shark Island WLS	1987	0.6
Andaman & Nicobar Islands	72	Shearme Island WLS	1987	7.85
Andaman & Nicobar Islands	73	Sir Hugh Rose Island WLS	1987	1.06
Andaman & Nicobar Islands	74	Sisters Island WLS	1987	0.36
Andaman & Nicobar Islands	75	Snake Island-I WLS	1987	0.73
Andaman & Nicobar Islands	76	Snake Island-II WLS	1987	0.03
Andaman & Nicobar Islands	77	South Brother Island WLS	1987	1.24
Andaman & Nicobar Islands	78	South Reef Island WLS	1987	1.17
Andaman & Nicobar Islands	79	South Sentinel Island WLS	1987	1.61
Andaman & Nicobar Islands	80	Spike Island-I WLS	1987	0.42
Andaman & Nicobar Islands	81	Spike Island-II WLS	1987	11.7
Andaman & Nicobar Islands	82	Stoat Island WLS	1987	0.44
Andaman & Nicobar Islands	83	Surat Island WLS	1987	0.31
Andaman & Nicobar Islands	84	Swamp Island WLS	1987	4.09
Andaman & Nicobar Islands	85	Table (Delgarno) Island WLS	1987	2.29
Andaman & Nicobar Islands	86	Table (Excelsior) Island WLS	1987	1.69
Andaman & Nicobar Islands	87	Talabaicha Island WLS	1987	3.21
Andaman & Nicobar Islands	88	Temple Island WLS	1987	1.04
Andaman & Nicobar Islands	89	Tillongchang Island WLS	1985	16.83

State	S. No.	Name of State/ Protected Area	Year of Establishment	Area (Km²)
Andaman & Nicobar Islands	90	Tree Island WLS	1987	0.03
Andaman & Nicobar Islands	91	Trilby Island WLS	1987	0.96
Andaman & Nicobar Islands	92	Tuft Island WLS	1987	0.29
Andaman & Nicobar Islands	93	Turtle Islands WLS	1987	0.39
Andaman & Nicobar Islands	94	West Island WLS	1987	6.4
Andaman & Nicobar Islands	95	Wharf Island WLS	1987	0.11
Andaman & Nicobar Islands	96	White Cliff Island WLS	1987	0.47
Andhra Pradesh	1	Coringa WLS	1978	235.7
Andhra Pradesh	2	Gundla Brahmeswaram WLS	1990	1194
Andhra Pradesh	3	Kambalakonda WLS	2002	71.39
Andhra Pradesh	4	Koundinya WLS	1990	357.6
Andhra Pradesh	5	Kolleru WLS	1953	308.55
Andhra Pradesh	6	Krishna WLS	1989	194.81
Andhra Pradesh	7	Nagarjuna Sagar-Srisailam WLS	1978	1401.81
Andhra Pradesh	8	Nellapattu WLS	1976	4.5892
Andhra Pradesh	9	Pulicat Lake WLS	1976	500
Andhra Pradesh	10	Rollapadu WLS	1988	6.14
Andhra Pradesh	11	Sri Lankamalleswara WLS	1988	464.42
Andhra Pradesh	12	Sri Penusila Narasimha WLS	1997	1030.85
Andhra Pradesh	13	Sri Venkateswara WLS	1985	172.35
Arunachal Pradesh	1	D'Ering Memorial (Lali) WLS	1978	190
Arunachal Pradesh	2	Dibang WLS	1991	4149
Arunachal Pradesh	3	Eagle Nest WLS	1989	217
Arunachal Pradesh	4	Itanagar WLS	1978	140.3
Arunachal Pradesh	5	Kamlang WLS	1989	783
Arunachal Pradesh	6	Kane WLS	1991	31

State	S. No.	Name of State/ Protected Area	Year of Establishment	Area (Km^2)
Arunachal Pradesh	7	Mahao WLS	1980	281.5
Arunachal Pradesh	8	Pakke (Pakhui) WLS	1977	861.95
Arunachal Pradesh	9	Sessa Orchid WLS	1989	100
Arunachal Pradesh	10	Tale WLS	1995	337
Arunachal Pradesh	11	Yordi Rabe Supse WLS	1996	397
Assam	1	Amchang WLS	2004	78.64
Assam	2	Barail WLS	2004	326.24
Assam	3	Barnadi WLS	1980	26.22
Assam	4	Bherjan-Borajan-Padumoni WLS	1999	7.22
Assam	5	Burachapari WLS	1995	44.06
Assam	6	Chakrasila WLS	1994	45.57
Assam	7	Deepor Beel WLS	1989	4.14
Assam	8	Dihing Patkai WLS	2004	111.19
Assam	9	East Karbi Anglong WLS	2000	221.81
Assam	10	Garampani WLS	1952	6.05
Assam	11	Hollongapar Gibbon WLS	1997	20.98
Assam	12	Lawkhowa WLS	1972	70.13
Assam	13	Marat Longri WLS	2003	451
Assam	14	Nambor WLS	2000	37
Assam	15	Nambor-Doigrung WLS	2003	97.15
Assam	16	Pabitora WLS	1987	38.81
Assam	17	Pani-Dihing Bird WLS	1995	33.93
Assam	18	Sonai Rupai WLS	1998	220
Bihar	1	Barela Jheel Salim Ali Bird WLS	1997	1.96
Bihar	2	Bhimbandh WLS	1976	681.99
Bihar	3	Gautam Budha WLS	1976	138.34
Bihar	4	Kaimur WLS	1982	1342
Bihar	5	Kanwarjheel WLS	1989	63.11
Bihar	6	Kusheshwar Asthan Bird WLS	1994	29.17
Bihar	7	Nagi Dam WLS	1987	1.92
Bihar	8	Nakti Dam WLS	1987	3.33
Bihar	9	Pant (Rajgir) WLS	1978	35.84

State	S. No.	Name of State/ Protected Area	Year of Establishment	Area (Km^2)
Bihar	10	Udaipur WLS	1978	8.87
Bihar	11	Valmiki WLS	1978	545.15
Bihar	12	Vikramshila Gangetic Dolphin WLS	1990	50
Chandigarh	1	City Bird WLS	1998	0.029
Chandigarh	2	Sukhna Lake WLS	1986	25.98
Chhattisgarh	1	Achanakmar WLS	1975	551.55
Chhattisgarh	2	Badalkhol WLS	1975	104.45
Chhattisgarh	3	Barnawapara WLS	1976	244.66
Chhattisgarh	4	Bhairamgarh WLS	1983	138.95
Chhattisgarh	5	Bhoramdev WLS	2001	351.24
Chhattisgarh	6	Sarangarh-Gomardha WLS	1975	277.82
Chhattisgarh	7	Pamed Wild Buffalo WLS	1985	262.12
Chhattisgarh	8	Semarsot WLS	1978	430.35
Chhattisgarh	9	Sitanadi WLS	1974	553.36
Chhattisgarh	10	Tamor Pingla WLS	1978	608.51
Chhattisgarh	11	Udanti Wild Buffalo WLS	1985	237.27
Dadra & Nagar Haveli	1	Dadra & Nagar Haveli WLS	2000	92.16
Daman & Diu	1	Fudam WLS	1991	2.18
Delhi	1	Asola Bhati (Indira Priyadarshini) WLS	1992	27.82
Goa	1	Bhagwan Mahavir WLS	1967	133
Goa	2	Bondla WLS	1969	7.95
Goa	3	Cotigaon WLS	1968	85.65
Goa	4	Dr. Salim Ali Bird (Chorao) WLS	1988	1.78
Goa	5	Madei WLS	1999	208.48
Goa	6	Netravali WLS	1999	211.05
Gujarat	1	Balaram Ambaji WLS	1989	542.08
Gujarat	2	Barda WLS	1979	192.31
Gujarat	3	Gaga (Great Indian Bustard) WLS	1988	3.33
Gujarat	4	Gir WLS	1965	1153.42
Gujarat	5	Girnar WLS	2008	178.8
Gujarat	6	Hingolgadh WLS	1980	6.54

State	S. No.	Name of State/ Protected Area	Year of Establishment	Area (Km²)
Gujarat	7	Jambughoda WLS	1990	130.38
Gujarat	8	Jessore Sloth Bear WLS	1978	180.66
Gujarat	9	Kachchh (Lala) Great Indian Bustard WLS	1995	2.03
Gujarat	10	Kachchh Desert WLS	1986	7506.22
Gujarat	11	Khijadiya Bird WLS	1981	6.05
Gujarat	12	Marine (Gulf of Kachchh) WLS	1980	295.03
Gujarat	13	Mitiyala WLS	2004	18.22
Gujarat	14	Nal Sarovar Bird WLS	1969	120.82
Gujarat	15	Narayan Sarovar Chinkara WLS	1995	442.91
Gujarat	16	Paniya WLS	1989	39.63
Gujarat	17	Porbandar Bird WLS	1988	0.09
Gujarat	18	Purna WLS	1990	160.84
Gujarat	19	Rampara Vidi WLS	1988	15.01
Gujarat	20	Ratanmahal Sloth Bear WLS	1982	55.65
Gujarat	21	Shoolpaneswar (Dhumkhal) WLS	1982	607.7
Gujarat	22	Thol Lake WLS	1988	6.99
Gujarat	23	Wild Ass WLS	1973	4953.71
Haryana	1	Abubshehar WLS	1987	115.3
Haryana	2	Bhindawas Lake WLS	1986	4.12
Haryana	3	Bir Shikargarh WLS	1987	7.67
Haryana	4	Chhilchhila Lake WLS	1986	0.29
Haryana	5	Kalesar WLS	1996	54.06
Haryana	6	Khaparwas WLS	1991	0.83
Haryana	7	Morni Hills (Khol-Hi-Raitan) WLS	2004	48.83
Haryana	8	Nahar WLS	1987	2.11
Himachal Pradesh	1	Bandli WLS	1962	32.11
Himachal Pradesh	2	Chail WLS	1976	16
Himachal Pradesh	3	Chandratal WLS	2007	38.56
Himachal Pradesh	4	Churdhar WLS	1985	55.52
Himachal Pradesh	5	Daranghati WLS	1962	171.5

State	S. No.	Name of State/ Protected Area	Year of Establishment	Area (Km^2)
Himachal Pradesh	6	Dhauladhar WLS	1994	982.86
Himachal Pradesh	7	Gamgul Siyabehi WLS	1962	108.4
Himachal Pradesh	8	Kais WLS	1954	12.61
Himachal Pradesh	9	Kalatop-Khajjiar WLS	1958	17.17
Himachal Pradesh	10	Kanawar WLS	1954	107.29
Himachal Pradesh	11	Khokhan WLS	1954	14.94
Himachal Pradesh	12	Kibber WLS	1992	2220.12
Himachal Pradesh	13	Kugti WLS	1962	405.49
Himachal Pradesh	14	Lippa Asrang WLS	1962	31
Himachal Pradesh	15	Majathal WLS	1954	30.86
Himachal Pradesh	16	Manali WLS	1954	29
Himachal Pradesh	17	Nargu WLS	1962	132.3731
Himachal Pradesh	18	Pong Dam Lake WLS	1982	207.59
Himachal Pradesh	19	Rakchham Chitkul (Sangla Valley) WLS	1989	304
Himachal Pradesh	20	Renuka WLS	2013	4
Himachal Pradesh	21	Rupi Bhaba WLS	1982	503
Himachal Pradesh	22	Sainj WLS	1994	90
Himachal Pradesh	23	Sech Tuan Nala WLS	1962	390.29
Himachal Pradesh	24	Shikari Devi WLS	1962	29.94
Himachal Pradesh	25	Shimla Water Catchment WLS	1958	10
Himachal Pradesh	26	Talra WLS	1962	46.48
Himachal Pradesh	27	Tirthan WLS	1992	61
Himachal Pradesh	28	Tundah WLS	1962	64
Jammu & Kashmir	1	Baltal-Thajwas WLS	1987	210.5
Jammu & Kashmir	2	Changthang WLS	1987	4000
Jammu & Kashmir	3	Gulmarg WLS	1987	180
Jammu & Kashmir	4	Hirapora WLS	1987	110
Jammu & Kashmir	5	Hokersar WLS	1992	13.75
Jammu & Kashmir	6	Jasrota WLS	1987	25.75
Jammu & Kashmir	7	Karakoram (Nubra Shyok) WLS	1987	5000
Jammu & Kashmir	8	Lachipora WLS	1987	80
Jammu & Kashmir	9	Limber WLS	1987	26

State	S. No.	Name of State/ Protected Area	Year of Establishment	Area (Km^2)
Jammu & Kashmir	10	Nandni WLS	1981	33.34
Jammu & Kashmir	11	Overa-Aru WLS	1987	425
Jammu & Kashmir	12	Rajparian (Daksum) WLS	2002	20
Jammu & Kashmir	13	Ramnagar Rakha WLS	1981	31.5
Jammu & Kashmir	14	Surinsar Mansar WLS	1981	55.5
Jammu & Kashmir	15	Trikuta WLS	1981	31.77
Jharkhand	1	Dalma WLS	1976	193.22
Jharkhand	2	Gautam Budha	1976	121.14
Jharkhand	3	Hazaribagh WLS	1976	186.25
Jharkhand	4	Kodarma WLS	1985	177.35
Jharkhand	5	Lawalong WLS	1978	211.03
Jharkhand	6	Mahuadanr Wolf WLS	1976	63.26
Jharkhand	7	Palamau WLS	1976	752.94
Jharkhand	8	Palkot WLS	1990	182.83
Jharkhand	9	Parasnath WLS	1984	49.33
Jharkhand	10	Topchanchi WLS	1978	12.82
Jharkhand	11	Udhwa Lake Bird WLS	1991	5.65
Karnataka	1	Adichunchunagiri Peacock WLS	1981	0.84
Karnataka	2	Arabithittu WLS	1985	13.5
Karnataka	3	Attiveri Bird WLS	1994	2.22
Karnataka	4	Bhadra WLS	1974	492.46
Karnataka	5	Bhimgad WLS	2010	190.42
Karnataka	6	Biligiri Rangaswamy Temple (B.R.T.) WLS	1987	539.52
Karnataka	7	Brahmagiri WLS	1974	181.29
Karnataka	8	Cauvery WLS	1987	1027.53
Karnataka	9	Chincholi WLS	2012	134.88
Karnataka	10	Dandeli WLS	1987	886.41
Karnataka	11	Daroji Bear WLS	1992	82.72
Karnataka	12	Ghataprabha Bird WLS	1974	29.79
Karnataka	13	Gudavi Bird WLS	1989	0.73
Karnataka	14	Gudekote Sloth Bear WLS	2013	38.48
Karnataka	15	Jogimatti WLS	2015	100.48
Karnataka	16	Malai Mahadeshwara WLS	2013	906.19

State	S. No.	Name of State/ Protected Area	Year of Establishment	Area (Km^2)
Karnataka	17	Melkote Temple WLS	1974	49.82
Karnataka	18	Mookambika WLS	1974	370.37
Karnataka	19	Nugu WLS	1974	30.32
Karnataka	20	Pushpagiri WLS	1987	102.96
Karnataka	21	Ranebennur Black Buck WLS	1974	119
Karnataka	22	Ranganathittu Bird WLS	1940	0.67
Karnataka	23	Ramadevara Betta Vulture WLS	2012	3.46
Karnataka	24	Rangayyanadurga Four-horned antelope	2011	77.24
Karnataka	25	Sharavathi Valley WLS	1974	431.23
Karnataka	26	Shettihalli WLS	1974	395.6
Karnataka	27	Someshwara WLS	1974	314.25
Karnataka	28	Yadahalli Chinkara WLS	2015	96.36
Karnataka	29	Talakaveri WLS	1987	105.01
Karnataka	30	Thimlapura WLS	2016	50.86
Kerala	1	Aralam WLS	1984	55
Kerala	2	Chimmony WLS	1984	85
Kerala	3	Chinnar WLS	1984	90.44
Kerala	4	Chulannur Peafowl WLS	2007	3.42
Kerala	5	Idukki WLS	1976	70
Kerala	6	Kottiyoor WLS	2011	30.38
Kerala	7	Kurinjimala WLS	2006	32
Kerala	8	Malabar WLS	2010	74.215
Kerala	9	Mangalavanam Bird WLS	2004	0.0274
Kerala	10	Neyyar WLS	1958	128
Kerala	11	Parambikulam WLS	1973	285
Kerala	12	Peechi-Vazhani WLS	1958	125
Kerala	13	Peppara WLS	1983	53
Kerala	14	Periyar WLS	1950	427
Kerala	15	Shendurney WLS	1984	100.32
Kerala	16	Thattekad Bird WLS	1983	25
Kerala	17	Wayanad WLS	1973	344.44
Lakshadweep	1	Pitti (Bird Island) WLS	1995	0.01

State	S. No.	Name of State/ Protected Area	Year of Establishment	Area (Km²)
Madhya Pradesh	1	Bagdara WLS	1978	478
Madhya Pradesh	2	Bori WLS	1977	485.72
Madhya Pradesh	3	Gandhi Sagar WLS	1981	368.62
Madhya Pradesh	4	Ghatigaon WLS	1981	511
Madhya Pradesh	5	Karera WLS	1981	202.21
Madhya Pradesh	6	Ken Gharial WLS	1981	45.2
Madhya Pradesh	7	Kheoni WLS	1982	122.7
Madhya Pradesh	8	Narsighgarh WLS	1978	59.19
Madhya Pradesh	9	National Chambal WLS	1978	435
Madhya Pradesh	10	Noradehi WLS	1984	1194.67
Madhya Pradesh	11	Orcha WLS	1994	44.91
Madhya Pradesh	12	Pachmarhi WLS	1977	417.78
Madhya Pradesh	13	Kuno WLS	1981	344.68
Madhya Pradesh	14	Panna (Gangau) WLS	1979	68.14
Madhya Pradesh	15	Panpatha WLS	1983	245.84
Madhya Pradesh	16	Pench WLS	1975	118.47
Madhya Pradesh	17	Phen WLS	1983	110.74
Madhya Pradesh	18	Ralamandal WLS	1989	2.35
Madhya Pradesh	19	Ratapani WLS	1978	823.84
Madhya Pradesh	20	Sailana WLS	1983	12.96
Madhya Pradesh	21	Sanjay Dubari WLS	1975	364.59
Madhya Pradesh	22	Sardarpur WLS	1983	348.12
Madhya Pradesh	23	Singhori WLS	1976	287.91
Madhya Pradesh	24	Son Gharial WLS	1981	41.8
Madhya Pradesh	25	Veerangna Durgavati WLS	1997	23.97
Maharashtra	1	Amba Barwa WLS	1997	127.11
Maharashtra	2	Andhari WLS	1986	509.27
Maharashtra	3	Aner Dam WLS	1986	82.94
Maharashtra	4	Bhamragarh WLS	1997	104.38
Maharashtra	5	Bhimashankar WLS	1985	130.78
Maharashtra	6	Bor WLS	1970	61.1
Maharashtra	7	Chaprala WLS	1986	134.78
Maharashtra	8	Deulgaon-Rehekuri WLS	1980	2.17
Maharashtra	9	Dhyanganga WLS	1997	205.23

State	S. No.	Name of State/ Protected Area	Year of Establishment	Area (Km²)
Maharashtra	10	New Maldhok Bird (Gangewadi) WLS	2012	1.98
Maharashtra	11	Gautala-Autramghat WLS	1986	260.61
Maharashtra	12	Ghodazari WLS	2018	159
Maharashtra	13	Great Indian Bustard WLS	1979	366.73
Maharashtra	14	Isapur WLS	2014	37.803
Maharashtra	15	Jaikwadi WLS	1986	341.05
Maharashtra	16	Kalsubai Harishchandragad WLS	1986	361.71
Maharashtra	17	Karnala Fort WLS	1968	4.48
Maharashtra	18	Karanja Sohal Blackbuck WLS	2000	18.32
Maharashtra	19	Katepurna WLS	1988	73.63
Maharashtra	20	Koka WLS	2013	100.138
Maharashtra	21	Koyana WLS	1985	423.55
Maharashtra	22	Lonar WLS	2000	1.17
Maharashtra	23	Malvan Marine WLS	1987	29.122
Maharashtra	24	Mansingdeo WLS	2010	182.59
Maharashtra	25	Mayureswar Supe WLS	1997	5.15
Maharashtra	26	Melghat WLS	1985	778.75
Maharashtra	27	Nagzira WLS	1970	152.81
Maharashtra	28	Naigaon Peacock WLS	1994	29.89
Maharashtra	29	Nandur Madhameshwar WLS	1986	100.12
Maharashtra	30	Narnala Bird WLS	1997	12.35
Maharashtra	31	Nawegaon WLS	2012	122.76
Maharashtra	32	New Bor WLS	2012	60.7
Maharashtra	33	New Nagzira WLS	2012	151.33
Maharashtra	34	Painganga WLS	1986	324.62
Maharashtra	35	Phansad WLS	1986	69.79
Maharashtra	36	Pranhita WLS	2014	420.06
Maharashtra	37	Radhanagari WLS	1958	351.16
Maharashtra	38	Sagareshwar WLS	1985	10.87
Maharashtra	39	Sudhagad WLS	2014	77.128
Maharashtra	40	Tamhini WLS	2013	49.62
Maharashtra	41	Tansa WLS	1970	304.81

State	S. No.	Name of State/ Protected Area	Year of Establishment	Area (Km²)
Maharashtra	42	Thane Creek Flamingo WLS	2015	16.91
Maharashtra	43	Tipeshwar WLS	1997	148.63
Maharashtra	44	Tungareshwar WLS	2003	85
Maharashtra	47	Umred-Kharngla WLS	2012	189.3
Maharashtra	48	Wan WLS	1997	211
Maharashtra	45	Yawal WLS	1969	177.52
Maharashtra	46	Yedsi Ramlin Ghat WLS	1997	22.38
Manipur	1	Yangoupokpi Lokchao WLS	1989	184.4
Manipur	2	Khongjaingamba Ching WLS	2016	0.412
Meghalaya	1	Baghmara Pitcher Plant WLS	1984	0.02
Meghalaya	2	Narpuh WLS	2014	59.9
Meghalaya	3	Nongkhyllem WLS	1981	29
Meghalaya	4	Siju WLS	1979	5.18
Mizoram	1	Dampa WLS	1985	500
Mizoram	2	Khawnglung WLS	1992	35
Mizoram	3	Lengteng WLS	1999	60
Mizoram	4	Ngengpui WLS	1991	110
Mizoram	5	Pualreng WLS	2004	50
Mizoram	6	Tawi WLS	1978	35.75
Mizoram	7	Thorangtlang WLS	2002	50
Mizoram	8	Tokalo WLS	2007	250
Nagaland	1	Fakim WLS	1980	6.41
Nagaland	2	Puliebadze WLS	1980	9.23
Nagaland	3	Rangapahar WLS	1986	4.7
Odisha	1	Badrama WLS	1962	304.03
Odisha	2	Baisipalli WLS	1981	168.35
Odisha	3	Balukhand Konark WLS	1984	71.72
Odisha	4	Bhitarkanika WLS	1975	525
Odisha	5	Chandaka Dampara WLS	1982	175.79
Odisha	6	Chilika (Nalaban) WLS	1987	15.53
Odisha	7	Debrigarh WLS	1985	346.91
Odisha	8	Gahirmatha (Marine) WLS	1997	1435

State	S. No.	Name of State/ Protected Area	Year of Establishment	Area (Km^2)
Odisha	9	Hadgarh WLS	1978	191.06
Odisha	10	Kapilash WLS	2011	125.5
Odisha	11	Karlapat WLS	1992	147.66
Odisha	12	Khalasuni WLS	1982	116
Odisha	13	Kothagarh WLS	1981	399.5
Odisha	14	Kuldiha WLS	1984	272.75
Odisha	15	Lakhari Valley WLS	1985	185.87
Odisha	16	Nandankanan WLS	1979	14.16
Odisha	17	Satkosia Gorge WLS	1976	745.52
Odisha	18	Simlipal WLS	1979	1354.3
Odisha	19	Sunabeda WLS	1988	500
Puducherry	1	Oussudu WLS	2008	3.9
Punjab	1	Abohar WLS	1988	186.5
Punjab	2	Bir Aishvan WLS	1952	2.64
Punjab	3	Bir Bhadson WLS	1952	10.23
Punjab	4	Bir Bunerheri WLS	1952	6.62
Punjab	5	Bir Dosanjh WLS	1952	5.18
Punjab	6	Bir Gurdialpura WLS	1977	6.2
Punjab	7	Bir Mehaswala WLS	1952	1.23
Punjab	8	Bir Motibagh WLS	1952	6.54
Punjab	9	Harike Lake WLS	1982	86
Punjab	10	Jhajjar Bacholi WLS	1998	1.16
Punjab	11	Kathlaur Kushlian WLS	2007	7.58
Punjab	12	Nangal WLS	2009	2.9
Punjab	13	Takhni-Rehampur WLS	1992	3.82
Rajasthan	1	Bandh Baratha WLS	1985	199.5
Rajasthan	2	Bassi WLS	1988	138.69
Rajasthan	3	Bhensrodgarh WLS	1983	229.14
Rajasthan	4	Darrah WLS	1955	80.75
Rajasthan	5	Jaisamand WLS	1955	52
Rajasthan	6	Jamwa Ramgarh WLS	1982	300
Rajasthan	7	Jawahar Sagar WLS	1975	153.41
Rajasthan	8	Kailadevi WLS	1983	676.38
Rajasthan	9	Kesarbagh WLS	1955	14.76
Rajasthan	10	Kumbhalgarh WLS	1971	608.58

State	S. No.	Name of State/ Protected Area	Year of Establishment	Area (Km²)
Rajasthan	11	Mount Abu WLS	1960	326.1
Rajasthan	12	Nahargarh WLS	1980	50
Rajasthan	13	National Chambal WLS	1979	274.75
Rajasthan	14	Phulwari Ki Nal WLS	1983	692.68
Rajasthan	15	Ramgarh Vishdhari WLS	1982	252.79
Rajasthan	16	Ramsagar WLS	1955	34.4
Rajasthan	17	Sajjangarh WLS	1987	5.19
Rajasthan	18	Sariska WLS	1955	219
Rajasthan	19	Sawaimadhopur WLS	1955	131.3
Rajasthan	20	Sawai Man Singh WLS	1984	103.25
Rajasthan	21	Shergarh WLS	1983	98.71
Rajasthan	22	Sitamata WLS	1979	422.94
Rajasthan	23	Tal Chhapar WLS	1971	7.19
Rajasthan	24	Todgarh Raoli WLS	1983	495.27
Rajasthan	25	Van Vihar WLS	1955	25.6
Sikkim	1	Barsey Rhododendron WLS	1998	104
Sikkim	2	Fambong Lho WLS	1984	51.76
Sikkim	3	Kitam Bird WLS	2005	6
Sikkim	4	Kyongnosla Alpine WLS	1977	31
Sikkim	5	Maenam WLS	1987	35.34
Sikkim	6	Pangolakha WLS	2002	128
Sikkim	7	Shingba Rhododendron WLS	1984	43
Tamil Nadu	1	Cauvery North WLS	2014	504.334
Tamil Nadu	2	Chitrangudi Bird WLS	1989	0.48
Tamil Nadu	3	Gangaikondam Spotted Dear WLS	2013	2.88
Tamil Nadu	4	Indira Gandhi (Annamalai) WLS	1976	841.49
Tamil Nadu	5	Kalakad WLS	1976	223.58
Tamil Nadu	6	Kanjirankulam Bird WLS	1989	1.04
Tamil Nadu	7	Kanyakumari WLS	2002	457.78
Tamil Nadu	8	Karaivetti Bird WLS	1999	4.54
Tamil Nadu	9	Karikilli Birds WLS	1989	0.61
Tamil Nadu	10	Kodaikanal WLS	2013	608.95

State	S. No.	Name of State/ Protected Area	Year of Establishment	Area (Km²)
Tamil Nadu	11	Koonthankulam-Kadankulam WLS	1994	1.29
Tamil Nadu	12	Megamalai WLS	2009	269.11
Tamil Nadu	13	Melaselvanoor-Keelaselvanoor WLS	1998	5.93
Tamil Nadu	14	Mudumalai WLS	1942	217.76
Tamil Nadu	15	Mundanthurai WLS	1977	567.38
Tamil Nadu	16	Nellai WLS	2015	356.73
Tamil Nadu	17	Oussudu Lake Bird Sanctuary	2015	3.32
Tamil Nadu	18	Point Calimere WLS	1967	17.26
Tamil Nadu	19	Pulicat Lake Bird WLS	1980	153.67
Tamil Nadu	20	Sakkarakottai Bird Sanctuary	2012	2.3
Tamil Nadu	21	Sathyamangalam WS	2008, 2011	1411.61
Tamil Nadu	22	Srivilliputhur Grizzled Squirrel WLS	1988	485.2
Tamil Nadu	23	Theerthangal Bird Sanctuary	2010	0.29
Tamil Nadu	24	Udayamarthandapuram Lake WLS	1991	0.45
Tamil Nadu	25	Vaduvoor Birds WLS	1991	1.28
Tamil Nadu	26	Vedanthangal Lake Birds WLS	1936	0.3
Tamil Nadu	27	Vellanadu Blackbuck WLS	1987	16.41
Tamil Nadu	28	Vellode Birds WLS	1997	0.77
Tamil Nadu	29	Vettangudi Birds WLS	1977	0.38
Telangana	1	Amrabad (Nagarjunasagar-Srisailam) WLS	1978	2166.28
Telangana	2	Eturnagaram WLS	1953	806.15
Telangana	3	Kawal WLS	1965	892.23
Telangana	4	Kinnersani WLS	1977	635.41
Telangana	5	Lanja Madugu Siwaram WLS	1978	29.81
Telangana	6	Manjeera Crocodile WLS	1978	20
Telangana	7	Pakhal WLS	1952	860
Telangana	8	Pocharam WLS	1952	130

State	S. No.	Name of State/ Protected Area	Year of Establishment	Area (Km^2)
Telangana	9	Pranahita WLS	1980	136.03
Tripura	1	Gumti WLS	1988	389.54
Tripura	2	Rowa WLS	1988	0.858
Tripura	3	Sepahijala WLS	1987	13.453
Tripura	4	Trishna WLS	1988	163.078
Uttar Pradesh	1	Bakhira WLS	1990	28.94
Uttar Pradesh	2	Chandraprabha WLS	1957	78
Uttar Pradesh	3	Dr. Bhimrao Ambedkar Bird WLS	2003	4.27
Uttar Pradesh	4	Hastinapur WLS	1986	2073
Uttar Pradesh	5	Jai Prakash Narayan (Surhatal) Bird WLS	1991	34.32
Uttar Pradesh	6	Kaimur WLS	1982	500.73
Uttar Pradesh	7	Katerniaghat WLS	1976	400.09
Uttar Pradesh	8	Kishanpur WLS	1972	227
Uttar Pradesh	9	Lakh Bahosi Bird WLS	1988	80.24
Uttar Pradesh	10	Mahavir Swami WLS	1977	5.41
Uttar Pradesh	11	National Chambal WLS	1979	635
Uttar Pradesh	12	Nawabganj Bird WLS	1984	2.25
Uttar Pradesh	13	Okhala Bird WLS	1990	4
Uttar Pradesh	14	Parvati Aranga WLS	1990	10.84
Uttar Pradesh	15	Patna WLS	1990	1.09
Uttar Pradesh	16	Pilibhit WLS	2014	602.798
Uttar Pradesh	17	Ranipur WLS	1977	230.31
Uttar Pradesh	18	Saman Bird WLS	1990	5.26
Uttar Pradesh	19	Samaspur Bird WLS	1987	7.99
Uttar Pradesh	20	Sandi Birds WLS	1990	3.09
Uttar Pradesh	21	Shekha Bird WLS	1852	0.25
Uttar Pradesh	22	Sohagibarwa WLS	1987	428.2
Uttar Pradesh	23	Sohelwa WLS	1988	452.47
Uttar Pradesh	24	Sur Sarovar Bird WLS	1991	4.03
Uttar Pradesh	25	Turtle WLS	1989	7
Uttar Pradesh	26	Vijai Sagar WLS	1990	2.62
Uttarakhand	1	Askot WLS	1986	600
Uttarakhand	2	Binsar WLS	1988	47.07

State	S. No.	Name of State/ Protected Area	Year of Establishment	Area (Km^2)
Uttarakhand	3	Govind Pashu Vihar WLS	1955	485.89
Uttarakhand	4	Kedarnath WLS	1972	975.2
Uttarakhand	5	Mussoorie WLS	1993	10.82
Uttarakhand	6	Nandhaur WLS	2012	269.96
Uttarakhand	7	Sonanadi WLS	1987	301.18
West Bengal	1	Ballavpur WLS	1977	2.02
West Bengal	2	Bethuadahari WLS	1980	0.67
West Bengal	3	Bibhuti Bhusan WLS	1980	0.64
West Bengal	4	Buxa WLS	1986	267.92
West Bengal	5	Chapramari WLS	1976	9.6
West Bengal	6	Chintamani Kar Bird WLS	1982	0.07
West Bengal	7	Haliday Island WLS	1976	5.95
West Bengal	8	Jorepokhri Salamander WLS	1985	0.04
West Bengal	9	Lothian Island WLS	1976	38
West Bengal	10	Mahananda WLS	1976	158.04
West Bengal	11	Pakhi Bitan Bird WLS	2016	14.09
West Bengal	12	Raiganj WLS	1985	1.3
West Bengal	13	Ramnabagan WLS	1981	0.14
West Bengal	14	Sajnakhali WLS	1976	362.4
West Bengal	15	Senchal WLS	1976	38.88
West Bengal	16	West Sunderban WLS	2013	556.45

(*Source*: National Wildlife Database, Wildlife Institute of India)

Conservation Reserves

Conservation reserves and community reserves in India are terms denoting protected areas of India which typically act as buffer zones to or connectors and migration corridors between established national parks, wildlife sanctuaries and reserved and protected forests of India. Such areas are designated as conservation areas if they are uninhabited and completely owned by the Government of India but used for subsistence by communities and community areas if part of the lands are privately owned. These protected area categories were first introduced in the Wildlife (Protection) Amendment Act of 2002 – the amendment to the Wildlife Protection Act of 1972. These categories were added because of reduced protection in and around existing or proposed protected areas due to private ownership of land, and land use.

List of Conservation Reserves (May, 2019)

State	S. No.	Name of State/ Protected Area	Year of Establishment	Area (Km²)
Gujarat	1	Chharidhandh Con R	2008	227
Haryana	1	Bir Bara Ban Con R	2007	4.19
Haryana	2	Saraswati Plantation Con R	2007	44.53
Himachal Pradesh	1	Darlaghat Con R	2013	0.67
Himachal Pradesh	2	Shilli Con R	2013	1.49
Himachal Pradesh	3	Shri Naina Devi Con R	2013	17.01
Jammu & Kashmir	1	Ajas (WL) Con R	1945	1
Jammu & Kashmir	2	Ajas Con R	1945	48
Jammu & Kashmir	3	Bahu Con R	1981	19.75
Jammu & Kashmir	4	Boodh Karbu Con R	1981	12
Jammu & Kashmir	5	Brain-Nishat Con R	1945	15.75
Jammu & Kashmir	6	Chatlam, Pampore (WL) Con R	1970	0.25
Jammu & Kashmir	7	Gharana (WL) Con R	1981	0.75
Jammu & Kashmir	8	Hokera (Ramsar Site) (WL) Con R	1945	13.75
Jammu & Kashmir	9	Hygam (WL) Con R	1945	7.25
Jammu & Kashmir	10	Jawahar Tunnel Con R	1981	18
Jammu & Kashmir	11	Khanagund Con R	1945	15
Jammu & Kashmir	12	Khimber/Dara/Sharazbal Con R	1945	34
Jammu & Kashmir	13	Khiram Con R	1945	15.75
Jammu & Kashmir	14	Khonmoh Con R	1945	67
Jammu & Kashmir	15	Khrew Con R	1945	50.25
Jammu & Kashmir	16	Kukarian (WL) Con R	1981	24.25
Jammu & Kashmir	17	Malgam (WL) Con R	1970	4.5
Jammu & Kashmir	18	Manibugh (WL) Con R	1970	4.5
Jammu & Kashmir	19	Mirgund (WL) Con R	1970	4
Jammu & Kashmir	20	Naganari Con R	1981	22.25
Jammu & Kashmir	21	Nanga (WL) Con R	1981	15.25
Jammu & Kashmir	22	Narkara (WL) Con R	1991	3.25
Jammu & Kashmir	23	Norrichain (WL) Con R	1981	2
Jammu & Kashmir	24	Panyar Con R	1945	10
Jammu & Kashmir	25	Pargwal (WL) Con R	1981	49.25

State	S. No.	Name of State/ Protected Area	Year of Establishment	Area (Km^2)
Jammu & Kashmir	26	Sabu Con R	1981	15
Jammu & Kashmir	27	Sangral-Asa Chak (WL) Con R	1981	7
Jammu & Kashmir	28	Shallabugh (WL) Con R	1945	16
Jammu & Kashmir	29	Shikargah Con R	1945	15.5
Jammu & Kashmir	30	Sudhmahadev Con R	1981	142.25
Jammu & Kashmir	31	Thein (WL) Con R	1981	19
Jammu & Kashmir	32	Tsomoiri (Ramsar Site) (WL) Con R	1981	120
Jammu & Kashmir	33	Wangat/Chatergul Con R	1945	12
Jammu & Kashmir	34	Zaloora, Harwan Con R	1970	25.25
Karnataka	1	Aghanashini Con R	2011	299.52
Karnataka	2	Ankasamudra Birds Con R	2017	0.9826
Karnataka	3	Banakapur Con R	2006	0.5627
Karnataka	4	Basur Amruth Mahal Kaval Con R	2011	7.37
Karnataka	5	Bedthi Con R	2011	57.3
Karnataka	6	Hornbill Con R	2011	52.5
Karnataka	7	Jayamangali Blackbuck Con R	2007	3.2328
Karnataka	8	Kappathagudda Con R	2015	178.72
Karnataka	9	Magadi Kere Con R	2015	0.54
Karnataka	10	Melapura Bee Eater Bird Con R	2015	0.0318
Karnataka	11	Puttenahalli Lake Birds Con R	2015	0.15
Karnataka	12	Shalmale Riparian Bio-system Con R	2012	4.89
Karnataka	13	Thimlapura Con R	2016	17.38
Karnataka	14	Thungabhadra Otter Con R	2015	20
Karnataka	15	Ummathur Con R	2017	6.08
Maharashtra	1	Anjneri Con R	2017	5.69
Maharashtra	2	Bhorkada Con R	2008	3.49
Maharashtra	3	Kolamarka Con R	2013	180.72
Maharashtra	4	Mamdapur Con R	2014	54.46
Maharashtra	5	Muktai Bhavani Con R	2014	122.74

State	S. No.	Name of State/ Protected Area	Year of Establishment	Area (Km^2)
Maharashtra	6	Toranmal Con R	2016	93.42
Punjab	1	Beas River Con R	2017	0
Punjab	2	Rakh Sarai Amanat Khan Con R	2010	4.95
Punjab	3	Ranjit Sagar Con R	2017	18.65
Punjab	4	Ropar Wetland Con R	2017	2.11
Rajasthan	1	Bir Jhunjhunu Con R	2012	10.4748
Rajasthan	2	Bisalpur Con R	2008	48.31
Rajasthan	3	Gogelao Con R	2012	3.58
Rajasthan	4	Gudha Bisnoiyan Con R	2011	2.32
Rajasthan	5	Jawai Bandh Leopard Con R	2013	19.7858
Rajasthan	6	Jor Beed Gadwala Bikaner Con R	2008	56.47
Rajasthan	7	Rotu Con R	2012	0.7286
Rajasthan	8	Shakambhari Con R	2012	131
Rajasthan	9	Sundha Mata Con R	2008	117.4892
Rajasthan	10	Umedganj Bird Con R	2012	2.7247
Sikkim	1	Sling Dong Fairreanum Orchid Con R	2008	0.06
Tamil Nadu	1	Tiruppadaimarathur Con R	2005	0.028
Tamil Nadu	2	Suchindrum-Theroor-Managudi Con R	2015	4.85
Uttarakhand	1	Asan Wetland Con R	2005	4.444
Uttarakhand	2	Jhilmil Jheel Con R	2005	37.835
Uttarakhand	3	Naina Devi Himalayan Bird Con R	2015	111.92
Uttarakhand	4	Pawalgarh Con R	2012	58.25
West Bengal	1	Deul Con R	2017	10.5
West Bengal	2	Garpanchkot Con R	2017	1340.34
West Bengal	3	Hijli Con R	2017	15.5
West Bengal	4	Mukutmanipur Con R	2017	43.7
West Bengal	5	Tekonia Con R	2017	5.87

(*Source*: National Wildlife Database, Wildlife Institute of India)

Community Reserves

Conservation reserves and community reserves in India are terms denoting protected areas of India which typically act as buffer zones to or connectors and migration corridors between established national parks, wildlife sanctuaries and reserved and protected forests of India. Such areas are designated as conservation areas if they are uninhabited and completely owned by the Government of India but used for subsistence by communities and community areas if part of the lands are privately owned. These protected area categories were first introduced in the Wildlife (Protection) Amendment Act of 2002 – the amendment to the Wildlife Protection Act of 1972. These categories were added because of reduced protection in and around existing or proposed protected areas due to private ownership of land, and land use.

List of Community Reserves (May, 2019)

State	S. No.	Name of State/ Protected Area	Year of Establishment	Area (Km^2)
Karnataka	1	Kokkare Bellur Com R	2007	3.12417
Kerala	1	Kadalundi Vallikkunnu Com R	2007	1.5
Meghalaya	1	Aruakgre Com R	2014	1
Meghalaya	2	Baladingre Com R	2013	0.5
Meghalaya	3	Balsri Adingi Com R	2017	0.456
Meghalaya	4	Bandarigre Com R	2013	0.67
Meghalaya	5	Chandigre Com R	2013	0.37
Meghalaya	6	Chenggni Com R	2017	1.74
Meghalaya	7	Chimanpara Com R	2016	0.102
Meghalaya	8	Chimitap Com R	2017	22.8
Meghalaya	9	Dallengggittim Com R	2017	0.2217
Meghalaya	10	Dambuk Allonge Com R	2017	36.6
Meghalaya	11	Dambuk Jongkol Com R	2017	2.648
Meghalaya	12	Dangkipara Com R	2014	0.025
Meghalaya	13	Daribokgre Com R	2013	1.73
Meghalaya	14	Dumitdikgre Com R	2013	0.7
Meghalaya	15	Dura Kalkgre Com R	2013	0.6
Meghalaya	16	Eman Asakgre Com R	2013	0.3
Meghalaya	17	Gokagre Com R	2017	0.1789
Meghalaya	18	Halwa Ambeng Com R	2017	0.941
Meghalaya	19	Jaksongram Com R	2017	0.555
Meghalaya	20	Jirang Com R	2014	2

State	S. No.	Name of State/ Protected Area	Year of Establishment	Area (Km²)
Meghalaya	21	Ka Khloo Langdoh Kur Pyrtuh Com R	2014	0.154
Meghalaya	22	Ka Khloo Pohblai Mooshutia Com R	2014	0.335
Meghalaya	23	Ka Khloo Thangbru Umsymphu Com R	2014	0.196
Meghalaya	24	Ka Lum Luwe Com R	2018	0.141
Meghalaya	25	Khloo Amrawan Com R	2015	1.29
Meghalaya	26	Khloo Blai Chyrmang Sein Raij Kongwasan Chyrmang Kwai Com R	2013	0.07
Meghalaya	27	Khloo Blai Ka Raij U Landoh longlang Com R	2016	0.15
Meghalaya	28	Khloo Blai Kongwasan Khloo Blai Chyrmang	2014	0.07
Meghalaya	29	Khloo Blai Sein Raij Tuber Com R	2014	0.965
Meghalaya	30	Kitmadamgre Com R	2014	0.7
Meghalaya	31	Kpoh Eijah Com R	2014	0.17
Meghalaya	32	Lawbah Com R	2014	2.1
Meghalaya	33	Lotnagar Com R	2017	0.456
Meghalaya	34	Lum Jusong Com R	2014	0.7
Meghalaya	35	Lumkohkriah Com R	2014	6.11
Meghalaya	36	Mandalgre Com R	2013	0.5
Meghalaya	37	Matchirampat Com R	2017	0.217
Meghalaya	38	Miewsyiar Com R	2014	0.87
Meghalaya	39	Mikadogre Com R	2013	0.01
Meghalaya	40	Mongalgre Com R	2014	0.2
Meghalaya	41	Nikwatgre Com R	2017	0.495
Meghalaya	42	Nongsangu Com R	2014	1
Meghalaya	43	Nongumiang Com R	2003	0.31
Meghalaya	44	Phudja-ud Com R	2014	1.2
Meghalaya	45	Raid Nongbri Com R	2014	0.7
Meghalaya	46	Raid Nonglyngdoh/Pdah Kyndeng Com R	2014	0.75
Meghalaya	47	Resu Haluapra Com R	2014	0.5
Meghalaya	48	Rewak Daburam Com R	2017	0.183
Meghalaya	49	Rewak Watregittim Com R	2017	0.098

State	S. No.	Name of State/ Protected Area	Year of Establishment	Area (Km^2)
Meghalaya	50	Rongalgre Com R	2016	0.165
Meghalaya	51	Rongcheng Com R	2017	2.356
Meghalaya	52	Rongma Paromgre Com R	2013	0.62
Meghalaya	53	Rongma Rekmangre Com R	2013	1.92
Meghalaya	54	Ryngibah Com R	2014	0.8
Meghalaya	55	Ryngud Com R	2014	5.22
Meghalaya	56	Sakalgre Com R	2013	1.22
Meghalaya	57	Sasatgre Com R	2013	0.6
Meghalaya	58	Selbalgre Com R	2013	0.2
Meghalaya	59	Siju Duramong-I Com R	2017	0.768
Meghalaya	60	Siju Duramong-II Com R	2017	25.1
Meghalaya	61	Taidang Com R	2017	1.214
Meghalaya	62	Thangkharang Com R	2014	1.11
Meghalaya	63	Thokpara Com R	2016	0.3
Meghalaya	64	Umsum Pitcher Plant Com R	2014	0.4
Meghalaya	65	Upper Dosogre Com R	2017	0.19691
Nagaland	1	Atoizu Com R	2015	4
Nagaland	2	Benreu Com R	2018	30
Nagaland	3	Bhumbak Com R	2018	6.5
Nagaland	4	Bonchu Com R	2009	9.05
Nagaland	5	Chemekong Com R	2015	29.175
Nagaland	6	Chishilimi Com R	2015	3.5
Nagaland	7	D. Khel, Kohima Village Com R	2015	3
Nagaland	8	Dihoma Com R	2015	2
Nagaland	9	Gariphema Com R	2018	2.65
Nagaland	10	Hukphang Com R	2018	3
Nagaland	11	Jotsoma Com R	2018	5
Nagaland	12	Kanjang Com R	2018	1
Nagaland	13	Kezoma Com R	2018	2.65
Nagaland	14	Khekiye Com R	2015	2.5
Nagaland	15	Khonoma Com R	2018	2.65
Nagaland	16	Khrieyalienuomaiko Com R	2018	2.65
Nagaland	17	Khrokhropfu – Lephori Com R	2009	6.15
Nagaland	18	Khudei Com R	2018	4.8
Nagaland	19	Khutur Com R	2018	4.89

State	S. No.	Name of State/ Protected Area	Year of Establishment	Area (Km²)
Nagaland	20	Khwuma Khel Jotsoma Com R	2018	3
Nagaland	21	Kidema Com R	2018	2.65
Nagaland	22	Kigwema Com R	2015	2.65
Nagaland	23	Kikruma Com R	2015	1.1
Nagaland	24	Kilo Old Com R	2018	2
Nagaland	25	Kiyelho Com R	2018	3
Nagaland	26	Litem Com R	2018	1.6
Nagaland	27	Lizuto Com R	2015	2.5
Nagaland	28	Longra Com R	2018	2.275
Nagaland	29	Longtang Com R	2018	5.8
Nagaland	30	Lotovi Com R	2018	1
Nagaland	31	Luzaphuhu Com R	2015	14
Nagaland	32	Mezoma Com R	2015	2.85
Nagaland	33	Morakjo Com R	2015	6.5
Nagaland	34	Mpai Namci CR	2018	20
Nagaland	35	Nerhema Perazatsa Com R	2018	20
Nagaland	36	Nerhema Yaoke Com R	2018	20
Nagaland	37	Nian Com R	2018	2
Nagaland	38	Noksen Com R	2018	1
Nagaland	39	Piphema "A" Com R	2018	1
Nagaland	40	Piphema "B" Com R	2018	2.8
Nagaland	41	Rangkang Com R	2018	5.15
Nagaland	42	Sakhabama Com R	2018	2.5
Nagaland	43	Sangdak Com R	2018	5.09
Nagaland	44	Scaly-Mopungchuket Com R	2009	15
Nagaland	45	Sitap Com R	2018	1.5
Nagaland	46	Tamlu Com R	2018	2
Nagaland	47	Tsekhwelu Com R	2015	8
Nagaland	48	Tsiepama Com R	2015	3.325
Nagaland	49	Tsuruhu Com R	2015	2.7
Nagaland	50	Tuophema Village Com R	2018	2.5
Nagaland	51	Viswema Com R	2018	2.65
Nagaland	52	Wakchin Chingla Com R	2018	30
Nagaland	53	Yali Com R	2018	14
Nagaland	54	Yangpi Com R	2018	3.0007

State	S. No.	Name of State/ Protected Area	Year of Establishment	Area (Km²)
Nagaland	55	Yaongyimchen Com R	2018	8
Nagaland	56	Yongshei Com R	2018	1.5
Nagaland	57	Yonyu Com R	2018	4.8
Punjab	1	Keshopur Chhamb Com R	2007	3.4
Punjab	2	Lalwan Com R	2007	12.67
Punjab	3	Siswan Com R	2017	12.95

(*Source*: National Wildlife Database, Wildlife Institute of India)

Wildlife Protection Act, 1972

Chapter	Particulars/Details	Section
Chapter I	Preliminary	1 to 2
Chapter II	Authorities appointed.	3 to 8
Chapter III	Hunting of Wild Animals.	9 to 17
Chapter IIIA	Protection of specified plants.	17A to 17H
Chapter IV	Protected areas: Wildlife Scantuary National Park Closed Areas.	18 to 38
Chapter IVA	Central Zoo Authority and Recognition of Zoo.	38A to 38J
Chapter IVB	National Tiger Conservation Authority.	
Chapter IVC	Tiger and Other Endangered species Crime Control Bureau.	
Chapter V	Trade and commerce in wild animals, animal articles and trophies.	39 to 49
Chapter VA	Prohibition of trade or commerce in trophies articles derived from certain animals.	49 to 49C
Chapter VI	Prevention and detection of offences.	50 to 58
Chapter VIA	Forfeiture of property derived from illegal hunting and trade.	
Chapter VII	Miscellaneous.	59 to 66

Important section

Section 9 – Totally prohibits hunting of wild animals included in schedule I to IV.

Section 18 (1)– Declaration/Notification of Wildlife Sanctuary by state government.

Section 35 (1)–Declaration/Notification of National Park by state government.

Section 38 (1)–Declaration/Notification of Wildlife Sanctuary or National Park by central- government.

Section 36 A – Conservation Reserves.

Section 36 C – Community Reserves.

Schedules

Schedule	Particulars/Details
Schedule I	Rare and threatened animals.
Schedule II	Special game.
Schedule III	Big game.
Schedule IV	Small game.
Schedule V	Vermin
Schedule VI	Threatened plant species.

Salient features of FSI report (2019):

Forest cover and tree cover:

1. The total geographic area of the country is 32,87,469 sq km.
2. According to the current assessment report the "total forest cover" of the country is 7,12,249 sq km (21.67%) of total geographic area of the country and there is an increase of 3,976 sq km forest cover.
3. In terms of density classes maximum area is covered by "moderate dense forest" (3,08,472 sq km) and minimum area by "very dense forest" (99,278 sq km).
4. The Forest cover in terms of area was maximum in Madhya Pradesh (77,482 sq km) followed by Arunachal Pradesh (66,688 sq km), Chhattisgarh (55,611 sq km), Odisha (51,619 sq km) and Maharashtra (50,778 sq km).
5. In terms of percentage maximum forest cover was found in states of Mizoram (85.40%), Arunachal Pradesh (79.63%), Meghalaya (76.33%), Manipur (75.46%) and Nagaland (75.31%).
6. Seventeen State's/UT's have more than 33% of the geographical area under forest cover
7. Ten State's/UT's have between 33 to 75% of the geographical area under forest cover
8. The total recorded tree cover area of the country is 95,027 sq km and there is an increase of 1, 212 sq. km.
9. The tree cover area is maximum in Mahatashtra (10, 806 sq. km), Madhya Pradesh (8, 339 sq. km), Rajasthan (8,112 sq. km) and Jammu & Kashmir (7, 944 sq. km).

Forest Cover of India

Class	Area (Km^2)	Percentage of Geographic area
Very Dense Forest	99, 278	3.02
Moderately Dense Forest	3,08,472	9.39
Open Forest	3,04,499	9.26
Total Forest Cover	7,12,249	21.67
Scrub	46,297	1.41
Non-Forest	25,28,923	76.92
Total Geographic Area	32,87,469	100.00

Growing stock

1. The "total growing stock" of wood in the country is 5,915.76 million cum and there is an increase of 93.38 million cum (1.6%).
2. The average per hectare growing stock is 55.69 cum

Mangrove cover:

1. According to the current assessment report "total mangrove area" in the country is 4975 sq km (0.15% of the total geographic area of the country), and has shown an increase of 54 sq km as compared to previous assessment
2. The main reason for the increase in mangrove cover is plantation in islands, along the river creeks and natural regeneration

Bamboo

1. There are 125 indigenous and 11 exotic species of bamboo found in India
2. The "Bamboo bearing area" of the country is 16 million ha.
3. The maximum bamboo bearing area lies in Madhya Pradesh (2 million ha) followed by Maharashtra (1.5 million ha), Arunachal Pradesh (1.49 million ha) and Odisha (1.18 million ha).

Carbon sink

1. According to the current assessment report the "Total Carbon stock" of the country is 7,124 million tonnes and there is an increase of 42.6 million tonnes as compared to previous report.

(*Source*: FSI-Report 2019)

Major Forest Groups of India (Champion & Seth Classification)

Tropical Forest	
Group 1	Wet evergreen forest
Group 2	Semi evergreen forest
Group 3	Moist deciduous forest
Group 4	Littoral and swamp forest
Group 5	Dry deciduous forest
Group 6	Thorn forest
Group 7	Dry evergreen forest
Montane Sub Tropical Forest	
Group 8	Broadleaved hill forest
Group 9	Pine forest
Group 10	Dry evergreen forest
Montane Temperate Forest	
Group 11	Montane wet temperate forest
Group 12	Himalayan temperate forest
Group 13	Himalayan dry temperate forest
Sub Alpine Forest	
Group 14	Sub alpine forest
Alpine Forest	
Group 15	Moist alpine scrub forest
Group 16	Dry alpine scrub forest

Non timber forest products

Fibres	**Species**
(a) Stem fibre	*Sterculia villosa, S. urens, Grewia optiva, Hardwickia binata, Butea monosperma*
(b) Leaf fibre	*Caryota urens, Musa* species, *Agave* species
(c) Flosses	*Ceiba pentandra, Bombax ceiba, Cochspermum reliogiosum*
(d) Canes	*Calamus, Daemonorops, Ceratolobus, Plectocomia*
(e) Coir	*Cocos nucifera*
Essential oils	
(a) Grass oil	Ginger grass oil- *Cymbopogon martinii var. sofia*
	Palmarosa oil- *Cymbopogon martinii var. motia*
	Citronella oil- *Cymbopogon nardus*
	Khus oil- *Vetiveria zizanoides*
	Lemon grass oil- *C. flexuous*

(b) Wood oil	Sandal oil- *Santalum album*
	Agar oil- *Aquilaria agallocha*
	Deodar oil- *Cedrus deodara*
	Pine oil- *Pinus* species
(c) Leaf oil	Eucalyptus oil- *Eucalyptus globulus*
	Citriodora oil- *E. citriodora*
	Camphor oil- *Cinnamomum camphor*
	Mint oil- *Mentha* species
	Wintergreen oil- *Gaultheria fragtissima*
	Patchouli oil- *Pogostemon patchouli*
	Pine needle oil- *Pinus roxburghii*
(d) Root oil	Costus oil- *Saussurea lappa*
	Valerian oil- *Valeriana jatamansi*
(e) Flower oil	Keora oil- *Pandanus tectorius*
Oil yielding trees	
	Tung oil- *Aleurites fordii*
	Indian butter tree- *Diploknema butyracea*
	Kokam butter- *Garcinia indica*
	Chaulmugra oil- *Hydnocarpus kurzii*
	Mahua butter- *Madhuca indica*
	Piney tallow- *Vateria indica*
Waxes	
	Carnuva wax- *Copernicia cerifera*
	Sepium sebiferum
Tannins	
(a) Wood tans	*Quebracho colorado*
(b) Bark tans	*Acacia nilotica*
	Terminalia arjuna
	Cassia fistula
	Cassia auriculata
	Ceriops roxburghiana
(c) Fruit tans	*Terminalia chebula*
	T. bellirica
	Emblica officinalis
(d) Leaf tans	*Anogeissus latifolia*
	Carissa spinarum
	Emblica officinalis

Dyes	
(a) Wood dyes	Cutch dye- *Acacia catechu*
	Brazilian dyes- *Caesalpinia sappan*
	Artocarpus dye- *Artocarpus heterophyllus*
	Santaline dye- *Pterocarpus santalinus*
(b) Bark dyes	Black dyes- *Acacia leucophloea, Acacia concinna.*
	Brown dyes- *Alnus nepalensis*
	Light reddish dyes- *Casuarina equisetifolia*
	Yellow dye- *Myrica esculenta*
	Fabric dye- *Terminalia alata*
(c) Fruit dye	Kamela dye- *Mallotus philippensis*
	Annatto dye- *Bixa orellana*
	Indigo dye- *Wrightia tinctoria*
(d) Flower dye	Dhak dye- *Butea monosperma*
	Saffron dye- *Crocus sativus*
	Toona ciliata (Yellow colour dye)
	Nyctanthus arbortristis (Orange colour dye)
	Mammea longifolia (Yellow colour dye)
(e) Root dye	*Berberis aristata* (Yellow dye)
	Punica granatum (Red dye)
	Rubia cordifolia (Red dye)
	Datisca cannabina (Yellow dye)
(f) Leaf dye	Henna dye- *Lawsonia inermis*
	Indigofera dye- *Indigofera tinctoria*
Gums	
	True Gum Arabic- *Acacia senegal*
	Indian Gum Arabic- *Acacia nilotica*
	Gum Kino- *Pterocarpus marsupium*
	Jhingan Gum- *Lannea coromandelica*
	Katira Gum- *Sterculia urens, Bombax ceiba*
	Salai Gum- *Boswellia serrata*
Resins	
(a) Hard resins	Pine resins- *Pinus roxburghii, P. wallichiana*
	True dammer- *Agathis loranthifolius*
	Black dammer- *Canarium strictum*
	Rock dammer- *Hopea odorata*

	White dammer- *Vateria indica*
	Lal dammer- *Shorea robusta*
	Green dammer- *Shorea tumbuggaia*
	Amber- *Pinus succinifera*
	Lacquer- *Rhus vernicifera*
	Sandarac- *Callistris quadrivalvis*
	Mastic- *Pistacia lentiscus*
	Shellac- *Laccifer lacca*
(b) Oleo resins	*Boswellia serrata*
(c) Gum resins	Gamboge- *Garcinia morella*
	Guggal- *Commiphora wightii*
	Asafoetida- *Ferula asafoetida*
Lac	
(a) Rangeeni	*Butea monosperma, Ziziphus mauritiana*
(b) Kusumi	*Schleichera oleosa*
Soap nut	*Sapindus mukrosii*
	Acacia concinna
Modified wood	
(a) Composite wood	Plywood, Laminated wood, Coreboards, Sandwich boards, Fibre boards, Particle boards.
(b) Improved wood	Impregnated wood Heat stabilized wood (Staybwood) Compressed wood Compregnated wood (Compreg) Heat stabilized compressed wood (Staypack)
Preservatives	
(a) Oil type	Creosote, Tar oil
(b) Water soluble leachable type	Zinc chloride, Zinc sulphate, Copper sulphate, Arsenic pentoxide, Borax
(c) Water soluble fixed type	Arsenic copper chromate, Acid cupric chromate, Chromated zinc chloride, Copper chrome boric, Zinc chrome boric
(d) Organic solvent type	Napthenic acid, Stearic acid, Abiet

Zeitfracht Medien GmbH
Ferdinand-Jühlke-Straße 7
99095 Erfurt, Deutschland
produktsicherheit@kolibri360.de